International Encyclopaedia of ENERGY RESOURCES

International Encyclopaedia of ENERGY RESOURCES

— *Volume 4* —

— *Edited By* —

Dr. M.P. Singh

Head of the Department of Forest Sciences

Birsa Agricultural University

Ranchi - 834 006 (Jharkhand)

2005

DAYA PUBLISHING HOUSE

Delhi - 110 035

ISBN 81-7035-357-2 (SET)
ISBN 81-7035-358-0 (Vol. 1)
ISBN 81-7035-359-9 (Vol. 2)
ISBN 81-7035-360-2 (Vol. 3)
ISBN 81-7035-361-0 (Vol. 4)

Published by : **Daya Publishing House**
1123/74, Deva Ram Park
Tri Nagar, Delhi - 110 035
Phone: 27383999
Fax: (011) 23244987
e-mail: dayabooks@vsnl.com
website: www.dayabooks.com

Showroom : 4762-63/23, Ansari Road, Darya Ganj
New Delhi - 110 002
Phone: 23245578, 23244987

Laser Typesetting : **Classic Computer Services**
Delhi - 110 035

Printed at : **Chawla Offset Printers**
Delhi - 110 052

PRINTED IN INDIA

Preface

This book is for people who still think that there are energy problems ahead, and who are not satisfied with an energy future dominated by coal and nuclear power.

Fortunately, the time has come when one can examine other alternatives seriously–in particular, energy conservation and renewable energy–without being relegated at once to the outer fringes. Southern California Edison, Pacific Gas and Electric, and numerous other progressive utilities, having embraced these alternatives, are a bit too formidable to be drummed out of the corps. The Office of Technology Assessment, the Gas Research Institute, and the Harvard Business School likewise are not thought of as outposts of a lingering counterculture.

Still, not enough people in the energy supply industries, in government, or in private agencies concerned with energy policy are aware of the full range of possibilities in energy conservation and in the numerous renewable energy technologies. For them, this book is an attempt to lay out, in an ordered fashion, the basis in fact, logic, and judgment, for an energy -efficient, renewable energy future.

One cluster of views about energy may be characterised as "supply oriented." The response to the oil shortage was quite expectedly framed in terms of increasing energy supply, especially in forms other than oil. The basic premises motivating supply-side arguments included the following assumptions:

Energy demand will grow with output. A healthy and growing economy requires a growing supply of energy. In early formulations the rate of energy supply growth was about equal to the growth rate of the economy.

While this volume does provide a great deal of information about energy sources, its more appropriate use is as a stimulus for discussion. Towards this end the book often raise more questions than they answer. The complexity of the ethical dilemmas posed by energy sources is such that we are still trying to identify what questions should be asked. The volume recognises that the first step in making the types of ethical decisions which are increasingly demanded of us is searching for the right questions. The book vary greatly in both length and orientation. This will make it useable by a wide range of groups–from students in the classroom, to professionals such as scientists, medical doctors, business people, lawyers, and ethicists, as well as environmental and other socially active groups. It is hoped that this volume will contribute in a better way to a society.

Dr. M.P. Singh

Contents

Preface v

— Volume 1 —

Introduction 1

1. Afghanistan 5
2. Algeria 9
3. Angola 21
4. Antarctica 31
5. Arab Maghreb Union 35
6. Argentina 47
7. Australia 58
8. Azerbaijan 69
9. Bahrain 80
10. Balkans Region 86
11. Baltic Sea Region 94
12. Bangladesh 109
13. Bolivia 118
14. Brazil 127
15. Brunei 143

16. Cameroon 149
17. Canada 157
18. Caribbean Fact Sheet 173
19. Caspian Sea Region 184
20. Caucasus Region 192
21. Central Asia: Kyrgyzstan Energy Sector 208
22. Central Asia: Tajikistan Energy Sector 212
23. Central Asia: Turkmenistan Energy Sector 216
24. Central Asia: Uzbekistan Energy Sector 221
25. Chad 227
26. Chile 234
27. China 244
28. Colombia 256
29. Congo-Brazzaville 266
30. Cote d'Ivoire 277
31. Economic Community of West African States (ECOWAS) 285
32. Ecuador 301
33. Egypt 311
34. Equatorial Guinea 321
35. France 329
36. Gabon 340
37. Germany 348

— Volume 2 —

38. Great Lakes Region: Burundi, Kenya, Rwanda, Tanzania, and Uganda 365
39. Greece 375
40. Horn of Africa 384
41. India 396
42. Indonesia 407
43. Iran 417
44. Iraq 432
45. Israel 447
46. Italy 459
47. Japan 470
48. Jordan 480

49. Kazakhstan 486
50. Kuwait 499
51. Libya 510
52. Malaysia 522
53. Mexico 530
54. Nigeria 543
55. North Central Europe 556
56. North Korea 568
57. North Sea 574
58. Norway 583
59. Oman 596
60. Pakistan 603
61. Panama 612
62. Paraguay 619
63. Persian Gulf Oil and Gas Exports Fact Sheet 625
64. Peru 633
65. Philippines 644
66. Portugal 653
67. Puerto Rico Fact Sheet 661
68. Qatar 666
69. Regional Indicators: Central America 674
70. Regional Indicators: European Union 687
71. Russia 694
72. Saudi Arabia 710
73. Singapore 723

— Volume 3 —

74. South Africa 729
75. South Asia Regional Overview 740
76. South China Sea Region 749
77. South Korea 756
78. Southeastern Europe 764
79. Southern Africa and the Southern African Development Community 776
80. Spain 790
81. Sudan 801

82.	**Syria**	**810**
83.	**Taiwan**	**819**
84.	**Thailand**	**826**
85.	**Turkey**	**833**
86.	**Ukraine**	**846**
87.	**United Arab Emirates**	**856**
88.	**United Kingdom**	**865**
89.	**United States of America**	**879**
90.	**Uruguay**	**901**
91.	**Venezuela**	**908**
92.	**Vietnam**	**921**
93.	**Yemen**	**931**
94.	**Monthly Energy Chronology 1996**	**937**
95.	**Monthly Energy Chronology 1997**	**978**
96.	**Monthly Energy Chronology 1998**	**1033**
97.	**Monthly Energy Chronology 1999**	**1083**

— Volume 4 —

98.	**Monthly Energy Chronology 2000**	**1113**
99.	**Monthly Energy Chronology 2001**	**1139**
100.	**Monthly Energy Chronology 2002**	**1179**
101.	**Monthly Energy Chronology 2003**	**1224**
102.	**Monthly Energy Chronology 2004**	**1265**
	Glossary	**1275**
	References	**1449**
	Index	**1457**

97

Monthly Energy Chronology 1999

The following information contains a listing of the major energy events that occurred in 1999.

January 1999

January 1

British Petroleum Company and Amoco Corporation complete their $53 billion merger. Chicago-based Amoco is the United States' fifth-largest oil company with roughly 9,300 gasoline stations. London-based British Petroleum, the world's third largest oil company, sells its products through a network of about 17,900 stations.

January 3

The Honduran government announces plans to sell off the state-owned electrical company and the country's four international airports in 1999. The National Electrical Energy Company, a 740-megawatt power supplier, could bring $500 million in the privatization sale.

January 4

Iraq and Jordan renew an agreement that will provide Jordan with Iraqi crude oil and refined petroleum products. Jordan's 1998 imports of Iraqi crude oil and refined petroleum products, including fuel oil and diesel, were estimated to have averaged about 100,000 barrels per day. This figure is expected to increase by 5 to 7 per cent in 1999.

January 5

Azerbaijan's state oil company SOCAR reports that it just exceeded its oil production target for 1998, while registering a shortfall in planned natural gas output. According to SOCAR, oil output in 1998 was about 180,000 barrels per day. Gas output reached nearly 198 billion cubic feet, 21 billion cubic feet below the planned figure.

January 6

Turkmenistan and Ukraine sign an agreement in Moscow that will allow the transit of about 705 billion cubic feet of Turkmen gas to Ukraine in 1999.

January 7

Several of Kuwait's opposition legislators demand that any future oil contracts with foreign companies conform to the country's constitution. The legislators are not opposed to the recent proposals to allow foreign investment in the operational aspect of Kuwait's upstream oil sector, but they want to make sure there is no violation of Kuwait's constitution, which bans the foreign ownership of national resources.

January 8

A new pipeline carrying crude oil from Azerbaijan's Caspian Sea fields across Georgia to the Black Sea port of Supsa begins operations. The new pipeline is expected to carry 50,000 barrels per day of oil in 1999 and marks another key development for the Caspian fields, which hold large reserves but have lacked an adequate pipeline system for exports.

January 9

Russia's Khabarovsk regional administration grants Exxon Neftegaz Limited the rights to construct a natural gas pipeline from the Sakhalin region in Siberia through Khabarovsk and on to China, according to Russia's Itar-Tass news agency.

January 11

Iraq rejects a proposal by Saudi Arabia to ease United Nations trade sanctions imposed on Iraq for its 1990 invasion of Kuwait. The initiative would recommend that Iraq be allowed to buy and sell all goods, except military equipment or materials that could be used for military purposes.

January 11

The Organization of Petroleum Exporting Countries (OPEC) releases figures showing that the December 1998 average price for the reference basket of seven crude oils used by OPEC as a gauge of world oil prices was the second lowest since July 1986. The December average was $9.69 per barrel, down from $11.19 per barrel in November 1998, and just a few cents above the July 1986 record low of $9.04 per barrel.

January 12

A coal mine explosion kills 16 workers and injures eight others, three seriously, in the northern province of Quang Ninh in Vietnam. Quang Ninh is Vietnam's main coal-producing area, yielding fuel both for domestic consumption and exports. Several deadly mine collapses have been reported in the province, mostly at private operations where support systems were poorly equipped.

January 13

As many oil-producing countries try to cut excess global production, Iraq announces plans to raise its oil output to 3 million barrels per day from its current 2.5 million barrels per day, and then to 3.5 million barrels per day within two years. Faleh al-Khayat, the Iraqi Oil Ministry's Director-General of Planning, says that the increases are contingent upon receiving spare parts for the country's ailing oil industry, which has been under United Nations trade sanctions for more than eight years.

January 13

A British oil worker kidnaped by Yemeni tribesmen is released unharmed. John Brooke was kidnapped by armed men from the Daham tribe who were demanding the release of a tribe member arrested three months ago on a murder charge. Brooke was taken at an oilfield run by the American Halliburton oil company in the Marib province, 105 miles southeast of the Yemeni capital of Sanaa.

January 14

Santa Fe Energy agrees to acquire Snyder Oil for about $450 million in stock in a deal combining the two mid-size independent oil companies into what would be called Santa Fe Snyder Corporation. The new company will have one-third of its oil and natural gas reserves overseas and two-thirds in the United States. Oil will represent about 60 per cent of the total reserves.

January 15

The United States proposes allowing Iraq to sell unlimited amounts of oil–but only if the proceeds go to buy food and other humanitarian supplies for the Iraqi people. The United Nations Security Council barred Iraq from freely exporting oil, its most valuable commodity, after its invasion of Kuwait in 1990. Concerned that sanctions were creating devastating hardships for Iraq's 22 million people, the Council agreed in 1995 to let Iraq sell limited amounts of oil to pay for humanitarian supplies.

January 18

More than 300 oil company officials attend a presentation by Brazilian petroleum regulators discussing the first oil and natural gas exploration blocks to be opened to foreign investors. Brazil is in the process of opening up to 92 per cent of its sedimentary basins to companies other than Petrobras, the country's state-owned oil company. Twenty-seven blocks will be part of the first round of competitive bidding, which is to occur in 1999.

January 20

The United Nations (U.N.) releases more than $81 million to Iraq to buy equipment to increase its supply of electricity. Iraq, which suffers frequent power shortages, as power plants fail and electrical demand rises, applied to buy the necessary generating equipment in 1998, when the oil-for-food program was expanded to allow Iraq to begin rebuilding its crumbling public services. U.N. trade sanctions and a more limited oil sales program had earlier prevented the Iraqis from replacing the old equipment.

January 25

Jordan's King Hussein appoints his eldest son, Abdullah, as his future successor. Hussein, who is suffering from cancer, passed over his brother Hassan, who had been Crown Prince for more than three decades. Abdullah, 36, is a career army officer.

January 25

Iranian President Mohammad Khatami approves a plan to restructure Iran's Oil Ministry. The plan focuses on decentralization of control by separating policy making from executive affairs, according to the official Islamic Republic News Agency.

January 31

Rebels fighting to oust the President of the Republic of the Congo abduct seven European forestry workers and cut electricity to the country's major seaport, Pointe-Noire.

January 31

According to U.S. officials, over the month of January, a series of almost daily U.S. missile air strikes on Iraq's air defenses has had a "grave impact" on President Saddam Hussein's ability to challenge allied enforcement of the no-fly zones over northern and southern Iraq.

February 1999

February 1

In southern Nigeria, a clash between Nigerian youths demanding work at a Shell oil terminal and Nigerian soldiers trying to break up the protest results in approximately 19 casualties. Shell temporarily shuts down the oil terminal as a result. The terminal handles more than 40 per cent of Nigeria's crude oil exports. Protests against Shell were first reported in the early 1990s. Nigerians have demanded employment in oil industry projects and development projects in their villages, and also compensation for use of their land, oil spills and other environmental damage. The number of protests has increased in recent years due to greater awareness of and discontent with the unequal distribution of wealth in the oil-based economy.

February 2

Arco Oriente Inc., a unit of U.S. Atlantic Richfield Company (ARCO) in Ecuador, temporarily halts operations in the Pastaza province, after about 1,200 residents in the area seize the company's oil installation. After several hours, the residents are forced to leave the company's compound by the Ecuadorian Army. The residents seized the installation in an effort to disburse around $4 million to finish construction of a road linking the cities of Puyo and Banos in Pastaza. The remote eastern jungle region contains about 200 million barrels of oil reserves.

February 3

A senior Iraqi oil minister says Iraq boosted its oil exports via its pipeline through Turkey by 107 per cent in 1998. The pipeline links northern Iraqi oil fields like Kirkuk with the Turkish terminal of Ceyhan on the Mediterranean. Under the U.N.-approved "oil-for-food" program, which allows Iraq to sell up to $5.2 billion of oil every six months, Iraq must pump the largest portion of its exports through Turkey.

February 4

Italy's ENI SpA and Russia's RAO Gazprom, the world's largest natural gas producer, agree to build a natural gas pipeline from Russia to Turkey at a cost of nearly $3 billion. Each project partner will hold a 50 per cent stake in the project. The proposed pipeline, called the Blue Stream project, is expensive by industry standards partly because it would run at great depth under the waters of the Black Sea. (Asian WSJ)

February 7

Jordan's King Hussein dies at the age of 63 from non-Hodgkins lymphoma. Hussein, who served as served as King of Jordan since May 2, 1953, had only a few days earlier appointed his eldest son, Abdullah, as his successor. Abdullah, 36, is a career army officer. Although containing almost no oil or gas resources of its own, Jordan is important to world oil markets because of its strategic location bordering Iraq, Saudi Arabia, Israel and Syria. Jordan also is an important trading partner of Iraq.

February 10

U.S. Energy Secretary Bill Richardson visits Saudi Arabia to discuss potential U.S. investment in the Kingdom's oil and gas sectors. Following his visit, Richardson says the Saudis are primarily interested in foreign investment in the natural gas sector and in the oil refining and marketing sectors, rather than in the upstream crude oil sector. Secretary Richardson's visit comes several months after a September 1998 meeting between several U.S. oil companies, Saudi Crown Prince Abdullah and Saudi Oil Minister Ali Naimi, in which Abdullah requested proposals from the companies on the development of Saudi oil reserves.,

February 11

Bangladesh and the United States sign three agreements aimed at helping Bangladesh reorganize its energy sector to avoid power shortages and to improve the extraction and sale of natural gas. The two sides will work together to assess Bangladesh's energy resources and to prepare a comprehensive energy policy. The United States and Bangladesh also will examine U.S. foreign investment options in Bangladesh's energy sector.

February 12

The U.S. government announces plans to purchase 28 million barrels of crude oil for placement in the federal Strategic Petroleum Reserve (SPR). The government announcement is aimed at helping small U.S. oil companies, which have been hurt due to the low oil price environment since late 1997.

February 15

Turkmenistan awards a $2.4 billion natural gas pipeline project to an American consortium including Amoco Corporation, and a new pipeline joint venture called PSG International, which is owned by affiliates of Bechtel Enterprises Inc. and General Electric Company's GE Capital Corporation. The consortium will construct a natural gas pipeline from fields in eastern Turkmenistan across the Caspian Sea through Azerbaijan and Georgia to the Erzurum in eastern Turkey. Issues the pipeline consortium will face include Turkmenistan's territorial dispute with Azerbaijan, and a determination by littoral states (those states touching the Caspian's shore) namely–Turkmenistan, Azerbaijan, Kazakhstan, Russia, and Iran–on whether the Caspian Sea is a sea or a lake. The outcome of these issues will impact conflicting territorial claims, including the ownership of oil fields and fishing rights.

February 18

Iraq announces that its section of a joint oil pipeline with Syria is almost ready. The pipeline links Iraqi oil fields located near the northern city of Kirkuk to Syria's Mediterrranean terminal at Banias. A spur off of the main pipeline leads to the Lebanese port at Tripoli. In August 1998, Iraq and Syria signed an agreement to reopen the pipeline, which has been closed for more than 16 years. The pipeline was originally designed to handle 1.4 million barrels per day, but Iraqi oil officials believe that a lack of storage capacity and pumps will enable Iraq to reach an initial export rate of only around 400,000 barrels per day. Under existing sanctions, Iraq needs to obtain special permission from the United Nations to export oil through Syria.

February 22

Kuwait Petroleum Corporation confirms a new find of sweet (*i.e.* low sulfur) oil in the Kra al-Marow oilfield, located northwest of Kuwait City. According to KPC, preliminary results from the new field indicate a test production rate of 1,800 barrels per day. Kra al-Marow, according to KPC, could be as large as Burgan, Kuwait's largest oil field.

February 22

Colombia's Cano Limon oil pipeline returns to normal operations after being shut down in a terrorist attack the previous week. The Cano Limon, with capacity to pump 175,000 barrels of oil per day, is a frequent target of leftist rebels who oppose foreign investment in the country's mineral resources. This marks the fourteenth such attack in 1999.

February 28

Turkish officials report that a U.S. bombing attack on Iraq damages communications at an oil-pumping facility in northern Iraq. Crude oil flows are temporarily disrupted on the line, which runs from Kirkuk in Iraq to Ceyhan in Turkey.

March 1999

March 4

Alliance Pipeline L.P., a consortium of Canadian and U.S. pipeline companies, starts construction of a high-pressure natural gas pipeline system that will stretch from northeastern British Columbia and northwestern Alberta across Saskatchewan to the Chicago area. The pipeline will be approximately 1,445 miles long, 42 inches in diameter, and have a natural gas capacity of 1.3 billion cubic feet per day. The pipeline is expected to begin service in the second half of 2000. The consortium includes affiliates of Coastal Corporation, Duke Energy, Enbridge, Fort Chicago Energy Partners L.P., Williams Companies, Unocal, and Westcoast Energy.

March 6

Sheik Isa bin Salman Al Khalifa, Emir of Bahrain since 1961, passes away unexpectedly, and his son, Sheik Hamad bin Isa Al Khalifa, is named as successor. Upon taking office, Sheikh Hamad pledges to follow his late father's pro-Western policies. Although Bahrain has minimal oil and natural gas reserves, it is strategically located in the heart of the Persian Gulf.

March 6

U.S. Secretary of Defense William Cohen and Saudi Arabia's King Fahd meet and reiterate their countries' commitments to maintaining friendly relations. Saudi Arabia, with one-quarter of the world's proven oil reserves, is an important source of crude oil imports for the United States, supplying 1.4 million barrels per day, or nearly 16 per cent, of U.S. crude oil imports in the first 10 months of 1998. Secretary Cohen also holds talks with Saudi officials regarding the continuing U.S.-Iraqi air battle in the no-fly zone over Iraq.

March 10

Reliant Energy, a large international energy services company based in Houston, agrees to spend $2.4 billion over the next seven years to acquire approximately 60 per cent of NV Energieproductiebedrijf UNA, one of the biggest power companies in the Netherlands. The transaction still must receive Dutch regulatory approval. The Netherlands, along with the rest of the European Community, is scheduled to begin opening up its electricity market to competition in 2007. The decision marks Reliant's first entry into the European electricity market.

March 15

Consortium members of Nigeria LNG Limited (NLNG), including Nigeria National Petroleum Corporation, Shell Gas BV, Elf Cleag Limited, and Agip International BV, confirm that a third liquefied natural gas (LNG) production train with a capacity of 2.9 million metric tons per year will be built, increasing NLNG's overall LNG processing capacity to 4.3 million metric tons per year. The third train is scheduled for completion by 2003, and more than 70 per cent of the LNG already has been sold under a long-term (21-year) contract to Spanish gas supply company Enagas. NLNG's first two LNG trains are expected to come on line in the summer of 1999.

March 16

El Paso Energy agrees to buy rival Sonat, for $3.9 billion in stock and $1.9 billion in assumed debt, creating the largest interstate natural gas pipeline company in the United States. The new company will be responsible for almost 25 per cent of all natural gas transportation in the United States.

March 17

DaimlerChrysler AG, the world's fifth largest automaker, and Ford Motor Company, the second largest, unveil plans to introduce hydrogen-fuel-cell cars to the mass market by 2004. Fuel-cell cars are run on electricity as a result of an electrochemical reaction between hydrogen and oxygen.

March 20

After three years of negotiations, Enron Corporation, one of the world's largest energy companies, and the Qatari government cancel plans to build a $4 billion liquefied natural gas (LNG) facility. Enron originally intended to ship about two million metric tons per year of Qatari LNG to India to fuel the Dabhol power plant and to sell the facility's remaining annual production to India and the Middle East. According to an Enron spokesman, the company has lined up other fuel suppliers for the Dabhol power plant, making the Qatar project less critical. Also, other suppliers have begun selling LNG in the Middle East and to India, making it more difficult for Enron to sell excess gas from the Qatar plant.

March 23

In an effort to raise oil prices, which fell sharply in late 1997 and stayed low through 1998 and into early 1999, OPEC and non-OPEC countries agree to cut oil output by a combined 2.104 million barrels per day, effective April 1, 1999, for one year. OPEC members have pledged to cut 1.716 million barrels per day, while several non-OPEC countries have pledged total reductions of 388,000 barrels per day. During 1998, due mainly to low oil prices, OPEC crude oil export revenues fell 30 per cent (to $100 billion) from the previous year.

March 24

Scientists at the Atlantic Richfield Company (Arco) announce the development of a new transportation fuel that emits less pollution than regular diesel fuel. In preliminary testing, the new fuel, called EC Diesel, emitted 15 per cent less particulates and 5 per cent less nitrogen oxide without reducing fuel economy, according to Arco. Further testing of the new fuel is to take place in California during the month of April. Arco began testing the new fuel in anticipation of additional regulatory policies in California requiring further reduction in fuel emissions.,

March 31

Arco agrees to be acquired by BP Amoco PLC for $26.6 billion in stock. If approved, the merger will create the largest oil producer in the United States and one of the largest energy companies in the world. The deal marks the fourth largest oil company merger since the onset of low oil prices in late 1997.

April 1999

April 1

Chile begins rationing its electricity across most of the country due to a severe drought in 1998 that nearly dried up reservoirs at several hydroelectric dams. Santiago, the Chilean capital, key copper mining zones in the northern part of the country, and parts of the central region of the country,have so far been spared cuts in electricity.

April 3

The Iraqi port of Mina al-Bakr, export terminal for roughly half of all Iraqi oil exports under the United Nations oil-for-food program, is reopened, following allied air strikes on Basra, in southern Iraq, the previous day. The attacks struck a communications facility and a radio relay station, which is used, in part, as one of three relay stations controlling the flow of crude oil from Iraq's southern oil fields to the port of Mina al-Bakr.

April 5

Following the arrival in the Netherlands of two Libyan suspects in the 1988 bombing of Pan American Flight 103 that killed 270 people, United Nations sanctions against Libya are suspended. The sanctions, imposed on March 31, 1992, initially included a ban on the sale of equipment for refining and transporting oil, but excluded oil production equipment. Sanctions were then expanded on November 11, 1993, to include a freeze on Libya's overseas assets, excluding revenue from oil, natural gas, or agricultural products.

April 9

Saga Petroleum SA, a Norwegian company, announces that its oil interests in Angola are up for sale. A Saga spokesman said that the company will concentrate on developing low-cost oil reserves in Libya and possibly Iran, as opposed to more expensive deepwater acreage off Angola.

April 12

The U.S. Deputy Ambassador to the United Nations, Peter Burleigh, becomes the first U.S. official to publicly reject a proposal that would allow direct foreign investment in Iraq's oil sector. The proposal was an attempt to allow Iraq to generate oil revenues closer to the target of $5.26

billion every 180 days under the United Nations oil-for-food program. So far, Iraq has not come close to the target ceiling since it was raised last year. However, Burleigh says that, if recent oil price increases of $5 to $6 per barrel are sustained, Iraq could "get to the $5.2 billion level for 6 months" by the end of the year. This would eliminate the need to enact "deep structural" changes in the oil-for-food program, according to Burleigh.

April 15

A consortium of British Gas and Royal Dutch/Shell bought a controlling 53 per cent stake in Cia. De Gas de Sao Paulo (Comgas), Brazil's largest natural gas distributor, for $1 billion, more than double the Brazilian government's minimum asking price. The British Gas–Royal Dutch/Shell consortium beat out a rival bid from a consortium led by Enron Corporation and Italy's Agip. The privatization of Comgas is the first state asset sale since Brazil devalued its currency in January.

April 15

The U.S. Department of Energy (DOE) announces that it will begin taking oil deliveries within the next few days under its plan to add 28 million barrels of oil to the U.S. Government's Strategic Petroleum Reserve (SPR) from federal oil royalty payments. In Phase 1 of the plan, the SPR is expected to acquire about 43,000 barrels per day over the next 3 months from oil companies operating in the Gulf of Mexico. Although about 50 per cent of the oil supplied in Phase 1 will be imported, domestic producers would still benefit from the entire acquisition since the oil market is international and fungible, according to a DOE official. Under Phase 2 of the program, the DOE expects to acquire about 100,000 barrels per day of royalty oil over a 6-month period.

April 17

An oil pipeline that transports oil from Baku, Azerbaijan, to Suspa, Georgia, is officially opened. This is the second pipeline dedicated to exporting Caspian Sea oil, but the first built since the Soviet Union disbanded in 1991. The other Caspian Sea oil pipeline, which runs through the Russian breakaway republic of Chechnya to the Russian port of Novorossisk, is often shut down. The new pipeline to Georgia has a capacity of 100,000 barrels per day.

April 20

The European Union follows the United Nations in suspending sanctions against Libya that prevented the supply of certain oil export and aviation-related goods and services.

April 23

The United Nations temporarily waives the requirement under the Iraq oil-for-food program that more than half of Iraqi oil exports must go through the pipeline to Turkey. As Iraq's oil production has increased, capacity through the Iraq/Turkey pipeline has not kept pace. Designed to handle up to 1.1 million barrels per day, the pipeline's capacity is now only about 900,000 barrels per day due to needed repair work. It is currently unclear as to when repair work on the second Iraqi oil pipeline into Turkey might allow it to become operational. So far, only about 45 per cent of Iraqi oil exports are going through the pipeline into Turkey.

April 24

The chief executive officer of Total, a major French oil company, meets with the Crown Prince of Saudi Arabia to discuss the possibility of future oil investments in Saudi Arabia. This follows separate meetings between either the Crown Prince or the Saudi Arabian oil minister with the chief executives of Exxon, Mobil, Marathon, and Conoco earlier in the month. All of these meetings are a result of an invitation last September by the Crown Prince for oil executives to submit proposals on areas of investment in the Saudi oil and natural gas sectors. However, Saudi Arabia has stated that it is mainly interested in downstream oil and gas joint ventures, ruling out upstream oil development at this time, since Saudi Arabia feels it is self-sufficient in this aspect of the industry.

April 24

The Kuwaiti oil minster announces that within the next 2 months, Kuwait is expecting to invite foreign oil companies to submit proposals for developing its northern oil fields. This would mark the first opening of the upstream sector in Kuwait since its oil industry was nationalized in the 1960s. However, the Kuwaiti oil minister made it clear that any foreign participation would be carried out in accordance with Kuwait's constitution, which bars foreign ownership of Kuwaiti natural resources. Total reserves of the fields being opened for development amount to 16 billion barrels. Kuwait hopes to increase production capacity from these fields to 1.3 million barrels per day from the current level of 590,000 barrels per day.

April 26

The European Union passes a ban of fuel shipments from its member countries to Yugoslavia as a means of putting economic pressure on Yugoslav President Slobodan Milosevic. The ban on shipments of crude oil and refined petroleum products is to begin on May 1. It does not include a blockade on shipments from non-European Union members, an issue that is under study by the North Atlantic Treaty Organization.

April 27

Azerbaijan's state-owned oil company Socar, signs 3 oil and natural gas production sharing agreements totaling an estimated $10 billion with Exxon, Mobil, and privately-held Moncrief Oil International. Exxon's agreement with Socar to develop the offshore Zafar/Mashal exploration block will amount to a total investment of $5 billion. Mobil's agreement to develop an offshore exploration block containing the Savalan, Dalga, Lerik-Deniz, and Janub prospects is worth about $4.5 billion. The agreement between Socar and Texas-based Moncrief is for an onshore block in the Lower Kura River Valley and accounts for the remainder of the investment.

April 28

The U.S. Department of Treasury's Office of Foreign Asset Control (OFAC), notifies Mobil that it has turned down Mobil's request for a license to swap crude oil it produces in Turkmenistan in exchange for Iranian oil. Mobil had hoped to be allowed to ship oil produced in Turkmenistan to northern Iranian oil refineries, while Iran, in turn, would provide Iranian oil from Iran's Persian Gulf export terminals to Mobil for shipment to global markets as payment. OFAC is responsible for enforcing U.S. unilateral sanctions against foreign countries. As a result of OFAC's denial of a swap arrangement with Iran, Mobil will have to continue exporting its Turkmenistan oil production across the Caspian Sea by barge to Azerbaijan, where it is then carried by rail or pipeline to Black Sea ports.

April 29

Spain's Repsol announces a bid to acquire the 85.01 per cent of Argentina's YPF that it does not already own. If the deal goes through, it would create the world's eighth largest energy group, with annual revenues of more than $26 billion. Repsol is offering a 25.4 per cent premium over YPF's closing price, which would amount to about $13.44 billion if all of YPF's shareholders take the offer. Repsol said the deal would only go through if it could obtain at least a controlling share (more than 50 per cent) of the company. In January, Repsol paid the Argentine government $2.01 billion for a 14.99 per cent share of YPF. Repsol is prepared to pay in cash, a requirement under YPF's company statutes for any buyer of 15 per cent of more of YPF.

April 30

Under new rules instituted by Ecuador's government, state-run Petroecuador will be allowed to have private partners, either foreign or domestic, in order to increase oil production. The rules state that the partnership contracts will have 20-year terms and be awarded to the companies or consortiums that offer the most participation for Petroecuador. As a result of the new rules, Ecuador is expecting investments of between $1 billion and $2 billion over the next four years.

May 1999

May 1

A European Union (EU) ban on oil shipments to Yugoslavia takes effect in an effort to deny Yugoslav forces access to fuel. The ban calls on EU member nations to cease shipping oil to Yugoslavia. In addition, U.S. President Clinton signs an executive order imposing a U.S. embargo on trade with Serbia, that exempts the Yugoslav republic of Montenegro. A broader blockade of Yugoslavia by NATO is still being discussed.

May 1

U.S. President Clinton unveils a plan to apply the same standard for tailpipe emissions to cars, light-duty trucks, and most sport utility vehicles (SUVs). Based on current nitrogen oxides (NO_x) emission levels, the proposed plan would result in a 77 per cent reduction for cars and a 95 per cent reduction for light-duty trucks and SUVs. The new standards would be phased in from the 2004 to 2007 model years. At the same time, the Environmental Protection Agency (EPA) proposes a rule that would require refiners to reduce gasoline sulfur content from a current average of nearly 330 parts per million (ppm) to 30 ppm. The new sulfur standard is being proposed in conjunction with the new tailpipe emission proposal since sulfur impedes catalytic converter efficiency, thus making it more difficult to reduce tailpipe emissions without reducing sulfur content in gasoline. Oil industry representatives have vowed to protest the proposed rule, claiming that it will cost refiners $3 billion to $6 billion. The EPA estimates that the cost of compliance for both the automobile and oil industries will be between $3.4 billion and $4.4 billion.

May 4

Japan signs a contract to lend $1 billion to the Korean Peninsula Energy Development Organization (KEDO), which is building two light-water nuclear reactors in North Korea as part of a 1994 agreement between the United States and North Korea. The money was delayed for several months as Japan was considering whether to continue funding some of the project after North Korea fired a missile through Japanese air space in 1998.

May 5

General Abdulsalami Abubakar, Nigeria's current leader, signs a new constitution designed to ensure a smooth transition from military rule to civilian rule when President-elect Olusegun Obasanjo assumes office on May 29. Dow Jones reported that General Abubakar hinted that the constitution would encourage decentralization of the federal government. Nigeria has been governed by the military for all but 10 of the nearly 40 years since it became independent.

May 6

South Korea's Ministry of Commerce, Industry, and Energy reports that the country's energy consumption in the first quarter of 1999 rose by 11.4 per cent compared to the same period in 1998. Oil consumption rose 11.5 per cent over the same period, averaging 2.17 million barrels per day. The Ministry reported that the increase in oil consumption was a result of improving consumer confidence in the South Korean economy.

May 9

Elf Aquitaine, a major French oil company, submits a proposal to the Saudi Arabian government to set up a new natural gas and electricity project worth over $1 billion. Elf is proposing to develop a natural gas field in central Saudi Arabia, and then to use the gas to fuel a new 1,500-megawatt electric power plant to be located near the gas field. This is the first firm proposal submitted to the Saudi Arabian government since Saudi Crown Price Abdullah's recent invitation to major energy companies to invest in Saudi Arabia's oil and gas sectors.

May 10

The Board of Argentine oil company YPF unanimously approved a $13.4 billion offer from Repsol, a Spanish company. Repsol, which already owns 14.99 per cent of YPF, made an all cash offer to purchase the remaining 85.01 per cent last month. The Board recommended to all shareholders to accept the Repsol offer. Two Argentine provinces, which own about five per cent of YPF's shares, remain concerned about Repsol's intentions for their regions.

May 10

Norsk Hydro, one of Norway's state-owned oil companies, offers to buy out privately-owned Norwegian oil and gas exploration company Saga Petroleum for more than $2.2 billion. Saga's board of directors will consider the offer before making a recommendation to their shareholders.

May 11

The Independent Petroleum Association of America (IPAA) votes to be "an interested party" to a petition before the U.S. International Trade Commission to investigate "dumping" of crude oil into the United States by Mexico, Venezuela, Iraq, and Kuwait. By being an "interested party", the IPAA will be able to file briefs in the case, without being the initiator of the petition, according to IPAA vice chairman Jerry Jordan. The petition will be filed on May 28 by a group called "Save Domestic Oil", which is led by Oklahoma Basic Energy Corporation and Continental Resources, another Oklahoma-based company.

May 12

The Caspian Pipeline Consortium (CPC) begins construction of a 981-mile pipeline that will carry crude oil from the Caspian Sea to the Russian port of Novorossisk for export to foreign markets. The pipeline's planned capacity is about 1.3 million barrels per day, and the CPC is expecting to load the first tanker in mid-2001.

May 13

Saudi Arabia notifies U.S. crude oil customers that state oil company Saudi Aramco will cut term-contracted volumes of crude delivered to them in June by 20 per cent to 25 per cent. Saudi Aramco had earlier notified European crude oil customers of cuts in volume averaging 35 per cent, with cuts for Asian customers averaging 17 per cent.

May 13

The U.S. special envoy for Caspian Sea energy issues, Ambassador Richard Morningstar, delivers a U.S. proposal to President Niyazov of Turkmenistan on the sharing of Caspian energy resources between Turkmenistan and Azerbaijan. The proposal was developed by a panel of U.S. advisors to the Turkmen-Azerbaijani commission on determining the sea boundary delineation between the two states. The same proposal was delivered to the government of Azerbaijan on May 12. Contents of the U.S. proposal are not made public.

May 17

The Environmental Protection Agency (EPA) states that it will not change its "Tier Two Plan" to cut gasoline sulfur content and tailpipe emissions, in response to a recent appellate court ruling that the EPA had overstepped its mandate in implementing some provisions of the Clean Air Act. Beginning in 2004, the Tier Two Plan would require refiners to cut gasoline sulfur content to an average of 30 parts per million, down more than 90 per cent from the current national average.

May 18

Acting Russian Fuel and Energy Minister Sergei Generalov states that Russia plans to maintain cuts in oil exports made in the first quarter of 1999 in support of efforts by the Organization of Petroleum Exporting Countries to raise oil prices. Russia had cut exports by 100,000 barrels per day.

May 18

Ecuador's Energy Minister, Rene Ortiz, announces plans to privatize all of Ecuador's downstream operations, including the sale of four refineries. State oil company Petroecuador currently operates the refineries.

May 20

China announces it is speeding up plans for the construction of a crude oil pipeline linking oil deposits in Kazakhstan to refineries in north and central China. The pipeline is now planned for completion in 2007. China National Petroleum Corporation has a stake in Kazakhstan's Uzen oilfield near the Caspian Sea.

May 21

The United Nations Security Council, in a unanimous vote, extends the "oil-for-food" program, under which Iraqi oil sales finance imports of food and medicine, for another six-month period. The resolution renews the program under its current terms, but promises a review of the program's $5.26 billion cap on exports if Iraq reaches the ceiling.

May 21

Venezuelan President Hugo Chavez states that a sharp increase in oil prices is not desirable. "If the price of a Venezuelan oil basket hovers around $13 to $14 per barrel, I will be happy," Chavez says.

May 24

Iraqi oil officials state that, despite a lack of spare parts, Iraq is capable of boosting output from its northern oil fields to 1.2 million barrels per day during the proposed sixth phase of the "oil-for-food" deal with the United Nations, from around 0.8 million barrels per day in the fifth phase.

May 25

Russian Prime Minister Sergei Stepashin signs a government decree doubling the tariff on oil exports to 5 euros (EUR) per metric ton. Earlier this year, the Russian government set out a schedule of tariffs based on export prices for Russian oil. The EUR 5 per ton tariff takes effect when the price of Russian oil rises above $12 a barrel.

May 27

Exxon and Mobil shareholders approve an $81.2 billion merger, in which Exxon will issue 1.32 shares for each share of Mobil's approximately 780.2 million shares outstanding. The merger still must receive regulatory approval from the U.S. government and the European Union. The chairmen of both companies state that they expect regulatory approvals to be obtained by the end of the third quarter of 1999.

June 1999

June 1

Sudan starts pumping oil through its pipeline linking the Heglig oil field in Western Kordofan province to Port Sudan on the Red Sea. The pipeline has a capacity of 250,000 barrels per day, and was financed by a consortium of Chinese, Malaysian, Canadian, and Sudanese firms.

June 2

Texaco terminates negotiations on the potential acquisition of the company by Chevron. Texaco states that Chevron's final proposal is "unacceptable" for reasons of complexity, feasibility, risk, and price. The talks had taken place as part of an overall trend toward consolidation among the major oil companies.

June 6

China National Petroleum Corporation (CNPC) announces plans to lay off 75,000 workers at five of its major refineries in northern China, reducing employment by 30 per cent overall and 50 per cent among managerial and administrative staff. The layoffs are part of CNPC's plan to make the refineries profitable by the end of 1999.

June 7

Indonesia holds its first-ever free elections. While there were minor irregularities, foreign election observers considered the elections generally free and fair. Forty-eight parties fielded candidates for the Parliament. Early returns show that no party will have a majority, so a coalition government will have to be formed.

June 9

A team of United Nations experts visits Iraq for talks with Iraqi Oil Ministry officials on purchases of equipment and supplies to repair Iraq's oil infrastructure. Iraq is allowed to spend up to $300 million on oil infrastructure repairs under the "oil-for-food" program's current six- month cycle.

June 10

BP Amoco announces its intention to sell several of its Canadian oilfields, accounting for about 30 per cent of the company's overall production in Canada. The sales are to be completed in the third and fourth quarters of 1999, and resulted from a strategic review by the company of its properties worldwide to prioritize the demand for investment capital.

June 12

Security forces in Chechnya, a breakaway province of Russia, take control of the main oil pipeline from Baku, Azerbaijan, to the Russian Black Sea port of Novorossiysk. The action follows attacks on the pipeline by renegade Chechen fighters.

June 14

The main Venezuelan oil workers union, Fedepetrol, states that it will not strike in the immediate future, as talks continue with Petroleos de Venezuela on its demands for wage increases. Venezuelan President Hugo Chavez had intervened in the negotiations, offering merit- based raises of between 3 per cent and 6 per cent to workers.

June 16

Brazil begins a two-day auction of 27 oil exploration blocks. The auction effectively ends the monopoly of the state-owned oil firm Petrobras over exploration and production in Brazil. Most of the blocks offered are deepwater offshore blocks, for which Brazil needs to raise foreign capital to speed development.

June 18

Texaco shuts down six oil platforms off the coast of Nigeria, cutting off a total of 50,000 barrels per day of production. Two of the platforms had been boarded by protesters in boats. Shell also is experiencing disruptions in Nigeria as a result of widespread civil unrest in the Niger Delta region.

June 24

Spain's Repsol announces the success of its buyout offer for YPF of Argentina. Repsol paid $13.5 billion for the outstanding 85 per cent of YPF shares.

June 24

Citing a lack of funds, the Jordanian government announces that it has shelved plans for construction of an oil pipeline from Iraq to Jordan. Jordan currently relies on a fleet of trucks to transport approximately 90,000 barrels per day of crude oil and refined petroleum products from Iraq.

June 25

The Yugoslav state-run petroleum company states that it can't provide enough petroleum products to satisfy demand due to unrepaired damage from NATO air strikes. Direct losses from the air campaign are estimated at $1.1 billion, mostly damage to refinery and storage facilities at Novi Sad and Pancevo.

June 26

Petroleum Development Oman, Limited announces the discovery of new oil deposits in commercial quantities in southern Oman. The new find is described as the most significant find in Oman in the last five years.

June 28

Russian oil firm Transneft announces that it will begin moving crude oil from Baku in Azerbaijan to the Russian Black Sea port of Novorossiysk by rail, temporarily bypassing the troubled pipeline through the breakaway province of Chechnya.

June 29

Brazil's Petrobras and Venezuela's Petroleos de Venezuela sign a letter of intent on forging a strategic alliance between the two firms. The agreement's stated aim is to be "the starting point to establish the economic viability of a joint company, which will be called Petroamerica."

June 30

The Independent Petroleum Association of America issued a statement declaring its intention to join as an "interested party" in the anti-dumping complaint filed by Save Domestic Oil, Incorporated. The complaint accuses Saudi Arabia, Iraq, Mexico, and Venezuela of dumping cheap oil on the U.S. market from the second quarter of 1998 through the first quarter of 1999.

July 1999

July 3

The Gulf Cooperation Council decides, at a meeting of foreign ministers, to establish a joint panel to represent the United Arab Emirates (UAE) in its territorial dispute with Iran over the islands of Abu Musa and the Tunbs. The move comes in response to tensions between Saudi Arabia and the UAE over Saudi Arabia's improving ties with Iran.

July 5

The Franco-Belgian oil company Total Fina makes a hostile takeover bid for the French oil firm Elf Aquitaine. Elf Aquitaine shareholders are offered a 4-for-3 swap of Elf Aquitaine shares for Total Fina shares, for a total value of $43 billion. While Elf Aquitaine was privatized in 1994, the deal would still require the approval of the French government, which retains a "golden share" in Elf Aquitaine permitting it to block a takeover.

July 6

Ehud Barak is sworn in as Prime Minister of Israel after a vote in the Knesset approving his government. Barak has promised to accelerate peace negotiations with Syria, Lebanon, and the Palestinians, which are considered a key factor in promoting stability in the Middle East.

July 7

The United Kingdom restores full diplomatic relations with Libya. Ties were severed between the two governments in 1984, after the killing of a British policewoman by a shot fired from the Libyan embassy. Meanwhile, the United States again restates its opposition to the immediate permanent lifting of United Nations sanctions against Libya. The air travel and arms embargoes were suspended on April 5, 1999, after Libya handed over two suspects in the 1989 Lockerbie bombing for trial in the Netherlands.

July 7

Phillips Petroleum announces the discovery of oil and gas in its Bohai Bay block 11/05 off the coast of southern China. The company intends to drill five more test wells in the near future, in addition to the two initial wells, which are capable of producing 3,000 to 5,000 barrels per day of crude oil. Phillips expects the drilling costs to be low, as water and reservoir depths are shallow. Analysts have estimated the reserves at 150 million barrels, but see it as a more significant development, as it May suggest that other, larger oil-bearing structures exist under the Bohai Bay. The China National Offshore Oil Company holds an option to buy a 51 per cent interest in any development of the field.

July 7

The Nigerian government announces the cancellation of 16 crude oil prospecting licenses, following an investigation by a panel probing awards of contracts by the previous military government. The canceled licenses were held by Nigerian firms, although some of the firms had foreign technical partners. Earlier in July, the Nigerian National Petroleum Corporation had revoked contracts previously awarded to foreign firms for crude oil lifting and refinery maintenance. The cancellations are effective in September 1999, at which time the government will conduct a review of the contracts.

July 9

Shell declares force majeure on all outstanding loadings of Bonny Light and Forcados crude oil from its facilities in Nigeria. The statement cites disruptions due to civil unrest by local residents demanding that the Niger Delta region receive a greater share of Nigeria's oil proceeds.

July 9

The United Nations Security Council issues a statement on Libya's handover of the two suspects wanted in the 1989 bombing of a Pan Am airliner over Lockerbie, Scotland, welcoming "significant progress" in Libyan compliance with earlier resolutions, but calling on Libya to "implement further" the remaining provisions of the resolutions dealing with compensation for the victims' families and cooperation with court proceedings in The Hague. The statement is issued after a resolution proposed by Namibia to lift the air travel and arms embargoes permanently was dropped under the threat of a United States veto. United Nations sanctions were suspended shortly after Libya handed over the two suspects.

July 11

Iran nominates its member of OPEC's Board of Governors, Hossein Ardebili, for the position of OPEC Secretary General. Ardebili ran unsuccessfully for Secretary General in 1994. Iran also states its support for the holding of an OPEC summit meeting in Caracas in 2000.

July 12

The United Nations Compensation Commission, charged with assessing Iraq's liability for damages resulting from the August 1990 invasion of Kuwait, issues a judgement of liability for $2.8 billion dollars to oil companies and oilfield service providers whose operations were disrupted. The largest share, $2.2 billion, is to go to the Kuwaiti Oil Company. American firms which are owed payments include Saudi Arabian Texaco, a subsidiary of Texaco, National-Oilwell, OGE Drilling Company, and the Halliburton Company.

July 12

BP Amoco reports a major natural gas find at Shah Deniz in Caspian Sea waters off Azerbaijan. The find could complicate plans to export natural gas from Turkmenistan to Turkey and Europe, if sufficient reserves are available without the expense of building an undersea pipeline across the Caspian.

July 12

President Mohammad Khatami of Iran calls for an end to student demonstrations in Tehran, which have entered their sixth day amid increasing violence. Reformist elements in Iran fear that the demonstrations will be used by hard liners to derail reform.

July 13

Schlumberger announces that it has agreed to spin off its offshore drilling operation, Sedco Forex, and merge it with Transocean Offshore, in a deal which will create the world's largest offshore drilling company. The merged company will be called Transocean Sedco Forex. Both companies said the spinoff was in response to the increasing costs and technological demands of offshore drilling, which is taking place in deeper waters than in previous years.

July 15

BP Amoco and Mobil announce the discovery of new crude oil reserves of approximately one billion barrels at the Crazy Horse deepwater field south of Louisiana in the Gulf of Mexico. BP Amoco is the operator, and holds 75 per cent equity, with the remaining 25 per cent held by Mobil.

July 15

Over 300 independent operators of Shell gas stations in Texas file suit against Shell Oil Company, accusing the firm along with refining companies Motiva Enterprises and Equilon Enterprises of using their dominant position in the marketplace to prevent independent operators from making a profit. According to the independent operators, Shell engaged in price discrimination, charging independent operators more than is charged to company-owned stations. A similar, but much smaller, lawsuit in Michigan resulted in a $2.4 million judgement against Sun Oil Company in favor of two independent dealers in December 1998.

July 16

BP Amoco announces that it intends to sell some $10 billion worth of assets and spend as much as $26 billion as part of a broad three-year initiative to finish the integration of British Petroleum and Amoco. Included in the sales will be $3 billion worth of refining assets, with the Alliance Refinery in Louisiana the first to be sold.

July 18

Elf Aquitaine launches a takeover bid for Total Fina, offering 190 euros (the European Union common currency) and three shares of Elf Aquitaine for every five shares of Total Fina. The offer follows by two weeks a hostile takeover bid for Elf Aquitaine by Total Fina, which is awaiting approval by French securities regulators.

July 23

Russia raises the export duty on crude oil by 5 euros per ton. The action comes amid a rash of petroleum product shortages in various regions of Russia, due largely to the rise in world crude oil prices, which makes it more lucrative for Russian oil companies to export their crude oil rather than refine and distribute petroleum products in Russia.

July 27

An advisory panel of the Environmental Protection Agency (EPA) releases a report urging a substantial reduction in the use of methyl-tertiary-butyl-ether (MTBE) as an air-pollution fighting fuel additive, due to concerns about pollution of groundwater by MTBE. MTBE is the major oxygenate which refiners began adding to gasoline as lead-based additives were phased out in the 1970's.

July 26

Venezuelan President Hugo Chavez' "Patriotic Pole" coalition sweeps elections for the new constitutional assembly, giving coalition members 119 of the 128 seats contested. The assembly is scheduled to convene in August 1999, and will have six months in which to draft a new constitution.

July 28

Italy's ENI announces that it has reached final agreements to invest $5.5 billion in oil and gas development in Libya. The projects to be funded will include a 372-mile pipeline which will carry Libyan natural gas to consumers in Sicily.

August 1999

August 4

The Russian Fuel and Energy Ministry orders Russian oil companies to sell all of their gasoline on the domestic market for the month of August 1999, due to domestic shortages. Russian refiners are also required to supply at least 70 per cent of their diesel and fuel oil output to domestic customers during August.

August 4

Mexico and Venezuela ratify the San Jose oil agreement, which provides for sales of crude oil to 11 Central American and Caribbean countries at discounted prices. When market prices for crude oil are below $15 per barrel, recipient states pay the market price. If prices rise above $15 per barrel, the excess is rebated to the beneficiaries as credits for imports of other products from the two oil-producing states.

August 5

Canadian Natural Resources and Penn West Petroleum make agreements for the purchase of British Petroleum-Amoco (BP Amoco) oil and gas production assets in Canada. Canadian Natural Resources will acquire the Bonnyville, Wabasca, and Nipisi properties for $1.06 billion, while Penn West Petroleum will acquire the Drayton Valley and Hoosier fields for $540 million.

August 6

The New York Mercantile Exchange (Nymex) makes an offer for the purchase of up to 70 per cent of London's International Petroleum Exchange (IPE). A similar offer from a group of energy companies failed to win the required 75 per cent approval of IPE's members last month.

August 9

The United States Department of Commerce dismisses a petition filed by Save Domestic Oil, Inc. under anti-dumping statutes. The petition alleged that Saudi Arabia, Venezuela, Mexico, and Iraq had sold crude oil to the United States at artificially low prices. The decision was based on the Department of Commerce's determination that "opposition to the petitions exceeded support." Majority support is defined as petitioner representation of at least 25 per cent of the domestic industry and support from at least 50 per cent of the industry expressing an opinion. Support from a majority in the affected industry is necessary under the law for Commerce to commence a formal investigation of an anti-dumping complaint.

August 10

British Petroleum-Amoco (BP Amoco) announces that it will have laid off 14,500 employees by the end of 1999 as part of its drive to cut costs. BP Amoco expects merger-related cost savings of $4 billion by 2001.

August 11

Chevron and Royal Dutch Shell sign an agreement with Benin, Togo, Ghana, and Nigeria to develop the West Africa Gas Pipeline. The $400 million project is scheduled for completion in 2002.

August 13

Texaco announces that it has shut in 50,000 barrels per day of production in Nigeria in response to civil unrest in the Niger Delta, which has prevented the company from delivering supplies to five of its offshore platforms.

August 13

The board of directors of Norway's state-owned oil firm Statoil formally recommends to the Oil Ministry that the firm be partially privatized.

August 16

United States Secretary of Energy Bill Richardson holds talks with Turkish President Suleiman Demirel on proposed oil and gas pipelines which would cross Turkish territory and terminate at the Turkish port of Ceyhan.

August 17

A major earthquake hits northwest Turkey, with its epicenter near the state-owned Tupras refinery. The resulting fire at the refinery causes damage which is expected to keep the 252,000-barrel-per-day facility idle for approximately 4 months.

August 18

Venezuela's Supreme Court rules in favor of several foreign oil companies in a longstanding dispute. At issue was the constitutionality of contract clauses involving exemptions from local taxes, arbitration of disputes, and subcontracting exploration projects. The clauses were challenged on constitutional grounds by current oil minister Ali Rodriguez, when he was serving as a senator in 1995, following the decision of a previous government to open Venezuela's oil sector to foreign investment.

August 18

French securities regulators extend the closing date for Elf Aquitaine's hostile takeover bid for TotalFina, in response to delays as a result of litigation over the regulators' earlier approval of the Elf Aquitaine offer. Elf Aquitaine and TotalFina are both currently engaged in competing hostile takeover bids for each other.

August 19

United States Secretary of Energy Bill Richardson visits Turkmenistan for talks on a proposed pipeline to carry Turkmen gas across the Caspian Sea, Azerbaijan, and Georgia to Turkey.

August 20

Indonesia's Minister of Mines and Energy, Kuntoro Mangkusubroto, announces that Indonesia's state oil firm, Pertamina, will have two years to restructure itself into an independent company. For five years afterward, Pertamina will have exclusive rights to production in certain designated zones. The change in policy is expected to open the Indonesian oil sector to increased competition. Pertamina has been shaken by a recent audit report from Price Waterhouse that found the company lost $6.1 billion between 1996 and 1998 due to poor management practices.

August 23

Alaska Governor Tony Knowles gives a speech outlining the State of Alaska's terms for approving the proposed merger between British Petroleum-Amoco (BP Amoco) and Atlantic Richfield Company (ARCO), stressing that continued competition should be a priority. Governor Knowles' administration has been concerned that the merger will have a negative effect on the State's economy. The two companies combined account for 72 per cent of Alaska's oil revenue.

August 24

Negotiators for the governments of Turkey and Azerbaijan begin two days of meetings in Washington on financing for the proposed oil export pipeline from Baku in Azerbaijan to Ceyhan on the Mediterranean coast of Turkey. The two delegations were to meet with the Overseas Private Investment Corporation (OPIC), the Export-Import Bank (Eximbank), and the World Bank.

August 27

The Russian oil ministry announces that the export duty for crude oil will be doubled, effective October 1, 1999, to 10 Euros (EUR) per metric ton. The move follows a shortage of petroleum products on the domestic Russian market.

August 28

Oil ministers from Mexico, Saudi Arabia, and Venezuela announce in a joint statement that they will continue to adhere to the oil production cuts which were agreed to in March 1999, and which are scheduled to remain in effect through March 2000.

August 31

Venezuelan President Hugo Chavez names Hector Ciavaldini as the new president of the state-owned oil monopoly, Petroleos de Venezuela. Ciavaldini replaces Roberto Mandini, who abruptly resigned the previous day. Mandini's resignation reportedly came as a result of policy disagreements with the Chavez administration, which has reversed the previous government's policy of investing in increased oil production capacity in order to secure greater market share.

September 1999

September 1

China National Petroleum Corporation announces plans to inaugurate a holding company, which will seek to have its shares listed on the New York Stock Exchange. An initial public offering of its stock is expected to attract $10 billion.

September 7

Indonesia's minister of energy and mines Kuntoro Mangkusubroto states that Indonesia intends to abrogate the treaty with Australia on joint development of oil and gas resources in the Timor Gap, between East Timor and the north coast of Australia. The issue of revenue sharing with Australia will be left for the new East Timorese government.

September 7

Lawyers representing Save Domestic Oil, Inc. file an appeal with the Court of International Trade in New York, seeking a reversal of the Department of Commerce dismissal of their anti-dumping complaint against four foreign oil producing countries. Save Domestic Oil is taking issue with the method used by the Department of Commerce to determine whether the petition had sufficient industry support to begin an investigation.

September 10

Chevron agrees to pay $95 million to settle allegations that it systematically underpaid royalties for crude oil produced from federal government lands. The settlement stemmed from a suit brought by two industry whistleblowers under the False Claims Act against a total of 18 firms which produce crude oil from federal lands.

September 10

Creditors of the Russian oil company Sidanco agree to liquidate the company, dealing a blow to BP Amoco's efforts to keep Sidanco, in which it owns a 10 per cent stake, united. BP Amoco alleges that the procedure used to make the decision to liquidate Sidanco was influenced by manipulation of the register of creditors by Sidanco's Russian rivals.

September 14

French oil companies Total Fina and Elf Aquitaine agree to merge, after a lengthy takeover battle, in a deal which will form the world's fourth largest oil company. The deal will give Elf Aquitaine shareholders 19 shares of Total Fina for every 13 shares of Elf Aquitaine. According to Total Fina's management, the merger will result in annual cost savings for the combined firm of $1.56 billion.

September 18

Russian forces begin air and artillery attacks against suspected guerrilla bases and infrastructure targets, including the Grozny refinery complex, in the breakaway region of Chechnya. The strikes come in the wake of a series of attacks by militants operating from Chechnya on the neighboring Russian region of Dagestan. Chechnya is along the route of the Baku-Novorossiysk crude oil pipeline, which has been closed during most of 1999 due to attacks.

September 20

Texaco announces the lifting of the state of force majeure it had declared August 13th on liftings of crude oil from the company's operations in Nigeria. Production had been shut in due to civil unrest in neighboring communities.

September 21

Brazil's state oil company, Petroleo Brasileiro S.A., announces the discovery of crude oil reserves in the Santos Basin estimated at between 600 million and 700 million barrels. The Santos Basin is located offshore from Rio de Janeiro state.

September 21

Senator Charles Schumer sends a letter to President Clinton and Secretary of Energy Bill Richardson requesting them to authorize the sale of crude oil from the Strategic Petroleum Reserve in response to rising oil prices. Senator Schumer's letter coincided with the ministerial meeting of the Organization of Petroleum Exporting Countries in Vienna.

September 22

The Organization of Petroleum Exporting Countries (OPEC), at a meeting of its member states' oil ministers, decides to maintain current production cuts until March 2000, despite the fact the crude oil prices have doubled since early 1999. In another development, OPEC announces that its current

Secretary General, Nigerian Rilwanu Lukman, will stay in office until March 2000. The announcement follows a vigorously contested race to succeed Lukman in the post, in which OPEC's three largest members, Saudi Arabia, Iran, and Iraq, had fielded candidates.

September 23

British Petroleum Exploration (Alaska), a subsidiary of BP Amoco, pleads guilty to a felony charge of illegally dumping hazardous waste on Alaska's North Slope. The company agreed to a $22 million settlement to resolve the charges.

September 23

In a joint statement, the foreign ministers of the five permanent members of the United Nations Security Council pledge to continue to work toward a consensus on a new policy toward Iraq. All five permanent members have accepted that on-site weapons inspections must resume in Iraq, though there is still disagreement over how much cooperation Iraq would have to give international inspectors before economic sanctions could be lifted.

September 23

The Senate approves a proposal to block the Clinton administration from increasing royalties on oil and gas from federal lands. The Department of the Interior has proposed a rule change which would tie royalties to market prices for crude oil.

September 24

China National Petroleum Corporation (CNPC) and Royal Dutch Shell sign a contract for the joint exploration and development of natural gas resources in northwest China's Ordos Basin. The $3-billion contract includes the development of the Changbei Block, where CNPC has found natural gas reserves estimated at 2.5 trillion cubic feet. Shell, as the operator, will cover the investment required to develop the field.

September 24

Italian Prime Minister Massimo D'Alema says he will evaluate the possibility of the Italian state-owned oil company ENI SpA merging with Total Fina and Elf Aquitaine once the latter two firms' merger is completed.

September 28

Iranian Oil Minister Bijan Zanganeh announces that the National Iranian Oil Company has discovered a new oilfield, Azadegan, with 26 billion barrels of crude oil in Khuzestan province. The discovery is the largest new find in Iran in the last three decades. Zanganeh expects the field to produce between 300,000 and 400,000 barrels per day of crude oil three to four years after development begins next year.

September 30

Japan suffers a serious nuclear accident at a uranium processing plant in Tokaimura, in which radiation is released after an apparent uncontrolled nuclear chain reaction. Three workers at the plant, operated by JCO, Inc., are injured. Japanese authorities issue a warning instructing 310,000 people in neighboring communities to stay indoors.

October 1999

October 4

The United Nations Security Council agrees to raise the monetary ceiling on Iraqi oil sales to $8.3 billion from $5.26 billion, guaranteeing the continuation of Iraqi production until the November 20 end date for the current six month extension of the "oil-for-food" program. The move is a one time adjustment, and does not bind the Security Council to continue a higher ceiling if the program is renewed for another six month term. The increase reflects the difference between previous monetary ceilings and actual Iraqi sales during previous phases of the program.

October 6

Exxon announces the commencement of production at the Balder field located in the Norwegian zone of the North Sea. The field is expected to reach a peak output of 100,000 barrels per day.

October 7

Brazilian state oil company Petrobras announces that it has broken off negotiations with Exxon for the development of the Albacora Leste offshore field. Petrobras is considering asking one of the other consortium partners to take over Exxon's stake and become the operator at the field, which is estimated to hold 600 million barrels of crude oil.

October 8

A spokesman for the Norwegian Oil and Energy Ministry announces that the Norwegian government is seeking to increase the country's crude oil production in 2000 by 20 per cent, to roughly 3.5 million barrels per day. According to the spokesman, if Norway were to decide to extend its 200,000-barrel-per-day production cut undertaken in support of the Organization of Petroleum Exporting Countries' (OPEC) March 1999 agreement, it would be subtracted from the new baseline amount of 3.5 million barrels per day.

October 8

Rem Vyakhirev, the chief executive of Russia's Gazprom, the world's largest producer and exporter of natural gas, announces his intention to attempt to forge a natural gas cartel in response to low prices in the European market. Other major exporters of natural gas to European countries include Algeria, the Netherlands, and Norway.

October 10

Two major Japanese downstream oil firms, Cosmo Oil and Nippon Mitsubishi Oil, announce a strategic alliance. While the move is short of a merger, it is expected to result in cost reductions for both firms by eliminating duplicate operations. The action follows a period of low refining margins and excess refinery capacity in the Japanese oil industry.

October 10

The French government announces its intention to retain its "golden share" in the merged Total Fina-Elf Aquitaine. The "golden share" allows the French government the right to block any merger plans or takeover attempt involving the company.

October 11

The Venezuelan newspaper El Nacional reports that the final draft of the new Venezuelan constitution includes a provision prohibiting privatization of the state-owned oil firm Petroleos de

Venezuela (PdVSA). A referendum on the proposed new constitution is scheduled to take place before the end of 1999.

October 13

Pakistan's military stages a coup, led by General Parvaiz Musharraf. According to a statement from Atlantic Richfield, the largest producer of oil in Pakistan, their operations are not expected to be affected.

October 13

U.S. House and Senate conferees agree on a six month moratorium on increases in oil royalty payments from firms lifting oil from federally-owned land. The moratorium has been opposed by the Clinton administration.

October 19

BP Amoco announces its decision to take the lead in promoting the construction of an oil export pipeline from Baku on the Caspian Sea to the Turkish Mediterranean port of Ceyhan. The company's support is seen as crucial, as it holds the largest stake in the Azerbaijan International Operating Company (AIOC), a consortium of foreign oil firms which will develop Azerbaijani reserves.

October 20

Saudi Oil Minister Ali Naimi says the Saudi government is still evaluating proposals submitted by foreign oil companies for investment in Saudi Arabia. He stresses the need to consider projects which "compliment the domestic industry, not replace it," such as petrochemical plants.

October 20

The Colombian government announces new measures to preserve the country's oil production level into the future. Companies holding concessions will be allowed to return unused portions of blocks back to the government, in exchange for an increased share of production in the portion they retain. The government also intends to solicit bids from foreign firms for some blocks currently operated by state-owned Ecopetrol.

October 25

The Federal Trade Commission approves the $6 billion merger between Sonat and El Paso Energy, which will create the largest natural gas pipeline company in the United States. The merged firm will be required to divest assets in some local markets to preserve competition.

October 28

The United Nations Security Council shelves a proposal to allow Iraq to double the amount, currently $300 million, it is allowed to spend under the "oil-for-food" program on spare parts for its oil industry. The issue of funding for repairs and improvements to Iraq's upstream oil infrastructure will now be considered when the next six-month renewal of the "oil-for-food" program is considered.

October 28

China National Petroleum Corporation (CNPC) announces the formation of a holding company, for which it intends to sell shares in an initial public offering (IPO) on the New York and Hong Kong stock exchanges to be scheduled in either January or February 2000. Earlier in the month, the China

National Offshore Oil Corporation (CNOOC) had announced the postponement of a similar IPO due to adverse market conditions.

October 29

Oklahoma Secretary of Energy Carl Michael Smith sends a letter to U.S. Secretary of Energy Bill Richardson urging him to resist calls for a sale of crude oil from the Strategic Petroleum Reserve (SPR). Smith argues that a sale of SPR oil would damage the recovery of U.S. domestic oil producers in the wake of low oil prices in 1998 and early 1999.

October 31

Isramco, an Israeli oil and gas exploration company, reports a discovery of natural gas in commercial quantities from an offshore test well in the Mediterranean. Previously, exploration in Israel had not discovered oil or gas in commercial quantities.

November 1999

November 1

The Russian government sells a 9 per cent stake in Russia's largest oil company, Lukoil, to a Cyprus-based firm, Reforma Investment. The Russian government had previously held a 27 per cent stake in Lukoil.

November 1

Atlantic Richfield and Triton Energy conclude a $5 billion agreement with the state-run oil companies of Thailand and Malaysia to sell natural gas from blocks jointly held by the two firms in the Gulf of Thailand. Deliveries of 390 million cubic feet per day are set to begin in 2002.

November 4

China and Germany sign a memorandum of understanding for the construction of a $2.7 billion petrochemical complex in China. The project will be a joint venture between China's state-owned petroleum firm Sinopec and BASF of Germany. The deal was signed during a visit to China by German Prime Minister Gerhard Schroeder.

November 6

China National Petroleum Corporation (CNPC) announces that it will be undertaking a reorganization of its corporate structure. The company's exploration, production, gas, refining, and petrochemical units will be transferred to a holding company. Its technical services and overseas operations units will remain outside the holding company. CNPC officials intend to offer shares in the holding company on the New York Stock Exchange in January or February 2000.

November 10

Alberto Calderon is appointed to head the Colombian state petroleum company, Ecopetrol, after the resignation of its previous president Carlos Rodado.

November 14

Royal Dutch Shell signs an $800 million development contract with the National Iranian Oil Company (NIOC) for two offshore fields in the Iranian sector of the Persian Gulf. The contract calls for an expansion in production at the Soroush field to 150,000 barrels per day and at the Nowruz field to 90,000 barrels per day. Under the "buy back" contract, Royal Dutch Shell will not have an ownership interest in the fields, but will recoup its investment plus a fixed rate of return. The move raises the possibility of United States sanctions under the Iran-Libya Sanctions Act (ILSA), which prohibits investment of more than $40 million per year by foreign companies in Iran's oil industry.

November 15

The United States and China conclude an agreement on Chinese entry into the World Trade Organization (WTO). Under the agreement, China will make a limited opening of its internal petroleum products market to foreign competition, allowing foreign firms to open 30 retail gas stations every three years after China's accession to the WTO. China also will abolish the existing oil import quota system in phases over four years.

November 17

California governor Gray Davis announces a lawsuit by the State of California against the federal government in an attempt to block extensions of offshore oil leases which eventually could expand oil production off the central California coast. California is demanding the right to scrutinize each of the 36 blocks prior to any federal government action. The move comes four days after the Department of the Interior unveiled a plan which allows oil companies to move forward with exploration in the areas.

November 18

The heads of state of Turkey, Azerbaijan, and Georgia sign an agreement to build a pipeline for the export of crude oil from the Caspian Basin. The 1,080-mile pipeline will begin at the Azerbaijani capital, Baku, and run through Georgia and Turkey to the Turkish port of Ceyhan. The project is expected to cost $2.4 billion, and the government of Turkey has offered guarantees that the cost of the Turkish segment of the pipeline will not exceed $1.4 billion. The signing ceremony took place during a visit to Istanbul by U.S. President Clinton for a summit of the Organization for Cooperation and Security in Europe (OSCE).

November 20

Iraq announces its rejection of the United Nations Security Council's two-week extension of the "oil-for-food" program, under which Iraq is permitted to make limited oil exports. The rejection produces a cutoff in Iraqi oil exports, which Iraq said was ordered to protest the failure of the Security Council to grant a six-month extension. Many analysts see the Iraqi move as coming at a particularly inopportune time for oil consumers, as world oil stocks have been falling in recent months.

November 25

The National Iranian Oil Company (NIOC) defers loadings of approximately 18 million barrels of crude oil exports scheduled for late November and early December, resulting in delays of as much as 11 days for some buyers. The deferment comes in response to an order from Iranian Oil Minister Bijan Zanganeh for strict control of exports to ensure compliance with Iran's OPEC production quota.

November 27

Iraqi Oil Minister Amer Mohammed Rashid states that Iraq is willing to accept a six-month extension of the United Nations "oil-for-food" program. He also urges other OPEC members not to raise output in response to Iraq's temporary halt in crude oil exports.

November 30

A spokesman for the Chinese Foreign Ministry states that if the dispute among several Asian countries over the potentially hydrocarbon-rich Spratly Islands cannot be resolved, China would consider the idea of joint development of the Islands with other claimants. However, the spokesman also reiterated China's claim to the islands.

November 30

The Federal Trade Commission (FTC) grants approval for the proposed merger between oil giants Exxon and Mobil. The $80 billion merger was approved by the FTC after the firms agreed to the largest divestiture of assets ever involved in a merger. The companies will sell over 2,400 retail outlets, mostly in the Northeast, Texas, and California, and a refinery in California.

December 1999

December 2

Independent refiner Tosco agrees to buy 1,740 gasoline stations from Exxon and Mobil in the Northeastern United States. The $840 million sale was one of the measures the Federal Trade Commission required to preserve adequate competition in the wake of the Exxon-Mobil merger.

December 2

BP Amoco wins approval from the State of Alaska for its planned takeover of Atlantic Richfield (ARCO). The Federal Trade Commission is still considering the matter.

December 4

Iraq's Minister of Petroleum, Amer Rashid Mohammed, states Iraq's rejection of a one week extension of the United Nations "oil-for-food" program approved by the Security Council a few hours earlier.

December 10

The California Air Resources Board approves a regulatory change that will halve the amount of sulfur allowed in gasoline sold in California from 30 parts per million to 15 parts per million, starting in 2003. The California limit would be half the national limit under a new rule proposed by the Environmental Protection Agency. The current federal sulfur limit for gasoline is 330 parts per million.

December 10

The United Nations Security Council passes a resolution extending the "oil-for-food" program by six months, under terms identical to the previous six-month phase. Iraq had previously stated a willingness to resume oil exports if a six month extension were to be granted.

December 14

Adrian Lajous, the director of Mexico's state-owned oil company Petroleos Mexicanos (Pemex), resigns. The move comes in the wake of delays and large cost overruns in the $10.5 billion Cantarell project, which seeks to increase Mexico's oil production capacity.

December 15

Chevron, BP Amoco, and Conoco agree to pay a total of $153 million to settle allegations that they underpaid royalties for oil produced on federal lands. The issue stems from a 1996 lawsuit filed by two former Atlantic Richfield executives.

December 16

ExxonMobil announces that it intends to cut 14,000 jobs, 13 per cent of its work force, in a round of layoffs stemming from its recent merger, as part of an effort to increase cost savings to $3.8 billion per year by 2002.

December 17

Turkmenistan signs a gas sales agreement with the Russian natural gas company Gazprom. Turkmenistan agrees to accept more than fifty per cent of the value of its payments from Gazprom in Russian foodstuffs and consumer goods, rather than cash.

December 17

The United Nations Security Council passes a resolution on returning weapons inspectors to Iraq. Under the resolution, sanctions could be suspended if Iraq were to cooperate with the inspectors over a period of nine months. Iraq has stated that it does not accept the resolution.

December 21

The Export-Import Bank drops a proposed $500 million loan to Russia's Tyumen Oil after Secretary of State Madeleine Albright exercises her statutory authority to block the transaction. The loan had been controversial in part because of Tyumen Oil's dispute with BP Amoco over the bankruptcy of Russian oil firm Sidanko, in which BP Amoco owns a major stake. BP Amoco and Tyumen Oil later settled the dispute on December 23.

December 21

The government of Kazakhstan signs an agreement with four foreign firms, including Texaco, for the construction of a pipeline from Bolshoy-Chagan to Atyrau, which will link into the export pipeline being developed by the Caspian Pipeline Consortium (CPC). The initial capacity of the pipeline will be 150,000 barrels per day, beginning in 2002.

December 21

President Clinton announces the publication of final rules on sulfur content in gasoline and environmental standards for light trucks and sport utility vehicles. The new sulfur content standard of 30 parts per million will become effective in 2004. Trucks and sport utility vehicles will be under the same emissions standards as automobiles by 2009.

December 23

Amid concerns over possible consumer hoarding of petroleum products in response to fears of supply disruptions caused by the "Y2K Bug," Secretary of Energy Bill Richardson advises drivers that they do not need to fill up their gasoline tanks or otherwise hoard fuel during the rollover to the year 2000, as there is no reason to believe there will be major disruptions that would affect the availability of fuels in the United States.

December 24

South Korea's Hyundai and Samsung sign a memorandum of understanding paving the way for the merger of their petrochemical business units. The merger comes as part of South Korea's drive to ease overcapacity in several key industries, including petrochemicals.

December 26

Iraqi officials state that Iraq intends to export the same amount of oil during the recently-begun seventh phase of the "oil-for-food" program as it had during the previous phase. Iraq's exports during the sixth phase averaged 2.1 to 2.2 million barrels per day. The country had suspended exports on November 24 to protest short extensions of the previous phase of the program, and had restarted exports in mid-December.

December 28

BP Amoco announces that it is withdrawing from a project to develop natural gas deposits in Ukraine, which BP Amoco said it had determined were below the minimum scale in which the company was willing to invest. BP Amoco's action will open the exploration blocks in central and eastern Ukraine to other firms, but is seen as a major blow to Ukraine's hopes of becoming less dependent on gas imports, due to the loss of BP Amoco's advanced deep-drilling technology.

December 31

The Panama Canal Zone reverts to Panamanian sovereignty at noon, after nearly a century of American control. More than a half-million barrels of crude oil and petroleum products transit the Canal each day.

December 31

After nearly two years of construction, ExxonMobil completes the Sable Offshore Energy Project, a $2 billion project to bring natural gas from fields offshore Nova Scotia to the northeastern United States. The fields are estimated to contain 3.5 trillion cubic feet of natural gas.

December 31

Russian President Boris Yeltsin makes a surprise announcement that he is resigning immediately. Vladimir Putin becomes Acting President, and presidential elections will be held within 90 days, with a date to be set by the State Duma. Russia is the largest exporter of energy in the world.

98

Monthly Energy Chronology 2000

The following information contains a listing of the major energy events that occurred in 2000.

January 2000

January 1

Energy companies and countries around the world report that they have passed into the year 2000 without significant problems from the "Y2K Bug." There was concern that the inability of some computers and embedded control systems to recognize the year 2000 could create serious problems.

January 6

Texaco announces that it has completed testing of the Agbami-2 well off the coast of Nigeria. The company states that the new field has crude oil reserves estimated at more than one billion barrels.

January 7

Statoil shuts in 390,000 barrels per day of crude oil production in response to severe weather in the North Sea. Including earlier moves by Norsk Hydro, Statoil, and Shell, a total of 1.27 million barrels per day of North Sea crude oil production is shut in due to weather.

January 13

The National Research Council, a body of the National Academy of Sciences, releases a report confirming that the rise in global temperatures is "undoubtedly real," deflating a key argument used by skeptics of global warming. The report does not directly address the causes of the warming trend.

January 13

The Environmental Protection Agency announces a fine of $30 million against refiner Koch Industries as part of a settlement of charges that the company contaminated lakes and streams with petroleum products which leaked from pipelines.

January 13

Indonesian President Abderrahman Wahid appoints a high level panel to oversee the restructuring of the state-owned oil company Pertamina and choose a new senior management team for the firm.

January 14

Chevron announces that it has agreed to pay $95 million to settle charges that it underpaid royalties to the federal government for oil produced on federal lands. The company says the move is not an admission of wrongdoing.

January 16

Takashi Fukuya, Japan's Minister of International Trade and Industry, meets with Saudi Arabia's Petroleum Minister Ali al-Naimi in an attempt to reach an agreement on the renewal of Japanese firm Arabian Oil Company's concession for crude oil production in the Saudi sector of the Neutral Zone. The ministers reportedly failed to reach agreement on an extension.

January 19

China discovers a large natural gas deposit in the Tarim Basin in the Xinjiang region of Western China. The country is considering the construction of a gas pipeline from the region to more heavily populated areas along the Pacific coast.

January 22

Iraq reaches an agreement on the continuation of oil supplies to Jordan. Under the agreement, Iraq will give Jordan $300 million worth of crude oil in 2000 free of charge, and Jordan will pay a maximum of $19 per barrel for any additional volumes imported.

January 25

Spot prices for number two heating oil close at $1.359 per gallon, after rising rapidly from $0.687 per gallon at the beginning of January 2000. The rapid rise takes place as winter storms hit the northeastern United States, increasing short term demand for heating oil. The spot price of West Texas Intermediate crude oil closes at $30.28 per barrel.

January 26

The United Nations Security Council reaches agreement on the appointment of Hans Blix of Sweden, the former head of the International Atomic Energy Agency (IAEA), to lead the new United Nations weapons inspection organization for Iraq. Iraq has indicated that it does not intend to accept the new Security Council resolution.

January 27

Senator Charles Schumer meets with Secretary of Energy Bill Richardson to press for a sale of oil from the Strategic Petroleum Reserve (SPR) in response to high oil prices. In particular, northeastern members of Congress have been concerned by the sharp rise in prices for heating oil in late January 2000 due to cold temperatures on the East Coast of the United States.

January 27

The Federal Trade Commission (FTC) decides to postpone a scheduled decision on whether to allow the proposed merger between BP Amoco and Atlantic Richfield, after the companies involved made a new offer to divest themselves of further assets.

January 27

Azerbaijan experiences a short-term domestic petroleum products shortage, which cripples the country's oil fired power plants. The government of Azerbaijan temporarily restricts oil shipments to the Russian port of Novorossisk and orders the import of a shipment of oil from neighboring Turkmenistan.

January 28

ExxonMobil's Australian unit, Mobil Oil Australia Limited, announces a compensation package to cover equipment damage and direct business losses arising from contaminated aircraft fuels. Around 5,000 small aircraft have been grounded in Australia due to concerns that engines could clog after consuming the fuels. Mobil has said the contamination resulted from a mistake in the production process, but has not admitted liability for financial losses incurred by its clients.

February 2000

February 2

The Federal Trade Commission (FTC) acts to block the proposed merger between BP Amoco and Atlantic Richfield, saying the merger would unduly restrict competition along the West coast of the United States.

February 3

The United States Navy seizes the Russian tanker Volgoneft-147 in the Persian Gulf. The vessel is transporting a cargo of smuggled Iraqi gasoil in violation of United Nations sanctions against Iraq.

February 8

Russia's second largest oil company, Yukos Oil, announces an agreement with state oil pipeline company Transneft to build a $1.7 billion oil pipeline from Siberia to China. The pipeline would run from Angarsk in Siberia to Beijing.

February 9

The Federal Energy Regulatory Commission (FERC) issues a group of policy changes which extend the deregulation of the interstate natural gas pipeline system begun under Order 636 in 1992. Among the changes is a lifting, for a trial period of 30 months, of the price ceiling on secondary market exchanges of short-term gas pipeline capacity. FERC's lifting of the ceiling is meant in part to encourage gas shippers to use longer-term contracts which would promote market stability.

February 11

Occidental Petroleum agrees to buy Altura Energy, the onshore U.S. oil exploration joint venture between BP Amoco and Shell, for $3.5 billion. Altura produces about 110,000 barrels per day of crude oil, mostly in West Texas and New Mexico.

February 14

The price of West Texas Intermediate crude oil closes on the New York Mercantile Exchange at $30.30 per barrel, the highest price (in nominal terms) since the Gulf War in 1991.

February 16

The United States announces sanctions against the Greater Nile Petroleum Operating Company (GNPOC), which is developing oilfields in Sudan. The sanctions prohibit American firms and individuals from doing business with GNPOC, but do not cover the foreign parent companies of GNPOC, which include China National Petroleum Corporation (CNPC), Petronas, and Talisman Energy.

February 16

President Clinton announces the release of $125 million in additional federal government assistance for low income households hit by high heating oil prices. On the same day, Secretary of Energy Bill Richardson addresses the New England Heating Oil Summit in Boston. The moves come after a dramatic spike in heating oil prices in the Northeastern United States. The price for number two heating oil at New York Harbor peaked at February 4th at $1.77 per gallon.

February 18

Iranian voters go to the polls to elect members of parliament. The election results in a landslide victory for reformist candidates allied with Iranian President Mohammed Khatami.

February 21

General Electric announces a breakthrough in the design of natural gas power generating plants, using steam instead of air to cool turbine blades. According to the company, the new design will use about 5 per cent less natural gas per unit of power generated than the best existing technologies.

February 22

Several hundred truck drivers form a convoy through downtown Washington in a protest against high diesel fuel prices. Trucking lobby groups such as the American Trucking Association have supported calls for a release of oil from the Strategic Petroleum Reserve. Senator Ben Nighthorse Campbell proposes removing the federal excise tax on diesel for one year.

February 24

BP Amoco announces that it plans to make $2.5 billion in investments in the development of natural gas reserves at the In Salah field in Algeria. Gas deliveries to Europe from the field are expected to start in 2003.

February 24

Saudi Arabia announces that it intends to invite representatives of companies which have submitted bids for natural gas and petrochemical investment in the country to talks which will be held in late March 2000, shortly after the end of the Muslim haj pilgrimage season.

February 26

Secretary of Energy Bill Richardson meets with Saudi Arabian Petroleum Minister Ali Naimi in Riyadh to discuss the recent rise in crude oil prices. The meeting is part of an overseas tour which includes stops in Mexico, Norway, Egypt, Israel, Saudi Arabia, and Kuwait. In a joint statement,

Naimi pledged to "continue to review oil supply and demand levels to ensure market stability, prevent oil price volitility, and avoid harming the world economy."

February 27

BP Amoco and Atlantic Richfield hold talks with the Federal Trade Commission (FTC) on their proposed merger, which the FTC has filed suit in a federal court to block. The two companies have reportedly offered major concessions, including a much larger sale of Alaskan North Slope production assets.

February 27

Columbia Energy agrees to be acquired by NiSource, ending an eight month long takeover battle. The deal is valued at $6 billion, will make NiSource the largest natural gas company east of the Rocky Mountains.

February 28

Arabian Oil Company, the Japanese firm which produces crude oil in the Neutral Zone of Kuwait and Saudi Arabia, announces the expiration of its concession for production in the Saudi half of the Zone. The firm's concession for the Kuwaiti portion of the Neutral Zone expires in 2003.

March 2000

March 2

Valero Energy Corporation agrees to purchase ExxonMobil's 160,000-barrel-per-day refinery at Benicia, California, and 340 retail gasoline stations for $895 million. The sale is part of the package of divestitures which Exxon and Mobil agreed to in order to secure approval for their merger from the Federal Trade Commission (FTC). Valero was selected despite a higher bid from Ultramar Diamond Shamrock Corporation, in the expectation that Valero posed fewer potential problems with approval by the FTC, as it has not previously sold gasoline in California and would introduce a new competitor into that market.

March 3

A strike by workers at Petroleos de Venezuela (PdVSA) officially begins, but fails to attract much support from union members. Only about 15 per cent of PdVSA workers fail to report for work, according to the company, and operations are not disrupted.

March 6

The United States Supreme Court overturns the State of Washington's law establishing state regulation of oil tankers, ruling unanimously that federal laws take precedence. The attempt to impose tougher regulatory standards came in the wake of the 1989 Exxon Valdez disaster in Alaska.

March 7

New York Mercantile Exchange front-month West Texas Intermediate crude oil futures contract closes at $34.13 per barrel, the highest level in nine years.

March 9

South Korea's Hyundai Corporation announces that it is to begin developing Libya's Elephant oilfield, discovered in 1997. The project is expected to produce 150,000 barrels per day when it reaches full output capacity in 2003.

March 14

The Clinton administration announces changes to the system for assessing royalties for oil production on federal lands. In the future, royalty calculations will be pegged closer to current spot market prices, instead of an arbitrary value at the wellhead. The changes are expected to generate an additional $67.3 million per year in revenues.

March 15

Phillips Petroleum announces that it has agreed to purchase Atlantic Richfield's assets in Alaska for $6.5 billion. The sale is being made in an effort to secure approval from the Federal Trade Commission (FTC) for the merger of Atlantic Richfield with BP Amoco. Earlier the same day, the FTC announced that it had suspended its antitrust lawsuit seeking to block the merger, citing progress in talks with the companies involved.

March 17

Secretary of State Madeleine Albright announces an easing of some United States economic sanctions against Iran. Purchases of several non-energy items produced in Iran, such as pistachio nuts and Persian rugs, now will be permitted. Sanctions dealing with Iran's oil and natural gas industries, however, remain in place.

March 21

Secretary of Energy Bill Richardson arrives in Algiers to meet with Algeria's oil minister, Chakib Khalil, as part of a final round of talks with oil producers about the need to increase production in advance of the OPEC ministerial meeting scheduled for March 27th. The visit is the first to Algeria by a cabinet level American official in over a decade. Earlier in the day, Secretary Richardson had visited Nigeria for talks with President Olesegun Obasanjo.

March 20

EPA Administrator Carol Browner announces that the Clinton Administration intends to push for a phase out of the use of methyl tertiary butyl ether (MTBE) as a gasoline additive. The administration wants Congress to pass legislation which would end the requirement for the use of MTBE in gasoline sold in some smog-prone urban areas, and instead require nationwide use of ethanol.

March 23

Vice Admiral Charles Moore, who oversees United States naval operations in the Persian Gulf, briefs the United Nations Sanctions Committee on the increased smuggling of Iraqi oil. Iraq is expected to earn in excess of $500 million from oil smuggling, and possibly up to double that amount, in the absence of strong action by Iran to prevent the use of its territorial waters by smugglers.

March 23

BP Amoco announces that it will purchase 20 per cent of the shares to be offered by PetroChina in its initial public offering (IPO) on the New York and Hong Kong stock exchanges on April 7. PetroChina is a unit of China National Petroleum Corporation China's largest oil and gas company. The announcement by BP Amoco provides a needed boost to the IPO, which had to be scaled back in size by more than half due to lack of interest on the part of many large institutional investors. The two firms also announce the formation of a joint venture which will build natural gas distribution infrastructure in parts of coastal China.

March 23

Russia's Lukoil announces the discovery of as much as 2.2 billion barrels of new oil reserves in the Russian sector of the Caspian Sea.

March 26

Vladimir Putin is elected president of Russia on the first ballot, winning 53 per cent of the popular vote. Putin took office as acting president in December 1999 after the resignation of Boris Yeltsin.

March 27

ExxonMobil files a lawsuit in an attempt to block the sale of Atlantic Richfield's assets in Alaska to Phillips Petroleum. ExxonMobil holds stakes in Prudhoe Bay assets currently operated by Atlantic Richfield.

March 28

After two days of meetings, oil ministers of the Organization of Petroleum Exporting Countries (OPEC) agree on an increase in oil production of 1.452 million barrels per day by its members, excluding Iran and Iraq. Iraq, has not been subject to OPEC production agreements while under U.N. Security Council sanctions. Iran, though not formally signing on to the agreement, stated its intention to raise its production in order to avoid loss of its market share. This would represent about a 1.7 million barrel per day increase in OPEC production targets, if Iran was included. Several major non-OPEC producers, including Mexico and Norway, also have indicated an intention to raise production.

March 30

The State Department announces that it is lifting its hold on an Eximbank loan to Russia's Tyumen Oil Company. The loan was blocked by the State Department in December due to a dispute between Tyumen and BP Amoco over the ownership of assets in Siberia's Chernegorneft oil field, but the two firms had since reached a settlement.

March 30

The United Nations Security Council votes to allow Iraq to import $1.2 billion in spare parts and other equipment for its oil industry this year under the "oil-for-food" program. This is an increase from the previous $600 million annual value allowed.

April 2000

April 5

The government of Iran announces that it has seized a tanker which was smuggling Iraqi oil through Iranian territorial waters. A spokesman for the United States Department of State welcomes the action.

April 6

PetroChina, a holding company which serves as a stock market listing entity for the China National Petroleum Corporation, launches an initial public offering (IPO) on the New York and Hong Kong stock exchanges. The IPO is valued at nearly $3 billion, scaled back drastically in size due to a lack of investor interest.

April 7

Tosco Corporation agrees to purchase the Wood River Refinery, located in Illinois from Equilon Enterprises, a joint venture between Texaco and Shell, for $420 million. Equilon officials say the sale is part of a shift to concentrate on the West Coast petroleum products market.

April 12

Several Chief Executive Officers (CEOs) of major United States oil companies meet with senior Saudi Arabian officials to discuss possible investments in natural gas and petrochemical projects. The firms represented at the meetings include Chevron, Conoco, ExxonMobil, Marathon Oil, Phillips Petroleum, and Texaco. The Saudi government announces, in conjunction with the meetings, a package of legal changes that will make Saudi Arabia more open to foreign investors. Complete foreign ownership will be allowed for some types of projects, and the maximum corporate tax rate for foreign enterprises will be reduced to 15 per cent.

April 14

BP Amoco receives approval from the Federal Trade Commission (FTC) for its $28 billion takeover of Atlantic Richfield Corporation (ARCO). As part of the approval, ARCO has agreed to sell its crude oil production operations in Alaska to Phillips Petroleum in a deal valued at $6.5 billion.

April 18

The China National Petroleum Corporation announces an agreement with Petroleos de Venezuela (PDVSA) to begin purchases of Orimulsion, a power-plant fuel made from bitumen from Venezuela's Orinoco region.

April 18

Amerada Hess Corporation announces an agreement with the Algerian state-owned oil company Sonatrach to develop three oil fields in central Algeria. The $555 million project will develop the El Gassi, El Agreb, and Zotti fields.

April 19

The United States Commerce Department reports a record $29.2 billion trade deficit for February 2000. This is due largely to the sharp increase in prices for crude oil imports, which added $1.3 billion to the monthly trade deficit.

April 24

The American Petroleum Institute (API) files a federal lawsuit seeking to overturn the Interior Department's new rules for royalty valuation of natural gas produced from federal government lands. The Independent Petroleum Association of America (IPAA), which represents independent oil and gas companies, also has filed a similar lawsuit.

April 25

Royal Dutch Shell agrees to pay a $2 million fine for transporting smuggled Iraqi oil aboard a Russian tanker. The tanker, Akademik Pustovoit, was detained by United States naval vessels enforcing United Nations sanctions against Iraq on April 5. Defense Department spokesman Kenneth Bacon states that Shell appeared to have acquired the Iraqi oil unwittingly, and would therefore be allowed to keep the cargo. The fine will go into a United Nations fund for the enforcement of sanctions.

April 28

Azerbaijan, Georgia, and Turkey sign a final governmental agreement in Washington on the planned Baku-Ceyhan pipeline, which would transport oil from the Caspian Sea region to Western markets through the Turkish port of Ceyhan. The agreement covers the issues of transit fees, security, and governmental liability involved in the project.

April 29

Iraq's oil ministry states that it expects to export more than $8.5 billion of oil in the current six-month phase of the United Nations "oil-for-food" program.

April 30

Total Fina Elf announces that it intends to invest $8 billion in Saudi Arabia's natural gas sector. The announcement comes following a meeting between Total Chairman Thierry Desmarest and Saudi Arabia's Oil and Gas Policy Committee, which includes Foreign Minister Saud al-Faisal and Petroleum Minister Ali Naimi. The Saudi government has been seeking foreign investment in natural gas and petrochemical projects.

April 30

The National Iranian Oil Company (NIOC) signs a 25-year agreement with a consortium of multinational firms for a set of studies dealing with its natural gas sector. The topics will include estimating reserves and assessment of Iran's prospects for both increased domestic use and export of natural gas. Firms involved in the agreement include BP Amoco, Royal Dutch Shell, British Gas, and Gaz de France, among others.

May 2000

May 2

ExxonMobil signs a production-sharing agreement with Qatar General Petroleum Corporation (QGPC) for the development of natural gas from Qatar's offshore North Field. The project is expected to eventually produce 1.75 billion cubic feet per day, much of which will be exported to the United Arab Emirates (UAE) as part of the Dolphin Project, a series of pipelines linking the gas grids of Qatar, the UAE, and Oman.

May 5

Indonesia's central government proposes three management options to the government of Riau province for the Penkanbaru field, Indonesia's largest. The options include the possibility of Riau taking over state oil company Pertamina's stake in the field. The move is part of an effort by the government of President Wahid to distribute oil and gas revenues more fairly among Indonesia's regions.

May 10

BP Amoco reports the discovery of a major natural gas field off the coast of Trinidad, which May hold 2 trillion cubic feet of gas. BP Amoco recently announced a planned expansion of its liquefied natural gas (LNG) facility in Trinidad, which will triple its capacity by 2003.

May 16

Several sources, including the Washington Post, report a major oil find at the Kashagan field offshore from Kazakhstan, with reserves reportedly greater than 8 billion barrels. If these early reserve estimates prove correct, the additional production volumes could boost chances for construction of the proposed Baku-Ceyhan pipeline.

May 16

Senate majority leader Trent Lott and other Republicans introduce legislation intended to boost United States domestic oil production. Among other actions, the bill would provide a tax credit of up to $3 per barrel for production from marginal wells during periods of low oil prices and open up the coastal portion of the Arctic National Wildlife Refuge (ANWR) to oil exploration.

May 17

Chevron announces that it is to acquire an additional 5 per cent stake in Tengizchevroil, the firm which operates the Tengiz field in Kazakhstan, in a deal valued at $450 million. The Kazakh government share in the field will fall to 20 per cent.

May 17

The Environmental Protection Agency (EPA) formally proposes a rule which, if finalized, would reduce allowable sulfur levels in diesel fuel by 97 per cent over the next five years. The move is opposed by major refiners.

May 19

BP Amoco receives approval from the European Union for its acquisition of Burmah Castrol, for $4.7 billion. After the purchase, BP Amoco will hold a 15 per cent market share for automotive lubricants in Europe, comparable to TotalFinaElf and ExxonMobil.

May 19

A federal appeals court denies a petition by six major oil companies to rehear a decision upholding a Unocal patent for phase two reformulated gasoline. Uncertainty over Unocal's patent rights has led some refiners to stay out of the market for reformulated gasoline, fearing that they May infringe on one of Unocal's five patents related to reformulated gasoline.

May 21

Russian President Vladimir Putin appoints Alexander Gavrin as Minister of Energy. Gavrin, a former Lukoil executive, replaces Victor Kalyuzhny. Putin also abolishes Russia's State Committee on the Environment, transferring its functions to the Ministry of Natural Resources, which licenses oil and gas development. (WSJ, WP)

May 22

The United States Supreme Court agrees to hear a case involving the power of federal agencies to issue regulations under broad grants of authority from Congress. The Environmental Protection Agency (EPA) is appealing a decision by a federal appellate court which invalidated new standards set by the EPA in 1997 for ozone and particulate matter emissions.

May 23

The Energy Information Administration releases a study of oil reserves in the Arctic National Wildlife Refuge (ANWR), which currently is off-limits to oil exploration. The study estimates that there are between 5.7 and 16 billion barrels of recoverable oil in the ANWR.

May 24

BP Amoco agrees to purchase the 18.1 per cent of Houston-based upstream oil and gas producer Vastar Resources which it does not already own. The deal is valued at $1.6 billion.

May 26

Devon Energy agrees to acquire Santa Fe Snyder Corporation for $2.35 billion in stock, plus the assumption of about $1 billion in debt. The deal will put Devon Energy among the largest independent oil and gas producers in the United States.

May 29

According to the Energy Information Administration, retail gasoline prices in the United States reach their highest level ever, with an average price of $1.538 per gallon. The previous peak was $1.529 per gallon, on March 20, 2000. These figures are not adjusted for inflation.

May 30

The Environmental Protection Agency (EPA) denies requests from petroleum marketers in Chicago and Milwaukee for a temporary exemption from a requirement to sell phase two reformulated gasoline, rejecting marketers' claims of supply shortages in those cities. Earlier, on May 18, the EPA had approved such a temporary waiver for St. Louis, based on a local supply shortage.

May 31

Iraqi Oil Minister Amir Muhammad Rashid states that Iraq does not intend to cooperate with the United Nations Monitoring, Verification, and Inspection Commission (UNMOVIC), the new body created to verify Iraq's destruction of prohibited weapons. In other comments, he also states that Iraq does not intend to sign more contracts with foreign oil companies to develop its oilfields until contracts previously awarded are implemented.

May 31

TransCanada Pipelines agrees to sell its stakes in several natural gas transmission assets in Argentina, Brazil, and Chile to TotalFinaElf for $440 million. The sales come as part of TransCanada's drive to divest non-core assets, which the company expects to result in revenues of $3 billion.

May 31

Oilfield services firms Baker Hughes and Schlumberger agree to create a venture which will unite the two companies' seismic survey units. The joint venture, Western Geco, will own seismic survey hardware, data processing operations, and the companies' archives of seismic survey data. The firm will be 70 per cent owned by Schlumberger and 30 per cent owned by Baker Hughes. Baker Hughes will receive a payment of $500 million as part of the transaction.

June 2000

June 6

The World Bank executive board votes to approve a loan of $193 million to support a project to build a crude oil pipeline from Chad to the coast of Cameroon. The countries will collect an estimated $2 billion in revenues from the project over a period of 25 years.

June 8

The Brazilian government conducts an auction of oil exploration and production concessions covering a total of 21 blocks, both onshore and offshore. The auction represents an important step in the opening of Brazil's oil industry to international competition and investment.

June 9

The United Nations Security Council passes a resolution extending the "oil-for-food" program, under which Iraq sells oil to finance imports of food and other civilian necessities, for a period of six months.

June 9

The United States and Mexico sign a treaty resolving the issue of economic rights over the deepwater "doughnut hole" area in the Gulf of Mexico between the two countries. The agreement is based on measuring distances from each country's coast, and gives the United States rights to 38 per cent of the area.

June 10

Syrian President Hafez Assad dies in Damascus. Syria's parliament votes to amend the constitution to lower the minimum age for a president, a move seen as facilitating a succession by his son, Bashar Assad.

June 15

The German government announces an agreement with utilities for the complete phaseout of nuclear power. Nuclear power plants will be closed after a lifespan of 32 years. Nuclear power supplies about one-third of Germany's electricity, and the phaseout plan May complicate Germany's plans to reduce fossil fuel consumption to curb greenhouse gas emissions.

June 15

The Department of Energy orders the release of 500,000 barrels of crude oil from the Strategic Petroleum Reserve (SPR). The oil is to be loaned to Citgo's refinery at Lake Charles, Louisiana, which has been cut off from it normal crude oil supplies by an obstructed waterway.

June 19

Husky Oil of Canada agrees to purchase Renaissance Energy in a $2 billion deal. The new entity created by the merger will be the second largest producer of oil and gas in Canada.

June 19

The Energy Information Administration (EIA) reports a one-week rise of five cents in the average price of regular gasoline, to $1.681. This is the seventh straight week of increasing prices. Gasoline prices in the Midwest are the nation's highest, at $1.874.

June 20

The Federal Trade Commission (FTC) launches a formal investigation into high gasoline prices in some areas of the Midwest, which have seen gasoline prices rise disproportionately in relation to the rest of the United States.

June 21

Oil ministers from the Organization of Petroleum Exporting Countries (OPEC), meeting in Vienna, agree to raise crude oil production quotas by a total of 708,000 barrels per day. OPEC's total production quota (excluding Iraq) will rise to 25.4 million barrels per day as of July 1, 2000. The next day, crude oil futures rise, with the New York Mercantile Exchange (NYMEX) August West Texas Intermediate contract closing June 22 at $32.19.

June 26

The United States Supreme Court orders the federal government to refund $156 million which Mobil Oil and Marathon Oil had paid for exploration rights off the coast of North Carolina. Plans to develop the area were derailed in 1990 by the Outer Banks Protection Act, which required extensive environmental analysis by the Department of the Interior before drilling would be permitted. The plaintiffs argued that the law amounted to an "open ended moratorium" on the companies' exploration plans.

June 27

The Wall Street Journal reports that PSG International, the joint venture company formed by GE Capital and Bechtel to build the Trans-Caspian Gas Pipeline (TCGP), has laid off most staff and closed offices associated with the project. The discovery of natural gas at the Shah Deniz field off Azerbaijan in 1999 has complicated negotiations on the project.

June 29

Norway's Oil and Energy ministry announces that it is rescinding its production cut of 100,000 barrels per day, which it had undertaken in cooperation with production cuts by the Organization of Petroleum Exporting Countries (OPEC), of which it is not a member. It is unclear whether the move will have a significant impact, as Norway's production cuts are subtracted from planned, not actual, quantities, and it is unclear whether Norway can meet its full planned production of 3.4 million barrels per day in the near term.

June 30

The Environmental Protection Agency (EPA) proposes a rule change which would make it easier for refiners to use ethanol in gasoline. The proposed change allows refiners to slightly increase the evaporative property in gasoline to accommodate ethanol, in exchange for the carbon monoxide reductions which result from ethanol use.

June 30

The Department of Justice and the State of Wisconsin file suit against Murphy Oil, alleging Clean Air Act violations at the company's refinery in Superior, Wisconsin.

June 30

Australia's Woodside Petroleum rejects an asset-equity swap deal proposed by Royal Dutch Shell, which would have resulted in Shell owning a majority interest in Woodside. Analysts have said the rejection by Woodside May result in a hostile takeover bid by Shell.

June 30

Shareholders of the Russian natural gas company Gazprom elect Dimitry Medvedev, a former senior government official, to take over as Chairman. Medvedev replaces former Prime Minister

Victor Chernomyrdin in the post. Gazprom provides the Russian government with almost one-fifth of its tax revenue and accounts for about 6 per cent of Russia's gross domestic product.

July 2000

July 3

News reports attributed to sources at the Organization of Petroleum Exporting Countries (OPEC) say that Saudi Arabia plans to increase production by 500,000 barrels per day in an effort to lower the price of the AOPEC Basket. Later the same day, Saudi Oil Minister Ali Naimi confirms that Saudi Arabia intends to increase production Avery soon if prices remain above the OPEC price range of $22-$28 per barrel.

July 5

Halliburton announces that it has received contracts from the Brazilian state-owned oil company Petrobras for the development of the Barracuda and Caratinga offshore oilfields. The contracts, with a total value of $2.5 billion, provide for the construction of two floating production, storage, and offloading units and the drilling of 51 wells.

July 5

Tosco Corporation agrees to sell its Avon refinery near San Francisco to Ultramar Diamond Shamrock in a transaction valued at $800 million.

July 11

Kuwait Petroleum Corporation announces plans to increase its crude oil production capacity to three million barrels per day, in a program of facility upgrades which will take approximately three years to complete.

July 12

The Kuwaiti parliament ratifies a treaty with Saudi Arabia resolving competing claims to offshore mineral rights. The two countries will share revenues from the Khafji, Dorra, and Hout oil and gas fields. The treaty will allow the two governments to begin negotiations with Iran to settle conflicting claims, which have again surfaced as Iran has begun drilling in the Dorra offshore gas field.

July 12

In a policy shift which will allow foreign investors a majority stake in the planned Xinjiang-Shanghai natural gas pipeline project, the Chinese government announces that it is ending its ban on foreign ownership of natural gas infrastructure. A tender for the project is planned for later this year and construction is to begin in 2001. The shift is seen as a further attempt by China to attract foreign capital to its energy sector.

July 13

BP Amoco agrees to sell its Alliance Refinery at Belle Chasse, Louisiana, to Tosco Corporation for $660 million. The sale will bring Tosco=s refining capacity to 1.35 million barrels per day. BP Amoco plans to sell its Alliance Pipeline to other buyers. BP Amoco currently owns 21 refineries, but has announced plans to sell off more refining capacity, including its interests in Singapore.

July 24

BP Amoco announces that it will launch a new brand identity worldwide, consolidating the acquisitions of Amoco, Burmah Castrol, and Atlantic Richfield. The company will drop the Amoco name, and will adopt a new green, white, and yellow logo based on the sun god Helios from Greek mythology.

July 25

Israeli-Palestinian peace talks at Camp David, Maryland, break off after two weeks of U.S.-mediated negotiations fail to produce an agreement.

July 25

A petroleum products pipeline catches fire in southeastern Nigeria, resulting in many deaths. Oil pipeline fires are common in the Niger Delta region, often resulting from attempts to steal petroleum products.

July 25

A large spill of diesel fuel takes place in Guanabara Bay near Rio de Janeiro in Brazil. The event is the third major oil spill in Brazil in 2000.

July 26

The Offshore Kazakhstan International Operating Consortium (OKIOC), which includes nine international oil companies, reports substantial flows of oil and natural gas from the first well drilled at the Kashagan field. OKIOC plans to drill a second test well at the western end of the structure, 25 miles away from the first well, before the end of 2000.

July 27

Qatar General Petroleum Corporation (QGPC) and ExxonMobil sign a memorandum of understanding with Kuwait Petroleum Corporation for exports of Qatari natural gas to Kuwait. The gas will come from Qatar=s offshore North Field.

July 27

Italy's ENI signs a deal with Iran worth $3.8 billion for the development of the country's South Pars gas field in the Persian Gulf. The project will take five years to become operational, and will eventually produce 530 million cubic feet of gas per day.

July 28

Spain's Repsol and Brazil's Petrobras agree to a $1 billion asset swap which will make Repsol the second largest oil refiner in Brazil. The two companies also agree to jointly produce electricity in Brazil.

July 30

Venezuelan President Hugo Chavez wins reelection with 60 per cent of the popular vote. His Patriotic Pole party also wins a controlling majority in the country's new unicameral legislature.

July 31

A Shell executive involved in the Trans-Caspian Gas Pipeline (TCGP) project warns that time is running out for reaching an agreement with the government of Turkmenistan. The Turkmen government has reportedly been asking for a pre-payment from the companies developing the pipeline, a demand they have not been willing to accept. In June 2000, consortium partner PSG International closed offices in the region associated with the TCGP project.

July 31

The Middle East Economic Survey (MEES) reports that Saudi Arabia has shortlisted eleven bidders who will be asked to submit formal proposals for integrated natural gas projects and downstream oil projects in August 2000. The firms reported to have been invited to bid include Royal Dutch Shell, Phillips Petroleum, Chevron, ExxonMobil, Texaco, Conoco, BP Amoco, ENI, Marathon Oil Canada, TotalFinaElf, Enron, and Occidental. (Enron and Occidental have submitted a joint bid.)

August 2000

August 1

ExxonMobil announces that it expects to attain cost savings of $4.6 billion as a result of its recent merger, up 65 per cent from its original estimate. Cost-cutting measures include a staff reduction of 19,000 from the original combined total of 123,000.

August 1

The Organization of Petroleum Exporting Countries (OPEC) officially tells member governments to cancel plans to raise production. On July 17th, with the price of the OPEC Basket above $28.00 since July 1, OPEC President Ali Rodriguez told members in a letter to prepare for an increase by the end of July, provided the OPEC Basket price stayed above the $28.00 ceiling in OPEC=s Aprice band. The price then fell below the threshold.

August 10

General Motors and ExxonMobil announce that they are jointly developing a gasoline processor for fuel cell-powered vehicles. The processor will create a stream of hydrogen derived from the gasoline which will run the fuel cell, which could be twice as efficient as a regular gasoline engine. General Motors says it plans to put fuel cell-powered vehicles on the market in significant numbers by 2004.

August 10

Venezuelan President Hugo Chavez meets with Iraqi President Saddam Hussein in Baghdad as part of a tour of members of the Organization of Petroleum Exporting Countries (OPEC). He is the first head of state to visit Saddam Hussein since the 1990 Iraqi invasion of Kuwait.

August 10

Shares of Petrobras, the Brazilian oil company, begin trading on the New York Stock Exchange, as the Brazilian government sells a 16.6 per cent stake in the majority state-owned firm, raising more than $4 billion.

August 16

ExxonMobil announces that it and four other refiners will ask the Supreme Court to review a lower court ruling upholding Unocal's patent covering reformulated gasoline. The Supreme Court will announce this fall whether it will hear the case.

August 17

The Wall Street Journal reports that, in a reversal, President Sapamurad Niyazov of Turkmenistan is again actively pursuing the Trans-Caspian Gas Pipeline. The shift in policy reportedly is communicated during a meeting with the Turkish ambassador in Ashgabat.

August 20

Ten people die as a result of a gas pipeline explosion outside of Carlsbad, New Mexico. The pipeline, which carries natural gas from West Texas to California, is owned by El Paso Energy.

August 22

PetroChina, China's largest oil and gas company, and the Hong Kong firm Hutchinson Whampoa announce the formation of a joint venture which will develop a business-to-business Web portal for oil and gas businesses in China. PetroChina says it expects the move to reduce its own procurement costs by 5 to 10 per cent.

August 23

The Energy Information Administration reports that crude oil stock levels in the United States have fallen to their lowest level since 1976. Crude oil for October delivery closes at $32.02 on the New York Mercantile Exchange (NYMEX), up 80 cents.

August 24

Enron announces an agreement with German regulators which will allow the company to begin supplying the municipal utilities of two German cities with natural gas. As with other members of the European Union (EU), Germany must open part of its natural gas market to foreign competition to fulfill its EU commitments.

August 24

Iraq's Deputy Prime Minister Tariq Aziz says Iraq will not cooperate with the United Nations Monitoring, Verification, and Inspection Commission (UNMOVIC), the body created by the United Nations to replace the former United Nations Special Commission on Iraq (UNSCOM).

August 25

President Clinton begins an official visit to Nigeria. In the first six months of 2000, Nigeria ranked as the fifth largest supplier of crude oil to the United States, accounting for about 10 per cent of American imports.

August 25

BP Amoco wins the right to retain its 10 per cent stake in the Russian oil company Sidanco, when Sidanco shareholders vote to allow the former rival to receive newly-issued shares in exchange for a subsidiary it lost in a disputed bankruptcy proceeding.

August 25

Florida Power and Light (FPL), the largest electric utility in the United States, files with the Florida Public Service Commission for approval to raise electricity tariffs in the state. The company cites increased costs for natural gas and fuel oil.

August 28

Futures contracts for heating oil on the New York Mercantile Exchange (NYMEX) reach their highest price since October 1990, trading as high as $1.00 per gallon before falling back to close at 99.9 cents per gallon. The price rise reflects low stock levels for heating oil in the United States, which have fallen almost 40 per cent in relation to the same period last year.

August 30

The Department of Energy awards contracts to create a two-million-barrel reserve of heating oil. The oil will be stored in privately owned facilities in Woodbridge, New Jersey, and New Haven, Connecticut.

August 31

PetroChina announces that it plans to eliminate 50,000 jobs annually over the next five years in a move to cut its operating costs. At the end of 1999, PetroChina had a staff of 480,000.

September 2000

September 2

Royal Dutch Shell announces its intention to invest in the development of natural gas reserves on Russia's Sakhalin Island in the northern Pacific ocean. Shell spokesmen say they expect Sakhalin's gas deposits to become a major source of supply for Northeast Asian countries.

September 8

Truck drivers in Britain begin a blockade of oil refineries to protest high fuel prices. The blockade follows a similar protest in France.

September 10

The Organization of Petroleum Exporting Countries (OPEC), at a meeting of OPEC oil ministers in Vienna, agrees to raise production quotas by 800,000 barrels per day (to 26.2 million barrels per day, not counting Iraq) in an attempt to push crude oil prices back under $28 per barrel. The quota increases become effective October 1.

September 12

Royal Dutch Shell announces that it plans to purchase 14 per cent of shares offered in the upcoming initial public offering (IPO) of the Chinese Petrochemical Corporation (Sinopec) for $430 million. BP Amoco had announced a purchase of $400 million in Sinopec equity the previous day. ExxonMobil is also reported to be planning a major purchase of Sinopec shares.

September 19

The Russian government sells off an 85 per cent stake in the oil company Onanko for nearly $1.1 billion to a company affiliated with Tyumen Oil. The Tyumen bid beat a lower bid from a consortium including Yukos Oil and Sibneft.

September 20

Oil prices close at $37.20 on the New York Mercantile Exchange (NYMEX), after trading as high as $37.80 during the day's trading session. The price spike comes amid an increase in tensions between Iraq and Kuwait. This level sets a new ten-year high for NYMEX crude oil.

September 20

The president of the China National Offshore Oil Company (CNOOC) announces that the firm is planning a $4 billion joint venture petrochemical plant in Guangdong province with Royal Dutch Shell. He also confirms that CNOOC still plans to move forward with an initial public offering (IPO) of stock on the New York and Hong Kong stock exchanges in the first quarter of 2001.

September 20

BP Amoco announces that it plans to begin supplying significant quantities of natural gas from Alaska's North Slope to the lower 48 states by 2007. The company says it is exploring options for utilizing the gas which include a pipeline, transportation as liquefied natural gas (LNG), and conversion of some of the gas into synthetic crude oil.

September 22

President Clinton authorizes the release of 30 million barrels of oil from the Strategic Petroleum Reserve (SPR) over 30 days to bolster oil supplies, particularly heating oil in the Northeast. The release will take the form of a "swap," in which crude oil volumes drawn from the SPR will be replaced by the recipients at a later date. Crude oil for November delivery falls four per cent, to $32.68, on the New York Mercantile Exchange (NYMEX).

September 23

President Clinton announces the release of $400 million in federal assistance for low-income households which rely on heating oil during the winter months. The amount is the largest allocation ever for the Low Income Heating Assistance Program (LIHEAP).

September 26

A summit of heads of government of the Organization of Petroleum Exporting Countries (OPEC) opens in Caracas, Venezuela. The summit is only the second OPEC meeting held at that level. The summit ends on a conciliatory note, with the communique calling for increased dialogue between OPEC and consuming nations.

September 27

Royal Dutch Shell announces that it intends to spend $1.3 billion on the development of the Na Kika offshore oilfield in the Gulf of Mexico. Production at the field is expected to begin in 2003 and peak at 100,000 barrels per day.

September 28

A federal grad jury indicts Koch Industries for alleged violations of environmental laws involving emissions of benzene at its refinery in Corpus Christi, Texas.

September 28

The United Nations Compensation Commission, which handles claims for reparations arising from Iraq's 1990 invasion of Kuwait, approves by consensus a $15.9 billion claim by Kuwait for compensation for lost oil production and damage to oil reserves and equipment. The proportion of revenues from Iraqi oil sales under the "oil for food" program which are used for payment of claims is reduced from 30 per cent to 25 per cent. Iraq condemns the decision, but states that it will not call a halt to oil exports, as had earlier been feared.

September 28

Crude oil prices drop after Saudi Arabia's Crown Prince Abdullah, at a summit of the Organization of Petroleum Exporting Countries (OPEC), states that Saudi Arabia is ready to supply the world market with whatever amount of oil is needed to stabilize prices. Crude oil futures on the New York Mercantile Exchange (NYMEX) close at $30.34 per barrel, down nearly four per cent.

September 28

The leader of Israel's Likud Party, Ariel Sharon, makes a controversial visit to the Temple Mount (known to Muslims as the Haram al-Sharif). Palestinians stage violent demonstrations in response.

October 2000

October 2

Gulf Canada Resources agrees to buy Crestar Energy for $1 billion in cash and stock. The deal will create North America's 11th largest independent oil and gas producer.

October 2

The United States Supreme Court rejects an appeal by ExxonMobil regarding $5 billion in damage awards against the company stemming from the 1989 Exxon Valdez oil spill in Alaska. ExxonMobil had argued that improper comments by a court bailiff to a juror in the case warranted a new trial.

October 12

The New Zealand Commerce Commission announces that it will block a proposed takeover of New Zealand's Fletcher Energy by Royal Dutch Shell. The transaction would have left Shell with control of about 80 per cent of New Zealand's oil and gas reserves.

October 12

Oil prices rise sharply on news of a terrorist attack on an American warship, the USS Cole, in the Yemeni port of Aden, as well as escalating violence between Palestinians and Israeli security forces. November crude oil on the New York Mercantile Exchange (NYMEX) rises $2.81 to close at $36.06 per barrel. Prices for Henry Hub natural gas hit a record high of $5.78 per million British thermal units (BTU) before falling back slightly to close at $5.63 per million BTU.

October 14

The Department of Energy finalizes contracts for crude oil to be withdrawn from the Strategic Petroleum Reserve as part of an effort to ease possible supply shortages during the upcoming winter heating season.

October 15

Chevron agrees to purchase Texaco for $35.1 billion in stock. The deal would create the fourth largest oil and gas company in the world, and follows a general trend toward consolidation among the major oil companies over the last two years. Analysts expect the merger, like other recent mergers, to face intensive antitrust scrutiny, especially as a combined ChevronTexaco would have a heavy share of both refining capacity and retail outlets on the west coast of the United States.

October 16

Venezuela's state-owned oil company, Petroleos de Venezuela (PdVSA), reaches an agreement on pay increases with its main labor union Fedepetrol, ending a four-day strike. The previous day, Venezuela's president Hugo Chavez had fired the president of PdVSA, Hector Ciavaldini, replacing him with Giacaipuro Lameda Montero, who had served as the director of Venezuela's government budget office.

October 18

Shares of Chinese state-controlled oil company Sinopec begin trading on the New York stock exchange. Several major multinational oil companies, including ExxonMobil, BP, and Royal Dutch Shell purchase Sinopec stock. The total value of the initial public offering (IPO) is $3.6 billion. The move to list shares on a foreign stock exchange comes in the wake of a similar IPO by the other major Chinese oil company, China National Petroleum Corporation (CNPC) earlier this year.

October 18

Saudi Aramco, Saudi Arabia's state-owned oil company, is placed under the direct control of the Supreme Council for Petroleum and Minerals Affairs. The move is seen by analysts as transferring more authority over the country's oil industry to the foreign minister, Prince Saud al-Faisal, who chairs the council.

October 18

ExxonMobil announces the beginning of construction on the Chad-Cameroon pipeline, which will bring oil from the landlocked central African country of Chad to a loading terminal on Cameroon's Atlantic coast. The project, partially funded by the World Bank and the United States Export-Import Bank (Eximbank), is scheduled for completion in 2003.

October 26

Cuban president Fidel Castro meets with Venezuelan president Hugo Chavez in Caracas. The two leaders sign an agreement which will provide Cuba with Venezuelan oil at discounted prices. (WSJ, NYT)

October 27

Russian president Vladimir Putin announces that Russia intends to resume supplies of natural gas to Yugoslavia, which had been cut off earlier this year as a result of Yugoslavia's failure to pay its debts to Gazprom. The announcement comes during a state visit by the new Yugoslav president Vojislav Kostunica to Moscow. Kostunica took office earlier in the month after winning an election against Slobodan Milosevic, and street demonstrations demanding that Milosevic honor the results. Western governments, including the United States, lifted economic sanctions against Yugoslavia in the wake of the transfer of power to a democratic government.

October 30

The president of the Organization of Petroleum Exporting Countries (OPEC), Venezuelan oil minister Ali Rodriguez, announces that the cartel will raise production quotas by 500,000 barrels per day, beginning November 1st. OPEC's action comes as a result of its "price band" mechanism, which triggers an increase in production quotas when the price of the OPEC Basket of crude oils closes over $28 per barrel for twenty consecutive trading days. Many analysts voice doubt as to whether the OPEC quota increase will lead to an actual increase in production of that magnitude, given the lack of spare production capacity of most OPEC members.

October 31

The United Nations Sanctions Committee approves an Iraqi request to be paid in Euros, rather than United States dollars, for oil exported under the "oil for food" program, which is part of the sanctions regime stemming from Iraq's 1990 invasion of Kuwait.

November 2000

November 3

Russia's Lukoil announces that it will purchase Getty Petroleum Marketing of the United States for $71 million. Lukoil eventually intends to switch Getty's 1,300 retail outlets in the Northeastern and Middle Atlantic states to the Lukoil brand name. The purchase represents the first takeover of a publicly traded American company by a Russian firm.

November 7

Saudi Arabia opens a border crossing point with Iraq to facilitate Saudi exports to Iraq under the United Nations "oil for food" program. The land border had been closed since the Iraqi invasion of Kuwait in 1990.

November 7

Amerada Hess Corporation agrees to purchase the UK's Lasmo Oil for $3.5 billion in cash and stock. The move will give Amerada Hess access to Lasmo's overseas oil production assets, and come as part of a general trend toward consolidation in the oil industry.

November 7

The United States holds presidential elections, with neither Al Gore nor George Bush being declared as the winner.

November 12

Oil ministers of the Organization of Petroleum Exporting Countries (OPEC), meeting in Vienna, announce a decision to put any further production increases on hold until their next meeting scheduled for January 17, 2001. The move effectively ends OPEC's "price band" mechanism, which called for automatic increases in production quotas of 500,000 barrels per day when the price of the OPEC Basket of crude oils remained over $28 per barrel for 20 consecutive trading days. OPEC also selects the Venezuelan oil minister, Ali Rodriguez, as its new Secretary General. He will formally take over from Nigeria's Rilwanu Lukman on January 1, 2001.

November 14

Turkmenistan's President Sapamurat Niyazov announces on state television that he has reached an agreement with Russia's Gazprom for the sale of more than one trillion cubic feet of Turkmen natural gas in 2001. The sale would account for the great majority of Turkmen gas exports, and is seen by many analysts as a blow to the proposed Trans-Caspian Gas Pipeline (TCGP).

November 15

BP Amoco announces plans to sell three of its smaller refineries in the United States as part of its plans to reduce its global refining capacity by 500,000 barrels per day. The refineries in Salt Lake City, Utah; Mandan, North Dakota; and Yorktown, Virginia have a combined capacity of 177,000 barrels per day. The company plans to complete the sales by mid-2001.

November 16

Iraq's State Oil Marketing Organization (SOMO) demands that companies lifting cargoes of Iraqi crude oil begin paying a fifty cent per barrel surcharge starting on December 1, 2000. The surcharge would be paid directly to the Iraqi government rather than being channeled into the account administered by the United Nations under the "oil for food" program, and would constitute clear violation of sanctions. The Iraqi move leads to concerns over a possible Iraqi cutoff of oil supplies beginning December 1.

November 17

New Zealand regulators approve the acquisition of Fletcher Energy by Royal Dutch Shell, after the company agrees to divest a number of Fletcher Energy assets to preserve competition. The deal the make Shell the largest oil and gas producer in New Zealand.

November 21

Press reports indicate that Iraq has begun pumping crude oil to Syria through a recently restored pipeline. A State Department spokesman says the United States embassy in Damascus has made an inquiry to the Syrian government about the reports.

November 22

Mexican President Vicente Fox names Ernesto Martens as the Minister of Petroleum in his new administration. Martens is to take office December 1. He was previously an airline executive.

November 22

Egypt announces that it is recalling its ambassador to Israel in protest of Israeli military retaliation against a terrorist attack the previous day in the Gaza Strip. Relations between Israel and Arab states have worsened considerably in the wake of Israeli-Palestinian clashes which started in late September 2000. Qatar had earlier closed a trade office in Israel.

November 22

The Caspian Pipeline Consortium (CPC) announces the completion of its oil pipeline connecting oilfields in Kazakstan to the Russian Black Sea port of Novorossiysk. While the pipeline itself is completed, some work remains on related service equipment. The pipeline will begin operation in mid-2001.

November 22

Caltex Indonesia halts work at 17 drilling rigs in Riau province due to a strike by more than 3,000 employees of Caltex subcontractors. While the action does not affect current production, it May delay planned increases in production in the first quarter of 2001.

November 26

The sixth Conference of Parties (COP-6) of the Kyoto Protocol in The Hague ends without an agreement between member states on implementing cuts in emissions of greenhouse gases. One of the main issues under negotiation at the conference was the possibility that member states could claim credit for "carbon sinks," forests and farmland which absorb carbon dioxide, as part of their overall commitment to reducing carbon dioxide emissions. Another main issue was "emissions trading," which would allow member states to purchase "emissions credits" from other member states whose carbon dioxide emissions were below their targets.

November 28

Canadian Prime Minister Jean Chretien wins reelection, with his Liberal Party winning 173 of the 301 seats in the Canadian House of Commons.

November 28

The Board of Directors of Australia's Woodside Petroleum rejects a takeover offer from Royal Dutch Shell, but agrees to present the offer to shareholders at its annual meeting in 2001. A takeover offer earlier in the year was also rejected.

November 30

Russia's Finance Minister Alexei Kudrin announces that an investigation by Russian tax officials has found widespread evidence of tax evasion by Russian oil companies, estimated to cost the Russian government $9 billion annually in lost revenues. Russian oil companies account for about 40 per cent of Russia's hard currency export revenues and 25 per cent of the Russian government's tax revenues.

November 30

Natural gas futures soar to a record high as forecasts for colder weather in the Northeast and Midwest threaten to boost demand at a time when supplies of natural gas in storage are at low levels. Natural gas closes at $6.59 per million British thermal units on the New York Mercantile Exchange (NYMEX), a rise of 40.8 cents.

December 2000

December 1

Iraq informs the United Nations Sanctions Committee that it intends to stand by its proposed oil prices for December, which the committee rejected earlier in the week because they were below fair market value. The Iraqi move is seen as part of its effort to force lifters of its oil to pay a 50 cent "surcharge" directly to Iraq. Iraq also suspends its oil exports.

December 4

California utilities are forced to cut off electricity supplies to some "interruptable" customers due to a supply shortage. California has suffered shortages and high wholesale electricity prices since May 2000. The immediate shortage stems, in part, from a reduction in electricity imports from the Pacific Northwest as a result of cold weather in the area. Other problems include: gas supply problems, low availability of hydroelectric and nuclear generating capacity, and high power demand.

December 5

The United Nations Security Council approves a six month extension to the Iraq "oil for food" program.

December 8

Press reports indicate that scheduled loadings of Syrian crude oil have increased by approximately 20 per cent for December 2000. The increase is suspected by some analysts to be due to possible supplies of Iraqi crude oil to Syrian refineries through a recently-refurbished pipeline linking the two countries.

December 12

The Chinese government announces that it intends to begin a limited opening of the country's retail petroleum products market to competition by foreign oil companies. The opening will initially take the form of a pilot program in one or two cities in 2001.

December 13

Iraq resumes exports of crude oil after a disruption of nearly two weeks due to a dispute over payments with the United Nations. Iraq had cut off exports when clients had refused to pay a surcharge directly to Iraq, which would violate terms of the "oil for food" program permitting Iraq to export oil while sanctions remain in effect, rather than to the account controlled by the United Nations Sanctions Committee. Earlier, on December 8th, the committee approved Iraq's revised pricing offer for December 2000, which oil traders said is still below market value relative to comparable grades of crude oil.

December 15

Natural gas prices soar 13 per cent after the National Weather Service issues a revised forecast predicting colder winter weather in the eastern United States. Henry Hub natural gas closes at $8.396 on the New York Mercantile Exchange (NYMEX), up 98.3 cents. Gas prices are running at record high levels on concerns over winter supplies, especially given colder weather this season.

December 16

Ukraine permanently shuts down the last reactor at its Chernobyl nuclear power plant, which gained notoriety for a major accident and radiation leak in 1986. The facility will still be the location of a major cleanup effort, as Ukraine tries to contain continuing radiation leakage from the containment structures around the reactors damaged in the accident.

December 19

Construction begins on a $3.5 billion pipeline linking oilfields in Chad to a loading terminal in Cameroon. Crude oil exports from the Chadian fields are expected to start in 2003-2004.

December 19

An Alabama jury returns a $3.5 billion judgement against ExxonMobil for alleged underpayment of royalties for offshore natural gas wells in Alabama waters. The bulk of the award is punitive damages. ExxonMobil says it plans to appeal.

December 21

Italy's ENI makes a $4-billion offer for the takeover of British independent oil company Lasmo Oil. Lasmo says it has agreed to the terms of the offer. The offer effectively outbids a lower offer from Amerada Hess which was made last month, and Amerada Hess says it will not raise its offer.

December 21

The Environmental Protection Agency (EPA) announces new regulations which will drastically reduce the allowable sulfur content in diesel fuel in the United States. The new diesel sulfur standard will be 15 parts per million (PPM). Oil industry trade groups have opposed the new standard.

December 25

Sinopec, one of the major Chinese oil companies, announces plans to lay off 27,000 employees in 2001. The layoffs are part of an overall program to trim costs and boost competitiveness in anticipation of China's expected entry into the World Trade Organization (WTO).

December 26

Heating oil prices in the United States rise more than 6 per cent in response to cold temperatures in the northeastern United States. Heating oil closes at 93.5 cents per gallon on the New York Mercantile Exchange (NYMEX).

December 26

Marathon Oil announces that it intends to acquire natural gas firm Penneco Energy for around $500 million. Penneco's operatios focus on coal bed methane in Wyoming and Montana.

December 27

Natural gas prices in the United States surge above $10 per million British Thermal Units (BTUs) first time ever in response to cold weather and stockdraws reported by the American Gas Association (AGA). Henry Hub natural gas closes at $9.978, after falling slightly from its intraday peak price.

December 27

Venezuelan President Hugo Chavez appoints Alvaro Silva Calderon to replace Ali Rodriguez as Minister of Petroleum. Calderon had previously served as a deputy minister. Rodriguez had recently been chosen as the new Secretary General of the Organization of Petroleum Exporting Countries (OPEC). Both will assume their new posts effective January 5, 2001.

December 27

Hunt Oil Company announces an unsolicited takeover offer for the 90.4 per cent of Canada's Berkeley Petroleum which it does not already own for $662 million. The move is seen as an attempt by Hunt Oil to become a larger player in the North American natural gas market in an era of rising demand. Berkeley Petroleum's management opposes the takeover.

December 28

AES Corporation purchases 61.6 per cent of the Chilean electricity generator Gener, S.A., for $841 million. Gener, Chile's second largest electricity generator, was one of only a few South American power firms which had not been purchased by foreign investors.

December 30

President Clinton orders the release of $300 million in emergency funds to help low-income Americans pay for high heating oil costs. Clinton says the months of November and December have been among the coldest on record.

December 31

Saudi oil minister Ali Naimi says that the Organization of Petroleum Exporting Countries (OPEC) will cut production when ministers meet in Vienna on January 17, 2001. Oil prices have fallen sharply in recent weeks, with the OPEC basket reaching $21.50 per barrel on December 25th, down one-third from highs reached in October 2000. Despite the recent decline, average oil prices for 2000 were the highest (not adjusted for inflation) in seventeen years.

99

Monthly Energy Chronology 2001

The following information contains a listing of the major energy events that occurred in 2001.

January 2001

January 2

A federal appeals court orders the Department of the Interior to reimburse Marathon Oil and ExxonMobil $78 million each for breaking leases for drilling rights off the North Carolina coast. The companies argued that new environmental laws effectively breached their drilling contracts.

January 5

The State of Texas sues ExxonMobil, alleging that the company had produced oil from state-owned land in Texas without paying royalties. The claim involves the Hawkins field in East Texas.

January 6

California power utility Southern California Edison announces that it will cut 1,450 jobs and reduce equipment maintenance in an effort to avoid bankruptcy. The company has been hit by rising prices for purchases of electricity, which is has not been able to pass on to its customers.

January 10

The White House announces that President Clinton will not designate the Arctic National Wildlife Refuge (ANWR) as a national monument prior to his departure from office. Environmentalist groups had been pressing for national monument status for the ANWR to prevent oil drilling.

January 11

Crude oil prices rise sharply on indications that the Organization of Petroleum Exporting Countries (OPEC) will cut output quotas at its next meeting scheduled for January 17th in Vienna. The New York Mercantile Exchange (NYMEX) contract for February crude closes at $29.48 per barrel, up $1.84, or about 7 per cent.

January 12

California narrowly avoids rolling blackouts as its power crisis again reaches a "Stage 3" alert. The state Department of Water Resources steps in to purchase power from suppliers in the Pacific Northwest, who had become unwilling to sell to the California Independent System Operator (ISO) due to its unstable financial condition.

January 13

Amerada Hess withdraws its offer to purchase Lasmo PLC, a British independent oil exploration and production company. Italy's ENI had offered a higher bid of $4 billion, and Amerada Hess said it would not match the higher offer.

January 15

Schneider Electric of France announces that it will take over another French utility, Legrand SA, in a deal valued at $6.85 billion. The move reflects a trend toward consolidation and cost-cutting as utilities in Europe prepare for competition.

January 16

Southern California Edison defaults on $600 million in payments owed to bondholders and electricity suppliers.

January 17

The Organization of Petroleum Exporting Countries (OPEC) agrees, at a meeting of ministers in Vienna, to reduce members' production quotas by 1.5 million barrels per day. The move comes in response to OPEC members' concerns about declining prices. Analysts expect the actual production cuts to total somewhat less than 1.5 million barrels per day, as some OPEC members had quotas above their actual production capacity.

January 17

California utilities impose rolling blackouts on large portions of the State as demand for electricity exceeds available supplies. The blackouts, which generally last one hour, are imposed to prevent an uncontrolled collapse of the electricity distribution system.

January 17

The Minerals Management Service (MMS), an agency of the Department of the Interior, sharply raises its estimates of oil and gas reserves in the United States deepwater outer continental shelf, due to recent drilling success in the Gulf of Mexico. The agency raises its estimates for recoverable natural gas by about 65 per cent and for recoverable oil by about 35 per cent.

January 18

The Department of Energy issues a rule which will tighten efficiency standards for air conditioners and heat pumps by 30 per cent. The rule is designed to help hold down future growth in electricity demand during peak usage periods.

January 20

George W. Bush is sworn into office as the President of the United States. Later in the day, the Senate votes to confirm Spencer Abraham as the new Secretary of Energy.

January 21

A small tanker carrying diesel fuel runs aground in Ecuador's Galapagos Islands and begins to leak, threatening an environmental disaster. This is later averted as winds push much of the oil slick out to sea.

January 22

President Bush names Curtis Hebert as chairman of the Federal Energy Regulatory Commission (FERC). Two seats on the five-seat body remain vacant.

January 29

Natural gas prices fall 13 per cent in a single day, on unseasonably warm temperatures in key urban markets. February Henry Hub natural gas closes at $6.293 per million British thermal units (Btu), down 96.3 cents. It had reached a high of $10.10 per million Btu on December 27, 2000.

January 29

President Bush names Vice President Richard Cheney to chair a White House task force which will oversee the new administration's efforts in devising a national energy policy.

January 31

Taiwan's legislature passes, by a 134 to 70 majority, a resolution calling for the resumption of work to complete the nation's fourth nuclear power plant. Construction on the $5.4-billion plant was halted in October 2000 after a new government under President Chen Shui-bian took office earlier in the year, promising to end the expansion of nuclear power. Earlier in the month, Taiwan's highest court ruled that the government's action in terminating construction had been legally improper.

January 31

The California Senate passes a bill to authorize the state government to spend up to $10 billion for the purchase of electricity, to be financed by the issuance of state bonds. The plan envisioned by the bill would have the state government enter into long-term power purchase agreements with suppliers, and resell the electricity at cost to private California utilities. Two of the largest California utilities, Southern California Edison and Pacific Gas and Electric (PG&E) have been unable to secure adequate supplies of electricity from other firms due to cash flow and credit problems. State intervention is intended to overcome these financial hurdles. Governor Gray Davis is expected to sign the bill into law later in the week.

January 31

Halliburton announces that it has agreed to sell its Dresser Equipment Group unit to two investment firms for $1.1 billion. Halliburton expects to realize a $300-million net gain from the transaction.

February 2001

February 2

California Governor Gray Davis signs into law a bill which authorizes the state government to sell up to $10 billion in bonds to finance purchases of electricity, in a bid to alleviate the state's electricity supply shortage. The state will enter into long-term contracts with suppliers, which are expected to bring a reduction in the rates paid over the last several months on the spot market.

February 5

El Paso Energy announces plans to invest $1.5 billion over the next five years on the construction of terminals to import liquefied natural gas (LNG) into the United States and Mexico. The company is considering three possible projects in the United States and one in the Bahamas which would supply Florida through an undersea pipeline, as well as two projects in Mexico. Enron also recently announced plans for a facility to import LNG into Florida through a terminal in the Bahamas.

February 8

Citing an energy crisis of "catastrophic proportions," a federal judge orders three major electricity suppliers to continue to supply electricity to California despite their concerns over the financial health of the state's two main investor-owned utilities.

February 12

Indonesia signs a $9-billion deal with Singapore to supply natural gas to the island nation over the next 20 years. The deal is to lead to the construction of a pipeline from gas deposits on the Indonesian island of Sumatra to Singapore, with supplies beginning in 2003. Indonesia already sells gas to Singapore from its offshore Natuna field, which flows through a separate pipeline.

February 12

Potomac Electric Power Company (PEPCO) agrees to acquire Connectiv, the parent company of Delmarva Power and Light and Atlantic Energy, for $2.2 billion. The combined company would serve more than 1.8 million customers in Delaware, the District of Columbia, Maryland, New Jersey, and Virginia. (DJ, WP)

February 12

A federal judge denies a request by Southern California Edison (SCE) to remove a rate freeze which has kept the company from recouping the costs imposed on it by soaring wholesale power prices from its customers.

February 13

Italy's ENI is chosen as the operator of Kazakhstan's giant offshore Kashagan oilfield, beating out ExxonMobil, Royal Dutch Shell, and TotalFinaElf. The decision comes one week after TotalFinaElf's purchase of BP's stake in the field for $400 million.

February 13

Taiwan's government orders the resumption of construction on the country's fourth nuclear power plant. Whether or not to build the plant has been a source of political discord over the last year, with President Chen Shui-Bian having been elected on a platform promising to cancel the project, but facing opposition from a majority in parliament.

February 14

Natural gas prices fall sharply on news that amounts in storage declined less than expected. Henry Hub natural gas closes on the New York Mercantile Exchange (NYMEX) at $5.518, down 50.1 cents.

February 14

Kuwait's Prime Minister appoints Adel al-Subeih as Minister of Petroleum.

February 16

United States and British aircraft strike Iraqi air defense targets near Baghdad.

February 16

Ecuador's government signs an agreement with a multinational consortium including Occidental Petroleum and Kerr McGee to build a $1.1-billion oil pipeline linking the country's Amazon jungle to an export terminal on its Pacific coast.

February 19

President George W. Bush visits Mexican President Vicente Fox in Mexico, his first foreign trip as president. Discussions with Fox are reported to include the oil and gas industry and electric power grid links between Mexico and the United States. Foreign investment in the energy sector is a controversial issue in Mexico, where the energy sector historically has been state-owned.

February 20

The United States Supreme Court declines to consider an appeal by five major oil companies against Unocal's patent on production of cleaner "reformulated" gasoline sold in California, allowing a lower court ruling in favor of Unocal to stand. The ruling May eventually have effects beyond the California market, as tighter environmental standards for fuels take effect across much of the country.

February 20

A federal jury returns a $500-million judgement against ExxonMobil for allegedly inflating wholesale fuel prices charged to independent retailers over a 12-year period from 1983 to 1994.

February 21

The Federal Energy Regulatory Commission (FERC) gives final approval for the construction of the Gulfstream Natural Gas System, a proposed $1.7-billion pipeline designed to deliver natural gas to growing markets in Florida.

February 22

For the first time in six weeks, California lifts all alerts on its electric power grid, due to the availability of more imported electricity and the return to service of power plants which had been closed for maintenance. With these developments, power reserves in the state reach 7 per cent.

February 23

The State of California and Southern California Edison (SCE) agree that the company will sell the state government its transmission lines for $2.76 billion. The utility will be allowed to issue bonds which will recover a "substantial" portion of its uncollected power costs.

February 23

Petroecuador declares force majeure on the shipment of 1.8 million barrels of oil, due to continuing protests in its inland oil-producing region. Local residents in the area began takeovers of the oilfields earlier in the week in order to press the Ecuadorean government for funding for local infrastructure and services.

February 23

Secretary of State Colin Powell begins a trip to the Middle East for consultations with regional leaders which will include stops in Egypt, Saudi Arabia, Israel, Gaza and the West Bank, Jordan, Kuwait, and Syria. It is his first foreign travel since taking office.

February 26

Enron signs a $1.3-billion energy management contract with pharmaceutical giant Eli Lilly. Enron will provide and manage the supply of electricity and natural gas at Eli Lilly facilities, as well as maintain energy assets and related infrastructure.

February 27

Shares of China National Offshore Oil Corporation (CNOOC) begin trading on the New York Stock Exchange. The company produces the majority of China's offshore oil.

February 27

The United States Supreme Court rejects a challenge to the regulatory authority of the Environmental Protection Agency (EPA) under the Clean Air Act. In a unanimous opinion, the Court rules that the EPA must consider only public health and safety in its decisions on acceptable levels of pollutant emissions, and is not required to conduct an analysis of financial costs and benefits.

February 28

The Environmental Protection Agency (EPA) announces that it intends to proceed with implementation of tighter restrictions on sulfur content in diesel fuel, which were proposed by the Clinton administration. The rule, which will require a reduction of 97 per cent in sulfur content by 2006, has been opposed by many in the refining industry.

February 28

The Federal Trade Commission (FTC) announces that it found no evidence of collusion among oil companies in relation to the summer 2000 surge in retail gasoline prices in the Midwest. The FTC will release its full report on its investigation in March 2000.

February 28

Calpine Corporation signs two contracts valued at up to $8.3 billion to sell electricity to the California Department of Water Resources, which was empowered to purchase power on the wholesale market due to the financial difficulties of the state's two largest investor-owned utilities. Deliveries of power under the contracts will begin July 1, 2001.

March 2001

March 2

California's largest utility, Pacific Gas and Electric, secures a $1 billion loan to pay its creditors and avoid bankruptcy.

March 4

Tests in recent days confirm the world's largest oil find in three decades in the Kashagan field in the Caspian Sea. Kashagan is a single reservoir at least 25 miles across, and two-and-a-half times the size of the nearby Tengiz field.

March 5

Governor Gray Davis of California announces that the state has reached forty separate deals worth $40 billion to buy power over the next ten years. This would provide an additional capacity of 8,886 megawatts.

March 5

The United States Supreme Court refuses to hear an appeal of the Environmental Protection Agency's new regulations that would reduce air pollution in the eastern half of the United States, paving the way for their implementation.

March 6

United States Secretary of Energy Spencer Abraham formally establishes the Northeast Home Heating Oil Reserve, a two million barrel government-owned reserve to be used in emergency circumstances.

March 7

Royal Dutch/Shell Group announces an unsolicited $1.8 billion takeover bid for Denver-based natural gas producer Barrett Resources Corporation.

March 7

Likud party leader Ariel Sharon is sworn in as Prime Minister of Israel.

March 8

United States Secretary of Energy Abraham attends the Hemispheric Energy Conference in Mexico, an annual meeting of Energy Ministers from 34 nations. He also meets with senior Mexican government officials in order to promote President Bush's "hemispheric energy policy."

March 8

Denver-based natural gas producer Barrett Resources Corporation rejects Royal Dutch/Shell Group's $1.8-billion takeover bid and invites other offers.

March 12

Russian President Vladimir Putin formally agrees to resume conventional arms sales to Iran and to complete a delayed nuclear power plant. These agreements also set out general principles for the Russia-Iran military relationship and principles for resolving competing claims over oil and gas deposits in the Caspian Sea.

March 12

Turkey signs a natural gas purchase deal with Azerbaijan that will deliver 233 billion cubic feet over 15 years. This adds momentum to United States and Turkish-backed pipeline plans from Baku to Ceyhan, Turkey.

March 13

United States President Bush, in a policy reversal, declares that his administration will not seek to regulate power plants' emissions of carbon dioxide, citing an Energy Department study that regulating these emissions could result in higher electricity prices.

March 13

ExxonMobil temporarily shuts down on-shore natural gas operations in the Indonesian region of Aceh due to escalating violence and risk to employees and contractors.

March 15

United States Secretary of Energy Abraham testifies before Congress that power blackouts in California this summer are "inevitable." However, he does not argue for any federal government action to avert this outcome.

March 15

The world's largest oil rig, located 80 miles offshore Brazil and operated by the Brazilian state oil company Petrobras, suffers three explosions. This one platform accounted for more than 5 per cent of Petrobras' total production. On March 20 Petrobras' Platform-36 sinks with 400,000 gallons of fuel and crude oil aboard.

March 17

OPEC (Organization of Petroleum Exporting Countries) decides to cut output by 4 per cent or 1 million barrels per day, effective April 1. The cut is aimed at preventing a price collapse in a time of weakening demand.

March 19

About 800,000 customers in California experience rolling blackouts in the wake of unseasonably warm weather that raised demand.

March 20

California experiences the second straight day of blackouts, with about 550,000 customers losing power in rolling blackouts.

March 21

BP and ENI of Italy agree to build a $2.5 billion gas liquefaction plant in Egypt with Egyptian Petroleum Corp. The facility will be built at Damietta on the Mediterranean with initial exports targeted to Southern Europe.

March 22

Mexican state oil company Pemex unveils a restructuring plan involving $3 billion in cuts for the company and possibly large job reductions.

March 26

Kazakhstan's Prime Minister opens an oil pipeline from the giant Tengiz field to the Russian port of Novorossiisk on Monday, giving the Central Asian producer its first direct link to international markets. The 900-mile pipeline will carry 600,000 barrels of oil per day by the end of the year, and eventually 1.5 million barrels per day.

March 27

A heavily guarded train carrying nuclear waste from France to storage in Germany sparks large protests across its route in Germany.

March 27

California regulators (State Public Utilities Commission) approve a 40 per cent -46 per cent rise in electricity prices. The panel defends this action by stating that it is the only way to avoid blackouts.

March 27

The Bush administration declares that it has no interest in implementing or ratifying the international climate treaty negotiated in Kyoto in 1997.

March 28

Oil workers in Venezuela, the world's third largest oil exporter, begin a strike but the state-run oil company PdVSA says vital operations continue under a contingency plan.

March 28

Crude oil prices drop after the United States Department of Energy reports a sharp increase in United States crude oil inventories. Inventories grew by 11.2 million barrels in the past week to a total of 301.5 million barrels. However, prices would go on to rise in the following week.

March 28

The two-day Arab Summit in Amman, Jordan ends with leaders agreeing to support the Palestinian cause and approving $240 million in aid for the Palestinians. However, no agreement was reached on a common policy toward Iraq.

March 29

President Bush announces that he is prepared to look elsewhere than the Arctic National Wildlife Refuge for oil and gas resources if opposition to drilling there is successful.

March 30

The oil workers' strike in Venezuela ends after weak support among workers and clashes with the Venezuelan National Guard.

March 30

The Western Hemisphere climate conference, a gathering of environment ministers in Montreal, ends with the United States declining to go along with a Latin American plan for industrialized countries to reduce their emissions.

April 2001

April 1

Qatar's Ras Laffan Liquefied Natural Gas Co. Ltd (RasGas II) signs an agreement with a consortium comprising Chiyoda Corp, Mitsui & Co. and Snamprogetti SpA to set up a third natural gas train at the RasGas plant. This sets in motion the largest-ever liquefied natural gas (LNG) export deal, with India to purchase 7.5 million tons per year for 25 years through PetroNet LNG.

April 2

The tentative $8-billion merger between FPL Group Inc. and Entergy Corp. collapses. This merger would have created the largest electric utility in the United States, with 6.3 million customers and 48,000 megawatts of generating capacity.

April 2

The Korean Ministry of Commerce, Industry, and Energy announces that it is splitting state-owned Korea Electric Power Company (KEPCO) into six privatized companies. According to the sell-off plan, foreigners will have the right to manage two of the five new companies, meaning that 30 per cent of the country's total generating capacity will be under the control of overseas investors. The Korea Power Exchange (KPX) has also been established to act as an intermediary for the trade of electricity between KEPCO and its subsidiaries. As a result, wholesale electricity prices will now be set by the market.

April 5

Governor Gray Davis of California, after previously opposing raising electricity rates, announces his support for a rate increase. The California Public Utilities Commission had already approved the biggest rate increase in State history the previous week, but Governor Davis' rate increase proposal would, in effect, be an alternative plan to that of the Public Utilities Commission.

April 6

California's largest utility, Pacific Gas and Electric, formally files for Chapter 11 bankruptcy protection. The utility is attempting to get relief from $9 billion in debt.

April 9

The Mozambican and South African governments sign an agreement finalizing a deal on a 900-km (559.2-mile), $1.35 billion gas pipeline from Mozambique's gas fields to the Gauteng industrial area in South Africa.

April 9

U.S. President George Bush's budget for the Department of Energy is released. It calls for a 50 per cent or $190 million cut in research programs for renewable energy sources. However, it adds $51 million for research on the use of hydrogen gas as an energy source and on advancing power transmission technology.

April 9

Governor Gray Davis announces that the State of California will buy Southern California Edison's transmission lines for $2.76 billion. This comes only days after California's largest utility, Pacific Gas and Electric, files for bankruptcy, and is an attempt to prevent the state's second largest utility from having to do the same.

April 10

E.ON AG of Germany announces plans to take over Powergen PLC of the United Kingdom in a $7.4-billion deal that would create the world's second largest electricity provider. E.ON is a conglomerate that derives half its market capitalization from non-utility businesses. Ownership of Powergen will give E. ON access to the U.S. market.

April 10

The front month gasoline futures price rises 3.4 per cent (3.39 cents per gallon) to its highest level since June 2000, following a fire that shut down El Paso Energy's Coastal Refinery in Aruba.

April 11

Shell Malaysia announces that it will invest $5.8 billion in Malaysia over the next five years to expand oil and gas operations in Malaysia, $2.6 billion of which will go towards exploration.

April 12

The administration of U.S. President Bush decides to retain rules from the previous administration regarding energy efficiency requirements for clothes washers and water heaters. The requirement is for the new machines to be 22 per cent more efficient by 2004 and 35 per cent more efficient by 2007.

April 12

ExxonMobil Corp. announces that it has made a major oil discovery in Indonesia. Its Mobil Cepu Ltd. unit made the discovery at the Banyu Urip No. 3 well, an onshore tract in the Central and East Java provinces, with estimated recoverable reserves in excess of 250 million barrels of crude oil.

April 12

Two studies are released by the U.S. Commerce Department's National Oceanographic Data Center and the Scripps Institution of Oceanography that show a direct connection between rising ocean temperatures and emissions of carbon dioxide and other greenhouse gases.

April 12

Data released by the International Energy Agency (IEA) show that the Organization of Petroleum Exporting Countries (OPEC) continued to produce oil above its target level in March. Output was reduced by only 120,000 bbl/d instead of the target cut of 770,000 bbl/d.

April 13

The U.S. Department of Energy announces that energy efficiency standards for central air conditioners will only need to be raised 20 per cent, instead of the 30 per cent proposed by the Clinton administration. U.S. Secretary of Energy Spencer Abraham comments that the 20 per cent standard is more realistic and affordable.

April 16

Ninety-two employees of Occidental Petroleum in Colombia are kidnaped in the eastern province of Aruaca. However, about 80 of them are released a few hours later after an army antiterrorism unit moves into the area. The rest are released on April 19.

April 16

Conoco's Killingholme refinery in the UK suffers an explosion, shutting down the refinery and killing two workers. The plant processes some 230,000 barrels of crude oil and other feedstocks per day, and its shutdown boosts NYMEX crude oil futures.

April 17

U.S. oil major Chevron announces that oil reserves in the Tengiz field in western Kazakhstan are about 368 million barrels according to its latest estimates. This is more than double previous estimates. The field, with Chevron as operator in a consortium consisting of Chevron, ExxonMobil, Kazakhstan's state oil company Kazakhoil and LukArco, a joint venture between BP and Russia's LUKoil, produced about 1.4 million barrels of crude in 2000.

April 17

A letter from U.S. Department of the Interior Secretary Gale Norton to Florida Governor Jeb Bush is released, stating that the Bush administration has decided to go ahead with plans to auction six million acres of potentially oil-and-gas-rich seabed in the Gulf of Mexico. The U.S. Department of the Interior estimates that the area contains 396 million barrels of oil and 2.9 trillion cubic feet of natural gas.

April 18

The Turkish Parliament passes a natural gas market bill that ends the state monopoly in natural gas and brings Turkish law in this area into compliance with European Union law.

April 18

A study is released by the Organization for Economic Cooperation and Development (OECD) predicting that, without strong government action, the emission of greenhouse gases will jump by one third in the next twenty years.

April 19

U.S. President George Bush states publicly that the U.S. government has no intention of removing economic sanctions on Iran and Libya. President Bush does not mention Iraq. According to the Washington Post, this statement comes after a draft of Vice President Cheney's energy task force report raises the possibility of lifting some of these restrictions. The Iran-Libya Sanctions Act expires in August, at which point the U.S. Congress will have to renew it for sanctions to continue.

April 22

The Summit of the Americas in Quebec ends, with all countries except Venezuela and Cuba (the only country not to attend) agreeing to establish a Free Trade Area of the Americas by the end of 2005. This would be the world's largest free trade zone, with combined output of more than $11 trillion. The United States, Canada, and Mexico agree to create a joint task force to look at ways to facilitate energy trade in the North American market.

April 23

The Australian Government rejects Royal Dutch/Shell Group's planned takeover of Woodside Petroleum Ltd, Australia's largest energy company. Australia invokes a rarely-used national interest veto and asserts that the government is ensuring development of the North West Shelf gas field, Australia's largest single ongoing natural resource development project.

April 23

U.S. major oil companies ExxonMobil and Conoco release earnings figures showing profits for the first quarter of 2001 of $5 billion and $616 million respectively. These figures represent more than a 50 per cent increase over the same period a year ago. Overall, companies in the Standard & Poor's energy sector are expected to post a 54 per cent increase in profits for the first quarter.

April 23

Russia reaches a credit agreement with the World Bank that involves a commitment to end subsidies for the coal industry in 2002. The World Bank is funding an $800-million modernization program for the Russian coal sector, as part of its support for structural reforms in the country.

April 24

Venezuela calls upon foreign companies operating in the country to cut their crude oil output in order for the country to meet its OPEC production quota. Most OPEC countries rely upon cuts in their own state-owned company's production to comply with quotas.

April 24

Tosco's Los Angeles-area refinery catches fire, halting production of about 5.5 million gallons of gasoline and other fuels per day. Gasoline prices on the Los Angeles spot market rise 7.5 cents per gallon to $1.425 per gallon.

April 25

A Federal Appeals panel dismisses a lower court's finding that strip mining of West Virginia mountaintops violates local environmental law, allowing the mechanized strip mining, that has largely replaced underground mining in Appalachia, to continue. Environmental groups plan to appeal this ruling.

April 25

The Federal Energy Regulatory Commission announces a plan to control California's electricity prices. Effective May 1, the regulators will be able to impose price controls on power generators when electricity reserves fall below 7.5 per cent.

April 26

The Directors of the Dabhol Electricity Plant of India, owned by Enron of the U.S., decide to halt electricity sales to the Maharashtra state electricity board. The state has declared its willingness to renegotiate the power purchasing agreement with Enron, but the company appears skeptical. The government would have to pay about $384 million to Enron if the project is terminated.

April 27

The Brazilian Government announces that it will invest $8 billion in the country's power sector by the end of 2002 in order to increase generating capacity by 6,000 megawatts. This comes as the possibility of power rationing looms due to a shortage of generating capacity.

April 27

Saudi Arabian Energy Minister Ali al-Naimi meets with a number of senior US officials, including Vice-President Dick Cheney and Secretary of Energy Spencer Abraham. Al-Naimi, in a statement before the meetings, says leading oil producers would not allow record-high gasoline prices to spin out of control.

April 30

U.S. Vice-President Dick Cheney previews the administration's energy plan in a speech in Toronto, Canada. Cheney, stating that conservation alone cannot solve America's energy needs, calls for increased domestic production of fossil fuels and increased usage of nuclear power to meet

America's energy demand. He also calls for construction of new coal and gas power plants, as well as upgrading and expanding of the country's transmission grid.

May 2001

May 1

Mexican President Vicente Fox announces plans to dissolve the board of Mexican state oil company Pemex and to instead create an eight-member advisory committee. This is a reversal from his previous policy to fill the board of Pemex with high-profile private-sector executives.

May 3

U.S. President George W. Bush orders federal agencies in California to cut their energy use and to take other appropriate actions to conserve energy. The Federal Government consumes 1.5 per cent of the nation's electricity.

May 3

ExxonMobil announces that it has made a major oil discovery offshore Angola, 217 miles northwest of Luanda. The discovery is located in Block 15 and is believed to contain 3.5 billion barrels of oil.

May 4

California Governor Gray Davis announces a $7 billion agreement by the State to buy electricity over 10 years from Sempra Energy. Governor Davis asserts that this will contribute to bringing reliable power to California at a lower price than the spot market.

May 4

A draft report of the Mitchell Commission is distributed to Israeli and Palestinian officials. The report says that Israel should freeze settlement construction, but does not recommend sending an international force to the region, as Palestinians have been demanding. The international commission, led by former U.S. Senator George Mitchell, was formed as part of a U.S.-brokered Israeli-Palestinian cease-fire agreement reached at an October 2000 summit at the Egyptian resort of Sharm el-Sheik.

May 7

The Williams Companies announces that it has acquired Barrett Resources for $2.5 billion in cash and stock. Barrett Resources is a natural gas exploration and production company. Williams builds power plants, transports natural gas, and trades electricity.

May 7

Valero Energy announces that it has reached an agreement to acquire Ultramar Diamond Shamrock for $6 billion. If approved by shareholders and regulators, the new company would be the second-largest petroleum refiner in the United States, with a refining of capacity slightly under 2 million barrels per day.

May 8

The State Planning Board of the People's Republic of China approves a plan to make a very large investment in the country's rural power grids. This includes maintenance and renovation as well as construction of new lines.

May 8

AES Corporation announces that it has suspended plans to invest $2-$2.5 billion in energy projects in Brazil. AES alleges that the Brazilian government's pricing policy jeopardizes the profitability of its operations there.

May 9

Four people are killed in the Aceh region of Indonesia when two bombs explode, damaging a natural gas pipeline and pumping station owned by ExxonMobil. ExxonMobil suspended on-shore operations, including shipments from the liquefied natural gas (LNG) terminal at Arun, on March 9, 2001, due to the possibility of violence. This has reduced Indonesia's export earnings from LNG.

May 9

BP Trinidad and Tobago announces the discovery of a one-trillion-cubic-foot natural gas field off the east coast of Trinidad, enough to power the entire country's electricity needs for about 50 years. More importantly, this represents another source for Trinidad and Tobago's liquefied natural gas exports and increases confidence in the hydrocarbon potential of Trinidad's east coast.

May 15

The California Public Utilities Commission approves a plan that calls for a $5.7 billion rate increase, much of it falling on residential customers. The plan is retroactive to March 27, 2001, with rates designed to reward those who conserve and to punish those who do not.

May 16

Norwegian Oil Minister Olav Akselsen announces that Norway's oil production has plateaued and that there are no plans to raise or lower production in concert with OPEC.

May 17

President Bush issues the administration's new energy policy. Among the plan's 105 specific recommendations are calls for reduced regulations to encourage more oil, gas, and nuclear production, tax incentives to boost coal output, and other tax incentives to promote conservation and alternative fuels. The plan also calls for increasing energy assistance to low-income households and for making the electricity grid more interconnected, both domestically and with Mexico and Canada.

May 17

The companies involved in the proposed Baku-Ceyhan oil pipeline, including the manager, BP, have found the initial cost estimate viable, and will proceed to the next phase of the project, a more detailed $150 million feasibility study. The proposed pipeline would loosen Russian control of Caspian Sea oil and serve as an alternative to the Bosporus as a transport route for oil from the Tengiz field.

May 17

BP and Shell say that they will build a $150 million, 100-mile natural gas pipeline in the Gulf of Mexico. The Okeanos pipeline will have the capacity to move as much as one billion cubic feet of gas per day from offshore production fields in ultradeep waters.

May 18

Brazil orders consumers and citizens to cut electricity consumption by 20 per cent to avoid blackouts. Rationing will begin on June 1, and includes surcharges for those who do not meet the cutback and rewards for those who do. This comes as a drought has reduced Brazil's hydroelectric generating capacity, which accounts for about 90 per cent of the country's total generation capacity.

May 18

Saudi Arabia selects the eight foreign companies to take part in its "Gas Initiative," three core venture gas projects that have an anticipated worth of $25 billion. They are: Core Venture 1: ExxonMobil (lead), Shell, BP, and Phillips; Venture 2: ExxonMobil (lead), Occidental and Enron (a joint bid); Venture 3: Shell (lead), TotalFinaElf, and Conoco. The Gas Initiative is the first major reopening of Saudi Arabia's upstream hydrocarbon sector since nationalization in the 1970s.

May 21

The price of Brent crude oil futures for delivery in early June peaks at $29.68 per barrel, the highest level in a year. This comes as concern increases over U.S. gasoline supplies, heightening Israeli-Palestinian tensions, a possible Iraqi halt in oil exports, and OPEC statements to the effect that a production quota increase in June is unlikely.

May 21

The Enron Corporation's power generating venture in India, the Dabhol Power Company, serves formal notice that it will terminate its power supply contract and pull out. The $2.9 billion Dabhol project represents the single largest foreign investment in India. The gas-fired plant already had a generating capacity of 740 megawatts and another 1,444 megawatts was scheduled to go on line in June.

May 23

Philippine Energy Secretary Jose Camacho announces that the country's power sector would require investments of around 382 billion pesos ($7.59 billion) for the next 10 years or an average of 38 billion pesos a year to ensure adequate power supply. According to many foreign investors, obtaining such a level of investment requires passage of the power reform bill before the Philippine Congress. The reform bill is aimed at privatizing debt-laden National Power Corp (Napocor) and restructuring the industry to bring down power rates. Napocor's debt represents over 12 per cent of the Philippines' national debt.

May 23

Shell announces that it has discovered a huge reserve of oil in Oil Mining Lease 118 offshore Nigeria. This is the same block where Shell is developing the 600-million barrel Bonga field. The discovery would appear to confirm the immense potential of Nigeria's deepwater offshore area.

May 29

California Governor Gray Davis meets with President Bush and informs him that he plans to file a lawsuit against the Federal Energy Regulatory Commission to force the federal government to impose price controls on wholesale electricity in California. President Bush reiterates his opposition to price caps.

May 29

Natural gas futures plunge 6 per cent to a 10-month low on speculation that growing U.S. inventories will help power plants meet summer demand for air-conditioning. The price for June delivery fell 23.5 cents, to $3.738 per million British thermal units on the New York Mercantile Exchange (NYMEX).

May 30

Iraqi Oil Minister Amir Mohammad Rasheed signs an oil and gas cooperation agreement with his Algerian counterpart, Chekib Khelil. The deal is thought to include an Algerian role in the development of the Touba oil field in southern Iraq and a new natural gas field in the Western Desert of Iraq.

May 31

The United States and Britain win Security Council approval of a one-month extension of the United Nations oil-for-food program. A vote on the new "smart sanctions" on Iraq proposed by the United States and Britain is delayed at least one month. Iraq demands the usual six-month extension, and says that it will cut off oil exports in response.

May 31

Ukrainian President Leonid Kuchma orders his new government to suspend the planned privatization of electricity generating and distribution companies. Privatization of electricity companies was one of the key terms laid down by the International Monetary Fund to secure loans to Ukraine under its currently frozen $2.6 billion aid program.

June 2001

June 1

Electricity rationing goes into effect in Brazil in response to electricity shortages caused by droughts in the heavily hydroelectric-dependent country. Three-quarters of consumers must cut consumption by 20 per cent for six months or face rolling blackouts and unscheduled power interruptions.

June 3

Iraq announces that it will halt crude oil exports in response to a United Nations Security Council resolution that extends the oil-for-food program by only one month, instead of the normal six-month period. The oil-for-food program affects revenues from Iraqi sales of about 2.1 million barrels per day. However, it has been reported Iraq will continue to sell several hundred thousand barrels per day to its neighbors through sales that are outside of the oil-for-food program. The Organization of Petroleum Exporting Countries (OPEC) announces that, if need be, it will make up for lost Iraqi production. Oil prices do not change greatly in response to either announcement.

June 3

The eight energy companies selected by Saudi Arabia on May 18 to take part in its Gas Initiative formally sign agreements to develop the projects. It is expected that the conversion of Saudi Arabia's power plants from oil to natural gas, which is part of the deal, will free up more crude oil for export.

June 4

An Omnibus Power Bill is passed in the Philippines. The effect is that the state-run National Power Corporation will be broken up and sold to private interests. Passage of this bill is seen as key to obtaining the necessary investment to maintain adequate power supply in the Philippines, and about $1 billion in development loans for the Philippines was contingent on its passage.

June 5

The Environmental Protection Agency (EPA) sets final health and safety standards for a proposed nuclear waste depository at Yucca Mountain in the Nevada desert. This is a key step in allowing construction of the facility, which is essential to efforts to rejuvenate the U.S. nuclear power industry. The following day, the EPA sets ground water standards for this same site.

June 5

OPEC ministers agree to leave the cartel's oil production quotas unchanged for at least a month, until a scheduled emergency meeting July 3. OPEC had been expected to leave the quotas unchanged until September, but Iraq's suspension of oil exports on June 3 created uncertainty.

June 5

Egypt and Jordan sign a 30-year agreement to export Egyptian natural gas to Jordan through a pipeline. The pipeline will have an initial capacity of 35.3 billion cubic feet per year, and the 156-mile pipeline will go from the Egyptian Mediterranean city of El-Arish to the Jordanian Red Sea coastal town of Aqaba. Construction is expected to take 18 months.

June 6

A report from the National Academy of Sciences on global warming that had been requested by the Bush Administration is released. The report affirms the view that global warming is a real problem, *i.e.*, that greenhouse gases are accumulating in the earth's atmosphere, and that air and ocean temperatures are rising.

June 6

Russian lawmakers in the lower house (Duma) approve a plan to import spent nuclear fuel (about 20,000 metric tons over ten years) from power plants in Europe and Asia. Russia would earn an estimated $20 billion for storing the radioactive material. However, about 90 per cent of the spent fuel Russia could import is from U.S.-designed nuclear reactors, and hence, cannot be transferred to a third country without the permission of the United States. The U.S. State Department states that it would want several assurances regarding reprocessing and transport security that Russia May be unwilling to agree to.

June 7

BP announces that it will build a new $600-million platform offshore Trinidad that is expected to double the company's production of natural gas there by 2004. BP currently produces one billion cubic feet per day in Trinidad.

June 7

Tony Blair is re-elected Prime Minister of Great Britain, as his Labour Party wins a large majority of seats in the House of Commons.

June 11

German Chancellor Gerhard Schroeder and leading energy companies sign an agreement that will shut down 19 nuclear power plants in Germany. The agreement limits nuclear plants to 32 years of operation, meaning that all nuclear plants in Germany could be shut down by 2021.

June 11

Saudi Arabia announces that it has seized ownership, effective June 7, of the 1.6-million barrel-per-day IPSA pipeline that had carried Iraqi crude oil to the Saudi Red Sea port of Mu'jiz prior to Iraq's invasion of Kuwait. The seizure includes pumping stations, storage tanks, and the maritime terminal. Saudi Arabia claims that the asset was confiscated as a result of aggressive Iraqi actions. Iraq insists that it still owns the pipeline.

June 14

President Bush meets with European leaders at a European Union (EU) gathering. After the meeting, European Commission President Romano Prodi announces that EU member nations will soon begin a concerted drive to ratify the 1997 Kyoto Protocol, an international pact to combat global warming that Bush has rejected.

June 14

The California Supreme Court rejects allegations that nine of the largest oil companies in the United States conspired to fix gasoline prices in California. This ruling May affect the California attorney general's ability to bring charges of antitrust behavior in the State's wholesale power market.

June 15

ExxonMobil and Qatar Petroleum sign a letter of intent for a natural gas to liquids (GTL) project that would be the largest in the world. The plant would have a production capacity of 80,000 to 90,000 barrels per day, and would use about 640 million to 720 million cubic feet of natural gas per day as feedstock. The project is expected to cost between $1.6 billion and $1.8 billion to construct.

June 16

The Iraqi Trade Minister, Mohammed Mehdi Saleh, states that Iraqi crude oil exports will not resume as long as the U.S.-British changes to the memorandum governing the oil-for-food program (*i.e.* "smart sanctions") are being considered.

June 18

The Federal Energy Regulatory Commission (FERC) votes unanimously to extend wholesale electricity price controls to ten western states that share their transmission grids with California. The order applies a proxy price based upon California-specific data to the whole area. That proxy price will be based on the operating costs of the least-efficient power plant operating in California during electricity emergencies.

June 18

Statoil, the Norwegian state-owned oil company, takes part in an initial public offering. The Norwegian government sells about 18 per cent of the company, raising about $2.9 billion, while retaining the remaining equity. Statoil is the ninth-largest publicly traded oil company in the world and controls vast oil and natural gas resources in the North Sea.

June 19

Hunt Oil Company of the United States agrees to pay $600 million to acquire Chieftain International of Canada. Most of Chieftain's assets are in the Gulf of Mexico, and this purchase boosts Hunt Oil's presence there.

June 19

Administrative workers for Shell in Nigeria go on strike, shutting down administrative offices and flow stations, with a production loss of 100,000 barrels per day, according to Shell. Downstream Shell workers will make a decision whether to strike on June 28.

June 19

Brazil's two-day auction of exploration and production rights ends, with 34 of 53 blocks offered by the National Petroleum Agency sold. About $240 million was raised from the sale, which exceeded expectations. Brazilian state-oil company Petrobras was the largest purchaser, obtaining 13 blocks on its own or as consortium leader.

June 21

A U.S. federal grand jury returns a 46-count indictment charging 13 Saudis and one Lebanese with the 1996 truck-bombing of the Khobar Towers apartments in Saudi Arabia that killed 19 Americans and injured nearly 400 others. The indictment also alleges Iranian government involvement in the bombing.

June 22

Electricity prices in the western United States fall to their lowest level in over a year in the wake of federal price controls and declining natural gas prices. The price for a megawatt-hour at the Palo Verde, Arizona transmission hub fell to $92, compared to $160 just before the FERC ruling of June 18.

June 24

Qatar signs a contract with Edison Gas of Italy to supply 3.5 million metric tons of liquefied natural gas (LNG) per year for 25 years, starting in 2005. Qatar aims to export 30 million metric tons of LNG per year by 2007.

June 27

Crude oil and gasoline futures prices on the New York Mercantile Exchange (NYMEX) collapse to their lowest levels in over a year. Gasoline futures for July delivery fall 6.35 cents, or 8.2 per cent, to 71.44 cents per gallon. Crude oil futures for August delivery fall $1.37, or 5.1 per cent, to $25.61 per barrel. This comes in the wake of reports that inventories are steady or rising for gasoline and crude oil.

June 28

The U.S. House of Representatives approves a measure banning new offshore drilling for oil and natural gas in the Great Lakes.

June 28

Brazilian power regulator Aneel sells licenses to build and operate eight hydroelectric plants, receiving $1.08 billion in concession fees from successful bidders. The new projects are expected to add 2,282 megawatts to the national grid.

June 29

The Energy Information Administration (EIA) releases a preliminary estimate showing that carbon dioxide emissions from energy sources in the United States rose by 41 million metric tons carbon equivalent, or 2.7 per cent, in 2000, much higher than the average annual growth rate of 1.5 per cent for the period 1990-2000.

June 30

ENI of Italy signs a $550 million contract to develop Iran's Darquain (Darkhovin) field, with expected production of 160,000 barrels per day. This deal May be seen as a test of the U.S. government's resolve to enforce sanctions against foreign companies investing in Iran's energy sector.

July 2001

July 2

U.S. Secretary of the Interior Gale A. Norton states that the Bush Administration will seek to let oil companies drill on about 1.5 million acres of the Gulf of Mexico out of the 6 million originally under consideration. This removes acreage closest to the shores of Alabama and Florida.

July 2

The United Nations (U.N.) Security Council, facing an almost certain Russian veto, agrees to postpone indefinitely a vote on the U.S.-led "smart sanctions" package for Iraq, despite support by the four other council members. Instead, it will extend, most likely through the end of the year, the program that allows Iraq to export oil and import food and other commodities under U.N. supervision.

July 3

At a meeting of its oil ministers, the Organization of Petroleum Exporting Countries (OPEC) agrees to maintain current production quotas. Ministers indicate that, if Iraqi oil returns to the market, they May cut production in response to maintain their desired level of prices.

July 5

Australia and East Timor sign an agreement to share royalties from oil and natural gas production in the Timor Sea, which separates the two countries. The deal supercedes the former agreement between Australia and Indonesia that divided royalties 50-50, with a new arrangement of 90 per cent for East Timor and 10 per cent for Australia. This agreement clears the way for $7.25 billion in proposed energy projects for the area and further downstream projects for Australia.

July 9

California State officials and power suppliers fail to reach agreement on disputed charges worth $8.9 billion. Chief Judge of the Federal Energy Regulatory Commission Curtis L. Wagner Jr. states that California consumers were overcharged by as much as $1 billion (going back to October 2000), but a group of energy companies offered refunds of $716.1 million, adding that these refunds should be offset by amounts they are still owed for previous power deliveries.

July 10

Amerada Hess agrees to acquire Triton Energy for $2.7 billion in cash. Both companies' boards have approved the transaction. Triton Energy is an international exploration and production company with major oil and natural gas assets in West Africa and Latin America. Triton's total proved reserves are estimated at 293.5 million barrels of oil equivalent. Amerada Hess' total proved reserves are estimated at 1.1 billion barrels of oil equivalent.

July 11

The Federal Energy Regulatory Commission (FERC) orders a consolidation of U.S. power-grid assets in order to foster large, competitive markets in the Northeast and Southeast. This is part of the FERC's goal to promote the development of four regional transmission organizations (RTOs) to oversee the national power grid, including an 11-state RTO in the west, including California.

July 11

Iraq resumes oil exports, ending a 5-week halt in protest of a U.S. and British-sponsored United Nations (U.N.) Security Council resolution that would have overhauled U.N. sanctions, after this resolution did not come to a vote. The oil-for-food program will be extended for five months.

July 12

The Russian government approves a plan to break up the state electricity monopoly RAO Unified Energy Systems (UES) into transmission and power-generating units, and to sell off some operations. Under the plan, UES' coast-to-coast transmission lines will be spun off into a separate company; the state will still own 52 per cent of UES.

July 12

A bomb attack shuts down pumping at Colombia's Cano Limon Pipeline, the country's second-largest crude oil export pipeline, just 19 hours after it had been returned to service after an earlier bombing. The 485-mile pipeline, which can transport 120,000 barrels per day to the port of Covenas, has been bombed over 100 times in 2001. In the previous week, the 110,000-barrel-per-day Colombia Pipeline was also bombed, shutting down production there. For the year-to-date, Colombia's oil exports have fallen 28.7 per cent compared to the previous year.

July 16

A 13-member panel appointed by the U.S. National Academy of Sciences recommends, in a draft report, that the federal government require automakers to improve the mileage of new vehicles. President Bush has said that the panel's findings will help him decide whether and how much to increase fuel economy standards.

July 16

BP agrees to buy a 51 per cent stake in German energy conglomerate E.On's Veba Oel petroleum station and oil unit in a deal valuing the business at about $5.56 billion. BP has the option to acquire the remaining 49 per cent. The deal would make BP the market leader in retail fuel sales in Germany.

July 17

The U.S. National Academy of Sciences releases a study of 39 research and development programs in energy efficiency and fossil fuel technology funded by the Department of Energy since 1978. The study determines that the programs yielded an economic return of $40 billion from a $13 billion investment.

July 18

Crude oil futures fall to their lowest levels in 14 months after data from the Energy Information Administration and the American Petroleum Institute show a larger-than-expected build in crude oil stocks. Crude oil for August delivery falls 68 cents per barrel to $24.89 on the New York Mercantile Exchange (NYMEX).

July 23

Following days of intense negotiations at the CoP-6 meeting in Bonn, an agreement is reached by 178 countries that would require industrialized countries to cut emissions of gases linked to global warming. The final product is a modified version of the Kyoto Protocol. The United States declines to participate in the agreement.

July 23

Endesa, Spain's largest power company, agrees to buy Eletrogen of Italy for $3.2 billion, making Endesa the third-largest electricity generator in Italy.

July 24

Mexican Energy Minister Ernesto Martens declares that maintenance work will reduce Mexico's crude oil exports by about 70,000 barrels per day for the next four months, meaning that no further Mexican production cuts will be needed to provide cooperation with expected Organization of Petroleum Exporting Countries (OPEC) production quota cuts. On July 9, Mexico had declared force majeure on export of Maya, its main export crude oil, until November due to maintenance on its oil and gas pipelines. (Reuters, OD)

July 24

An Iranian warship in the Caspian Sea threatens a BP oil exploration ship off the coast of Azerbaijan. This prompts BP to suspend exploration in the area. The two vessels were in the Araz-Alov-Sharg field 90 miles southeast of Baku. Iran claims the field is in Iranian waters. Caspian Sea region countries have been unable to agree on a division of the Sea.

July 25

Faced with declining oil prices, ministers of the Organization of Petroleum Exporting Countries (OPEC) agree to cut crude oil production quotas by about 4 per cent, or one million barrels per day. The cut will take effect September 1, and is aimed at maintaining the price of the OPEC basket of crude oils at around $25 per barrel. Crude oil futures for September delivery climbed 47 cents per barrel, to $26.78, on the New York Mercantile Exchange (NYMEX) after the announcement.

July 26

U.S. Environmental Protection Agency (EPA) Administrator Christie Whitman proposes significant changes in the regulation of power plant pollution. Five specific programs would be replaced with a single, flexible approach that includes expanded pollution credit trading.

July 26

Former Indonesian President Abdurrahman Wahid leaves the presidential palace and the country, ending a 2-day standoff and clearing the way for his successor, Megawati Sukarnoputri, the former vice-president, to take over. The National Assembly had voted on July 23 to remove Wahid from office and install Sukarnoputri in the presidency.

July 30

The flow of natural gas from Iran to Turkey through the Tabriz-Ankara pipeline, completed on July 26, is delayed. Turkey states that its state gas company, Botas, wants to conduct additional tests on the pipeline. The National Iranian Gas Company claims that the delays are a tactic by Turkey to avoid paying compensation to Iran for not being technically prepared on the agreed start-date for exports. Under the 23-year deal signed in 1996, Iran was to have begun exporting 106-billion cubic feet per day, increasing to 353-billion cubic feet per day from 2007 onward.

July 30

Energy authorities from Colombia and Ecuador sign an agreement in Quito to build a $50 million, 230-kilovolt interconnection to allow the countries to share electric power. The interconnection will link the substation of Jamondino in Pasto, Colombia with the Pomasqui substation in Quito. Ecuador will likely be a net importer from Colombia because of overcapacity in Colombia.

August 2001

August 1

A Phillips Petroleum-led coalition announces that it is postponing its investment commitment for the Bayu-Undan project, which involves construction of a $500 million pipeline to transport natural gas from the offshore field in the Timor Sea to Darwin, Australia for processing into liquefied natural gas (LNG). The Bayu-Undan field contains an estimated 3.4 trillion cubic feet of gas. The reason for the postponement reportedly concerns taxation uncertainties after oil and gas reserves in the Timor Gap were redistributed between Australia and East Timor.

August 1

The U.S. House of Representatives votes to allow oil drilling in the Arctic National Wildlife Refuge. It also votes to reject a proposal to substantially boost the gasoline mileage of sport-utility vehicles.

August 2

U.S. President George Bush signs an executive order calling on the federal government to purchase products that consume no more than 1 watt of electricity in their standby mode.

August 2

A three-day strike costing $1 million per day at Venezuela's Sincor Heavy Oil project ends. Sincor is a joint venture of TotalFinaElf, Statoil, and PdVSA that will convert the heavy oil into 31 to 32 degree API Zuata Sweet crude oil.

August 2

Crude oil futures prices for September delivery on the New York Mercantile Exchange rise $0.94 to $27.71 per barrel, their highest level in six weeks. This comes after a July 31 report by the American Petroleum Institute showing a decline in U.S. petroleum inventories for the first time in more than a month.

August 3

U.S. President George Bush signs into law the Iran and Libya Sanctions Act (ILSA) Extension Act of 2001. This Act provides for a 5-year extension of ILSA with amendments that affect certain of the investment provisions. ILSA sanctions foreign companies that provide new investments of over $40 million for the development of petroleum resources in Iran or Libya, or that violate existing United Nations prohibitions against trade with Libya. The law allows the president to waive sanctions against a foreign company if doing so is deemed to be in the U.S. national interest. U.S. companies are prohibited by U.S. law from engaging in any commercial or financial transactions with Iran or Libya.

August 7

Venezuelan President Hugo Chavez announces that Venezuela will have a new hydrocarbons law by mid-November. He states that the new law will promote the participation of domestic companies in the oil sector. The bill is expected to include taxation and royalty reforms.

August 8

The Energy Information Administration announces that petroleum demand in July averaged 19.7 million barrels per day, up from 19.3 million barrels per day in June, for the second-highest July figure ever recorded. Gasoline demand averaged 9 million barrels per day for July an all-time high.

August 8

U.S. Secretary of Energy Spencer Abraham announces that four programs to encourage the development of low-generation-cost fuel cells will receive $270 million in subsidies from the Department of Energy.

August 10

The United States and Great Britain reject a proposal by United Nations Secretary General Kofi Annan to permit the Iraqi government to use $1 billion per year to fund infrastructure improvements and to increase oil production capacity. It has been suggested that without infrastructure investment, Iraq's production could fall significantly over the next few years.

August 10

The State of California files a lawsuit against the U.S. Environmental Protection Agency over federal requirements that expensive anti-pollution additives be added to gasoline in the State. The additives raise the cost of gasoline by 3 to 5 cents per gallon, and California maintains that these additives are not necessary to maintain clean air standards.

August 13

Iraqi Vice-President Ramadan announces that Syria will soon hire contractors to build a new oil pipeline stretching from the Iraqi border to Syria. The pipeline would replace an old one that was shut down in 1982, but is reported to be operating.

August 13

An electricity interconnection between Venezuela and Brazil is inaugurated today by the presidents of the two countries at Santa Elena de Uairen, Venezuela. The transmission line will link this city to Boa Vista in Brazil. The power is generated by the 12,500-megawatt Guri hydroelectric plant in Venezuela.

August 13

BP signs an agreement with Algerian state oil and gas company Sonatrach for the development of the 7.1-trillion-cubic-foot in Salah natural gas field. The $2.5 billion project will split development costs 65/35. Production is expected to come on stream in 2004, and will peak at 141 billion cubic feet per year.

August 14

Devon Energy buys Mitchell Energy for $3.1 billion in cash and stock. The transaction, which was approved by the boards of trustees of both companies, also involves the assumption by Devon of $400 million of Mitchell's debt. This deal makes Devon the second-largest independent producer of natural gas in the United States.

August 14

The Environmental Protection Agency announces that it will not issue a reassessment of an air pollution rule affecting coal-fired power plants by August 17, as originally announced in the

President's energy policy plan speech in May. Instead, the decision, which is part of a plan to ease the regulatory burden on the energy industry without having an environmental rollback, will be delayed until September in order to be part of a more comprehensive set of policy options.

August 14

President Bush names Pat Wood III to be the new chairman of the Federal Energy Regulatory Commission, effective September 1, to succeed Curtis Hebert, who announced his resignation on August 6.

August 15

Rebels in southern Sudan attack oil field operations for the first time since operations there restarted two years ago. Oil production was interrupted for only 12 hours, according to Sudanese officials. Production from these fields is about 200,000 barrels per day.

August 20

A Superior Court Judge in San Jose, California signs a settlement agreement between the environmental group Communities for a Better Environment and five large oil companies (Shell, Chevron, Texaco, Equilon Enterprises, and Unocal) that will force the companies to clean up their sites that have been contaminated with the gasoline additive methyl tertiary butyl ether (MTBE). Arco, Tosco, and ExxonMobil are still in litigation.

August 24

The United States decides to support a modified British proposal to tighten procedures for pricing Iraqi crude oil. According to reports, Iraq is attempting to price its oil at artificially low levels, and favoring buyers willing to pay surcharges to secret accounts, thereby circumventing United Nations (U.N.) control over Iraqi oil revenue. Britain had proposed that the U.N. and Iraq set prices every 10 days, instead of the current 30 days, to make it more difficult for Iraq to exploit fluctuations in the market. The United States and Britain today agreed to 15 days instead of 10.

August 24

The retail price of gasoline in the United States is found to have risen by 6.25 cents per gallon nationwide over the previous two weeks, according the Lundberg Survey of 8,000 stations. This is the first reported rise in national average retail gasoline prices since May 18.

August 24

Iranian President Khatemi and Turkmen President Niyazov call for a suspension of oilfield development in disputed sectors of the Caspian Sea. Both countries claim areas being developed by Azerbaijan. The unresolved legal status of claims by the five littoral States (Azerbaijan, Iran, Kazakhstan, Russia, and Turkmenistan) to the Caspian is an impediment to hydrocarbon development in some areas.

August 27

Several dozen oil workers, including eight foreigners, are released by militants after being taken hostage during a raid on a drilling rig off the coast of Nigeria on August 23. The militants release the hostages and evacuate the rig after negotiations with company officials. Work was suspended on the rig, which is now being guarded by a security force.

August 28

The European Commission approves the $5.3 billion takeover of Italian power company Montedison by Italenergia, a holding company led by the Fiat Group. However, the Commission warns that, if Electricite de France, a minority stakeholder in Italenergia, were to gain control of Italenergia, the case would be open to reexamination.

August 28

National Grid PLC of the United Kingdom announces that it is the managing member of a new electricity-transmission alliance that will manage the electric-transmission systems of 10 companies in 11 American states. The new regional transmission organization (RTO) will be called Alliance Transco, and will have a gross book value of $12 billion. The RTO plans to start operating December 15, 2001.

September 2001

September 3

Santa Fe International agrees to buy rival oil and natural gas driller Global Marine for $3 billion in stock. Santa Fe is the smaller company, but is paying a 17 per cent premium for shares in Global Marine. The deal, which has been approved by both boards of directors, creates one of the world's largest drilling companies, GlobalSantaFe Corporation, from two mid-sized companies.

September 3

Libya's Foreign Minister announces that U.S. companies will be given one year to resume oil operations in the country before Libya decides whether their licenses should be revoked and given to other firms. Current U.S. sanctions forbid companies from operating in Libya due to previous Libyan involvement in terrorist acts against the United States and other people.

September 4

Devon Energy agrees to buy natural gas-producer Anderson Exploration of Canada for $3.4 billion. Under the merger plan, Devon will commence a tender offer for at least two-thirds of Anderson's outstanding shares while Anderson agrees not to solicit further offers while also paying a $135 million breakup fee. The deal gives Devon access to large undeveloped gas reserves in Canada. The transaction will increase Devon's proven reserves by 35 per cent, to about 2 billion barrels of oil equivalent. Natural gas production will rise to 2.2 billion cubic feet per day, making Devon the largest independent producer of natural gas in North America.

September 5

The New York Mercantile Exchange (NYMEX) begins trading of 15-day Brent futures contracts, which had previously been traded only on the International Petroleum Exchange in London. The price of Brent crude oil is an important international benchmark used in pricing formulas for up to two-thirds of the world's crude oil.

September 5

Enron Chairman and Chief Executive Kenneth Lay announces that the company will divest itself of $4-$5 billion in assets in the next two years. Enron, one of the world's largest energy companies, is restructuring itself and placing more emphasis on its trading operations.

September 6

Twelve crude-oil pipelines at China's largest refinery, Maoming, explode. About 20,000 people are evacuated and six workers are injured. The refinery's output is lowered by 73,000 barrels per day. Maoming is a unit of the China Petroleum and Chemical Corporation (Sinopec). The oil pipelines begin functioning at a reduced capacity on September 16.

September 7

BP decides to withdraw from negotiations with PetroChina for a stake in the construction of PetroChina's planned $4.8-billion, 2,485-mile natural gas pipeline from Xinjiang to Shanghai. ExxonMobil, Royal Dutch/Shell, and Gazprom are still in negotiations with PetroChina.

September 7

The U.S. Federal Trade Commission approves Chevron's bid to buy Texaco. Texaco must sell its Equilon Enterprises and Motiva Enterprises units in order to complete the $39-billion deal. The new company, ChevronTexaco, will have a market value of over $100 billion, assets of $83 billion, net proven reserves of 11.5 billion barrels of oil equivalent (boe), and daily production of 2.7 million boe.

September 10

Dominion Resources announces that it has agreed to buy Louis Dreyfus Natural Gas for $1.8 billion in cash and stock. Dominion Resources is a natural gas and electricity concern with 2.8 trillion cubic feet (Tcf) of gas reserves and 4 million electricity customers. Louis Dreyfus is an independent natural gas exploration and production company operating primarily in Texas and the Gulf of Mexico, with proven reserves of 1.8 Tcf of natural gas equivalent.

September 11

The largest terrorist attack in world history occurs as 2 hijacked airplanes crash into the twin towers of the World Trade Center in New York City, one hijacked plane crashes into the U.S. Department of Defense's Pentagon headquarters, and another hijacked plane crashes into a rural part of Pennsylvania. The World Trade Center is destroyed, and the Pentagon is heavily damaged. Over 5,000 people die and economic damage is estimated to be in the billions. Aviation is halted in the United States and all major trading markets (including energy) are closed for the remainder of the week. The U.S. government blames the attack on Osama Bin Laden's terrorist network.

September 13

Relative calm returns to world oil markets as U.S. retail gasoline prices return to normal levels and Brent crude oil futures fall back to $28.02 per barrel for October delivery after spiking to above $31.00 in the aftermath of the September 11 attacks. Also, energy trading by Houston energy companies resumes and limited commercial aviation starts.

September 16

The California State Legislature ends its session without a bailout for Southern California Edison, as Governor Gray Davis had requested. A bill that would have let Southern California Edison issue $2.9 billion in bonds was not allowed to come to a vote in the Senate. Southern California Edison, with 4.3 million customers, acquired a debt of $3.9 billion during the California power crisis in the winter of 2001.

September 17

Phillips completes its $7.36-billion stock acquisition of Tosco, after getting approval from the U.S. Federal Trade Commission. This makes Phillips the second-largest oil refiner in the United States, with a capacity to refine 1.7 million barrels per day.

September 17

Major trading markets in the United States, including the New York Stock Exchange and the New York Mercantile Exchange (NYMEX), reopen for the first time since September 11.

September 19

Enron invokes a clause in its Dabhol power plant contract, claiming that because the Maharashtra State Electricity Board has violated its power purchase agreement, the Maharashtra state government and the government of India are liable for $5 billion. India could avoid facing liabilities that could total $5 billion by choosing to settle the dispute by paying offshore sponsors and foreign lenders, according to Enron's Dabhol unit.

September 19

BP and Methanex of Canada conclude a deal to construct the world's largest methanol plant, in central Trinidad. The $400-million plant would begin production in 2003 and have a capacity to produce 5,000 metric tons (5,500 short tons) per day.

September 20

Iraq accuses Kuwait of excessive extraction of the joint al-Ratqa border oilfield. Iraq's foreign minister requests compensation from Kuwait.

September 20

PG&E, California's largest utility, files its bankruptcy plans. PG&E argues that it should reorganize into two utility companies because this would put it in a better position to borrow the money needed to pay off $13.2 billion in debt. Some consumer groups object to the plan, contending that it could lead to higher electricity rates because of reduced State oversight.

September 20

Duke Energy agrees to pay $3.5 billion in cash and stock and to assume $4-$5 billion in debt in acquiring Canadian Westcoast Energy. Duke Energy is the largest U.S. utility owner, and the deal will add U.S. and Canadian natural gas pipelines to its portfolio. The companies' combined natural gas assets will include 18,900 miles of transmission pipes, 58,700 miles of gathering pipeline, 16,500 miles of distribution pipes, 241 billion cubic feet of storage, and 84 processing facilities.

September 24

Crude oil and petroleum products futures fall to their lowest levels in nearly two years amid fears that a recession will reduce energy demand. At the New York Mercantile Exchange (NYMEX), crude oil set for October delivery falls $3.96 to $22.01 per barrel, and crude oil for November delivery falls $3.82 to $22.44 per barrel. Over the past six trading sessions crude oil and gasoline futures have fallen more than 26 per cent and heating oil futures have fallen nearly 29 per cent.

September 25

U.S. Nuclear Regulatory Commission Chairman Richard A. Meserve announces that a comprehensive review of nuclear plant security is underway in light of the September 11 terrorist attacks. One concern is that a fully-fueled airliner could penetrate the concrete dome enclosing nuclear reactors.

September 25

An explosion at the deepest coal mine in the United States, at Brookwood, Alabama, kills thirteen miners in the deadliest domestic mining accident since 1984. The blast occurs at Blue Creek Mine No. 5 when the ceiling in a shaft falls in over a battery charger, sparking naturally occurring methane. It could be weeks or months before production at the mine can resume, as fires continue to rage and the mine is being flooded by water in an effort to extinguish the fires.

September 26

Reliant Resources agrees to buy Orion Power Holdings for $2.9 billion in cash in addition to assuming about $1.8 billion of Orion's debt. Reliant has 20,000 megawatts of capacity in the United States and nearly 3,500 in Europe. Reliant expects the acquisition of Orion will enhance its position as a provider of wholesale power, natural gas, and energy services.

September 26

Federal Energy Regulatory Commission (FERC) Chairman Pat Wood III proposes that electric utilities that refuse to join multi-state regional power grids by December 15 would lose the ability to charge market-based rates and face delays in merger approvals. FERC is pressing for the creation of four super-regional grids to minimize the market power of dominant utilities.

September 27

At its two-day meeting in Vienna the Organization of Petroleum Exporting Countries (OPEC) decides to keep its production quotas unchanged at 23.2 million barrels per day, despite crude oil being at its lowest price levels since 1999.

October 2001

October 2

The California Public Utilities Commission approves a plan that saves Southern California Edison from bankruptcy by allowing the utility to pay off $3.3 billion in debt and eventually to resume normal operations. The plan would keep the current high rates in place for several years. The utility and its shareholders would pay off slightly less than half of the cost of retiring the debt, and the utility's customers would pay off the remainder under the plan. The plan is temporarily blocked for two weeks on October 30, pending a civil suit.

October 4

The U.S. Energy Information Administration (EIA) releases its Winter Fuels Outlook 2001/2002, which predicts lower gasoline, heating fuel, and electricity costs for this winter as compared with the previous one. The main reasons are adequate inventories and declining energy demand because of the slow economy.

October 4

Japan's National Police Agency tightens security around 34 nuclear plants, government buildings, and U.S. facilities in Japan in response to the September 11 terrorist attacks in the United States.

October 7

Crude oil resumes flowing through the trans-Alaska pipeline after workers welded shut a bullet hole that caused 260,000 gallons of oil to spill out. The pipeline, which carries about 17 per cent of the United States' oil production, had been shut down on October 4 after being pierced with a bullet in an apparent act of criminal mischief.

October 7

Air raids by the United States and its coalition partner, Great Britain, begin against Taliban and al-Qaeda targets in Afghanistan, after the Taliban refuse to hand over alleged terrorist mastermind Osama Bin Laden and his associates. The raids are intended to "..disrupt the use of Afghanistan as a terrorist base of operations and to attack the military capability of the Taliban regime," according to U.S. President George Bush.

October 8

The Federal Energy Regulatory Commission (FERC) orders four companies-Dynegy, Mirant, Williams, and Reliant Energy-to give refunds because the prices they charged for electricity in July exceeded federal limits. The amounts were not specified, but the order comes after FERC set price ceilings in June for ten western states that the companies did not abide.

October 8

Burlington Resources agrees to purchase Canadian Hunter Exploration for $2.08 billion. Canadian Hunter Exploration has estimated proven reserves, 90 per cent of which are in Canada, of 1.2 trillion cubic feet (Tcf) of natural gas and 6.2 million barrels of oil and natural gas liquids. Canadian Hunter Exploration also has two million acres that are undeveloped. This deal expands Burlington Resources' reserve base by 12 per cent, to 11.5 Tcf of natural gas equivalent.

October 8

Enron agrees to sell its Portland General Electric Utility to Northwest Natural Gas for $1.88 billion in cash and stock. Northwest Natural Gas also is expected to assume $1.1 billion in debt. Northwest Natural Gas will now have assets of $5 billion, more than 1.25 million customers, 2,000 megawatts of generation capacity, 26,000 miles of electric transmission and distribution lines, and 12,000 miles of natural gas distribution lines.

October 9

Royal Dutch/Shell announces that it will acquire Texaco's interests in two U.S. refining ventures, Motiva Enterprises and Equilon Enterprises, for $2.1 billion in cash, $1.4 billion in debt, and $300 million in pension liabilities. Texaco agreed to sell the companies as part of an agreement with the U.S. Federal Trade Commission in order to complete its merger with Chevron. Motiva will be wholly owned by Shell, and Equilon will be majority owned by Shell and minority owned by Saudi Refining. Motiva and Equilon own over 20,000 gasoline stations and eight refineries in the U.S.

October 9

The U.S. House of Representatives pass a measure urging the Bush administration to pump more oil into the Strategic Petroleum Reserve. The resolution, which is not legally binding, urges that the emergency stockpile be expanded to its full capacity of 1 billion barrels.

October 11

The Federal Energy Regulatory Commission (FERC) approves a plan to reopen the Cove Point, Maryland liquefied natural gas (LNG) terminal by April, 2002. The facility had been closed in the 1980s, as it was then uneconomic, but it May now serve as an alternative to Boston's Everett terminal (see October 16) and El Paso Energy will most likely make use of the facility (see October 18). The Cove Point terminal has connections to a number of important east coast pipelines.

October 15

The first tanker loading of the new $2.5-billion Kazakh-Russia Pipeline takes place. This is a trial run that informally inaugurates the pipeline. Initial capacity of the pipeline is expected to be 28.2 million metric tons per year (around 560,000 barrels per day). The Caspian Pipeline Consortium (CPC), led by ChevronTexaco, runs the pipeline.

October 16

The U.S. Coast Guard lifts a ban on liquefied natural gas (LNG) tankers entering Boston Harbor to makes deliveries to Distrigas' Everett LNG terminal that had been imposed on September 26 in response to the terrorist attacks of September 11. LNG regasified at the Everett terminal normally provides 15 per cent -20 per cent of the natural gas that heats homes and businesses in New England, with the percentage rising to 35 per cent on the coldest days. On October 26, the Mayor of Boston asks a federal court to prevent tankers from entering because he claims there are inadequate disaster response plans.

October 17

Norway's Labor Prime Minister Jens Stoltenberg announces that he will resign effective October 19, so that a new coalition government, headed by Kjell Magne Bondevik and consisting of the Conservative Party and the Christian People's Party, can take power. The incoming government has promised to cut taxes, rationalize state ownership, and boost privatization.

October 17

Venezuela's state-owned natural gas entity, Enagas, announces that it will increase its share of the $2.2-billion Cristobal LNG project from 30 per cent to 60 per cent, sharply reducing the shares of its partners ExxonMobil, Royal Dutch/Shell, and Mitsubishi. It is feared that this May jeopardize the future of the project.

October 18

Crude Oil for November delivery falls to its lowest level since August 1999 on the New York Mercantile Exchange (NYMEX). Light, sweet crude falls 50 cents per barrel to settle at $21.31 per barrel. Brent crude for December delivery closed at $20.36 at London's International Petroleum Exchange (IPE), down 37 cents per barrel. Poor economic prospects in the next few months, and OPEC's inability to respond so far are seen as factors contributing to the sliding prices of crude oil.

October 18

El Paso announces that it will buy 1.8 million tons per year of liquefied natural gas (LNG) from Norway's Snohvit LNG project, led by Statoil, in what will be Norway's first LNG project. Deliveries will begin in 2006 to Williams' Companies' Cove Point, Maryland terminal, though El Paso May make use of other terminals as well.

October 22

OPEC announces that its 10 members with output restrictions implemented only 539,000 barrels per day (bbl/d) of a promised one million bbl/d cut in September.

October 22

The natural gas transportation unit of Brazilian state energy company Petrobras invites private firms to submit requests for the expansion of the BrazBol imports route, a key move toward free access to the pipeline between Brazil and Bolivia. Moreover, Petrobras will not be able to acquire more than a 40 per cent share of new capacities under the rules recently imposed by the National Petroleum Agency (ANP). The Brazilian part of the pipeline is operated by TBG.

October 25

Some 34,000 Brazilian oil workers go on strike against Petrobras, dramatically reducing oil production and refining. Crude oil production falls by 60 per cent -70 per cent and natural gas production by 34 per cent. Eight of eleven refineries cease or reduce operations. Workers claim that Petrobras' $2 billion in profits in the first half of 2001 should be used to raise salaries. Workers end the strike October 29 after accepting a smaller wage hike proposal than they had demanded.

October 28

Mexican President Vicente Fox states that Mexico, despite pressure from OPEC to help reverse lower oil prices, would not cut or freeze Pemex's crude output for now. This will make it more difficult for OPEC's production quota cuts to have a lasting effect on world oil prices.

October 29

ExxonMobil announces that a consortium it leads will spend $4 billion over 5 years to develop large offshore oil and natural gas fields in Russia's far eastern Sakhalin region. The fields are estimated to contain 2.3 billion barrels of oil and 17 trillion cubic feet of natural gas. ExxonMobil will be the operator and own a 30 per cent interest in the fields. Sakhalin Oil and Gas Development of Japan will own 30 per cent, ONGC Videsh of India 20 per cent, Sakhalinmorneftegas-Shelf of Russia 11.5 per cent, and RN-Astra of Russia 8.5 per cent. The total investment could grow to $12 billion over the 30-40 year project life. This is the single largest foreign investment in Russia, as Russia continues to undertake market reforms.

October 29

Experts from OPEC member countries and from non-OPEC from non-OPEC countries Russia, Egypt, Norway, Mexico, Kazakhstan, and Angola, meet in Vienna to discuss the outlook for oil markets, but no recommendations for production cuts are made. Few details of the meeting are released.

October 30

U.S. retail gasoline prices fall to $1.27 per gallon, a drop of 26 cents per gallon since mid-September, and the lowest level since early-January 2000. This is despite the fact that gasoline demand is up 1.7 per cent for the past four weeks, to 8.6 million barrels per day compared to the same period a year ago, according to the U.S. Energy Information Administration.

October 30

The U.S. Federal Aviation Administration (FAA) releases a statement banning aircraft flying under 18,000 feet within a radius of 10 nautical miles of 86 sensitive nuclear sites. This follows a general terrorist threat warning issued by the FBI, in response to which Homeland Security Director Tom Ridge urges U.S. energy companies to be on high alert.

October 31

The U.S. Department of Commerce releases data that shows that the U.S. economy declined by 0.4 per cent for the quarter ending in September. This is the first contraction since the first quarter of 1993.

October 31

OPEC President (and Algerian Oil Minister) Chakib Khelil states that OPEC oil producers are prepared to cut supply to get weak oil prices back up to the group's $25-per-barrel target price. OPEC member Kuwait states that it will support any move that OPEC makes.

November 2001

November 2

The United States recalls its ambassador to Venezuela for consultations, following comments made recently by Venezuelan President Chavez about the United States war on terrorism.

November 2

Crude oil production begins at BP's Northstar field located off Alaska's north shore. Production of 65,000 barrels per day is expected by next quarter. Northstar is the first field to be successfully developed in federal waters off Alaska's North Slope, coming online 18 years after it was initially discovered. The Northstar production module is the largest ever built in Alaska, with 22 wells planned. Recoverable crude oil reserves are estimated at 175 million barrels.

November 6

Crude oil for December delivery on the New York Mercantile Exchange (NYMEX) falls to a two-year low after members of the Organization of Petroleum Exporting Countries (OPEC) warn that a downward price spiral could occur if major non-OPEC oil exporters do not reduce oil production. The NYMEX price settles at $19.92 per barrel, down 10 cents per barrel from the low of November 5, and the first time it has been under $20 per barrel since mid-1999.

November 6

ChevronTexaco signs an exploration and production sharing agreement with Bahrain. ChevronTexaco expects to drill its first well by the end of 2002. Bahrain recently gained sovereignty over the Hawar Islands through an International Court of Justice decision in March 2001, opening a formerly disputed area of the Persian Gulf to exploration.

November 7

The Federal Energy Regulatory Commission (FERC) orders California's electric grid operator (ISO) to begin sending utility bills to the State, rather than to utility companies, for more than $1.6 billion of power used to stabilize the electric system since the State's power crisis last year. California assumed payment responsibility for the State's cash-strapped utilities in January.

November 7

A U.S. Federal Appeals Court rules that a $5 billion punitive damage judgment against ExxonMobil stemming from the Exxon Valdez oil spill in Alaska is excessive. However, the court rejects ExxonMobil's claims that the plaintiffs are entitled to no punitive damages. Instead, the case returns to federal trial court to have the award reduced.

November 8

Brunei announces the creation of a national oil company. According to an official statement, the company will "consolidate and mobilize the petroleum industry of Brunei." Brunei is the third-largest oil producer and the fourth-largest liquefied natural gas (LNG) producer in Southeast Asia. Royal Dutch/Shell and TotalFinaElf are active in Brunei.

November 9

The U.S. Energy Information Administration (EIA) reports that energy-related carbon dioxide emissions rose 3.1 per cent in the year 2000, and have risen 14 per cent since 1990. EIA reports that the emissions increase between 1999 and 2000 was a result of strong economic growth and more use of fossil fuels due to cold weather and a drought that reduced hydroelectric power generation.

November 9

Enron, the world's largest electricity and natural gas trading company, agrees to an all-stock takeover by former competitor Dynegy. ChevronTexaco, a 27 per cent stakeholder in Dynegy, will immediately inject $1.5 billion cash into Enron, and an additional $1 billion into the combined entity. The merged company will be called Dynegy Inc., and Dynegy executives will occupy all top positions. The deal is expected to take at least six months to close.

November 10

An agreement is reached at talks in Marrakech, Morocco, on rules for implementation of the Kyoto climate change treaty. Rules for joint implementation projects, the Clean Development Mechanism, and funding for less developed countries are elaborated. The United States does not participate actively in negotiations or agree to the rules.

November 13

U.S. President George Bush orders that the Strategic Petroleum Reserve be filled to capacity over the next few years. The reserve has a capacity of about 700 million barrels of oil, and now contains about 545 million barrels of oil. The Strategic Petroleum Reserve is intended, in the short run, to smooth out price spikes and shortages caused by a supply disruption.

November 13

President Hugo Chavez of Venezuela announces new hydrocarbon laws that will increase production royalty payments to the government by as much as 80 per cent. These laws also reverse the policy of allowing foreign oil companies to hold majority partnerships with the state oil company PDVSA. There are concerns that these laws May reduce foreign investment in Venezuela's hydrocarbons sector.

November 14

At its meeting in Vienna, Austria, OPEC announces that it intends to cut its crude oil output quotas by 1.5 million barrels per day effective January 1, but only if non-OPEC producers cut their output by 500,000 barrels per day as well. The production cuts are an effort to steady or raise world oil prices, which have fallen markedly since September.

November 14

Mexico pledges to cut its crude oil exports by 100,000 barrels per day as of January 1, 2002, in order to strengthen world oil prices in concert with OPEC actions. Mexico is the world's seventh-largest crude oil producer and exported 1.6 million barrels per day in September 2001.

November 15

The NYMEX crude oil price for December delivery falls 11.6 per cent to $17.45 per barrel, after Russia appeared to reject OPEC's proposal to cut oil production. Over the last week the December delivery price has fallen 21 per cent. Oil prices have not been this low in over two years.

November 18

Phillips Petroleum and Conoco agree to merge into a new company to be called ConocoPhillips, which would be the third-largest oil and natural gas company in the United States, and the sixth-largest in the world, in terms of production. The company also would be the largest gasoline retailer in the United States and the fifth-largest refiner in the world. Combined total reserves of the new company would be 8.7 billion barrels of oil equivalent, and production would be 1.7 million barrels of oil equivalent per day. The new company expects to be able to compete more effectively with its larger rivals and to achieve significant cost savings. The new company will be based in Houston, Texas.

November 20

Non-OPEC oil exporter Oman indicates that it will cut production by about 3 per cent or 25,000 barrels per day, in order to reduce oil supply and cooperate with OPEC's contingent cuts.

November 22

Norway's Oil and Energy Minister announces that he has a mandate to reduce the country's current crude oil production of about 3.2 million barrels per day by as much as 200,000 barrels per day. The cut would be pro rata, meaning that all companies will have to participate in the output cut. The timing and extent of the actual cut will depend on the overall package of supply reduction agreed upon by OPEC and other large non-OPEC producers. Brent crude oil at the International Petroleum Exchange (IPE) rises $1.17 per barrel to $19.02 per barrel, on the news.

November 26

The European Commission approves German utility E. On's $13.5 billion takeover of U.K. utility Powergen (which also owns LG&E of the United States), creating one of the world's largest electric utilities. The takeover had previously been approved by American and British regulators.

November 26

The U.S. Energy Information Administration (EIA) reports in its Weekly Survey of Gasoline Prices that U.S. gasoline prices have fallen to their lowest levels since July 1999, the result of a supply glut. The average price for regular unleaded gasoline fell 4 cents per gallon over the last week to $1.127 per gallon, down 38 cents per gallon from a year ago and the lowest level for any week since early July 1999.

November 27

Iraq rejects a call by U.S. President George Bush to let United Nations weapons inspectors back into the country to determine whether it is building weapons of mass destruction. An Iraqi spokesman states that, before asking Iraq to allow weapons inspectors to return, the United Nations should lift the 11-year-old sanctions on Iraq and the West should abolish the no-fly zones in northern and southern Iraq.

November 28

Dynegy withdraws from its merger offer with Enron. Without an infusion of capital from a merger or other source, it will be difficult for Enron to continue payments to its creditors in light of the downgrade of Enron's debt rating to single-B-minus earlier in the day by Standard & Poor's. If Enron is unable to repay or refinance its debt, the company, formerly one of the largest energy companies in the world, May have to declare bankruptcy.

November 29

The United Nations Security Council unanimously approves a resolution extending the Oil-for-Food program in Iraq for another six-month period. This resolution allows Iraq to sell unlimited quantities of oil on the condition that the proceeds are used to buy food, medicine, and other humanitarian goods, and to pay war reparations. This resolution also calls on members of the Security Council to agree by May 31, 2002, on a list of "dual use" items that would require United Nations approval before Iraq could import them through the program.

December 2001

December 2

Enron files for Chapter 11 bankruptcy in the Southern District of New York for 14 affiliated entities, including Enron, Enron North America, Enron Energy Services, Enron Transportation Services, Enron Broadband Services, and Enron Metals & Commodity Corporation. Enron was formerly the world's largest electricity and natural gas trading company, and the seventh-largest publicly-traded energy company in the world. Enron also files a $10 billion lawsuit against Dynegy, alleging breach of contract, in connection with Dynegy's November 28 termination of its proposed merger with Enron.

December 3

Iran condemns a November 30 agreement between Azerbaijan and Kazakhstan on sharing the resources of the Caspian Sea, labeling it "provocative." Iran believes that only a comprehensive agreement involving all five states bordering the Caspian Sea should be negotiated.

December 5

Sempra Energy announces that it has signed a memorandum of understanding with a consortium led by Repsol, BP, BG, and Brindas of Argentina to bring up to 800 million cubic feet of liquefied natural gas (LNG) per day from Bolivia to Sempra's proposed Baja California LNG terminal where it would be regasified. The natural gas would then be piped to its final destinations in California and various areas of Mexico.

December 5

Bolivia gives authorization to Sociedad Transierra, an international consortium of energy companies, to build a 280-mile pipeline to Brazil. Sociedad Transierra is comprised of Brazil's state energy company Petrobras, Spain's Repsol-YPF, and France's TotalFinaElf. Construction of the pipeline is set to begin in mid-January and will transport 777 million cubic feet of natural gas per day to Brazil.

December 6

The High Court of Bangladesh blocks for at least three months a government move to export natural gas to India through a pipeline. Bangladesh has proven reserves of more than 12 trillion cubic feet of gas, enough to cover domestic requirements for 15 years, according to an unofficial estimate. But the country's opposition parties, including former Prime Minister Hasina's Awami League, and some local experts are opposed to the export of gas unless there is a surplus after estimated domestic needs for the next 50 years have been met.

December 6

The U.S. Energy Information Administration (EIA) predicts that, for the first time since 1991, U.S. petroleum demand will decline year on year. For 2001, EIA expects U.S. petroleum demand to average 19.69 million barrels per day, down 10,000 barrels per day from 2000. The decline in demand, especially for jet fuel since the September 11 attacks, has contributed to a buildup of U.S. oil inventories and has helped keep crude oil and petroleum product prices relatively low.

December 7

The California State Government makes its first payment of $404.8 million to the State's non-profit grid operator following an order by the Federal Energy Regulatory Commission (FERC) on November 7 that sent the grid operator's power bills to the State rather than to utilities. In January 2001, the State of California bought power on behalf of PG&E and Southern California Edison after the two companies had trouble meeting their fiscal obligations, but the State has since then been reluctant to pay the high rates charged by generators in that period.

December 10

In Venezuela, a nationwide, 12-hour strike takes place during which thousands of businesses close and millions of Venezuelans stay home in order to protest the economic policies of Venezuelan President Hugo Chavez. The strike is led by Fedecamaras, Venezuela's largest private business chamber.

December 12

The state-owned Turkish Petroleum Corporation (TPAO) announces that the company plans to explore for oil in the Kurdish-controlled area of Northern Iraq. On December 14, TPAO announces that it has signed a deal with the Iraqi Oil Ministry to drill 20 wells at the Khurmala field near Kirkuk, just south of Kurdish-controlled areas. Later in the month, TPAO denies having plans to explore Kurdish-controlled areas, but claims the contract with the Iraqi Oil Ministry's North Oil Company is UN-approved. A UN spokesman later denies this.

December 14

The U.S. Government announces that 2.9 million barrels of crude oil are scheduled to be added to the U.S. Strategic Petroleum Reserve in December, and a further 5.6 million barrels in January. This nearly triples the amounts scheduled in mid-November.

December 14

A new rule from the U.S. Department of Energy takes effect such that the U.S. government no longer must prove that the Yucca Mountain nuclear waste storage facility in Nevada would prevent radioactive contamination of the environment through its underground rock formations. Rather, a combination of advanced storage containers and natural geological barriers would be able to satisfy environmental standards for protecting the ground water and atmosphere from the release of radioactive materials.

December 17

The U.S. Energy Information Administration (EIA) reports that the average U.S. retail price of unleaded regular gasoline fell 3.6 cents per gallon over the last week to $1.059 per gallon, the cheapest level since mid-March 1999. The price is down 36 cents per gallon from a year ago and is at the lowest level since the week of March 22, 1999, according to EIA's weekly survey of more than 800 service stations.

December 18

Royal Dutch/Shell announces that it has completed the first successful production tests in a Brazilian offshore oil field in deep waters. Shell said a short-term production test at the BS-4 block in the Santos basin, some 124 miles off the Rio de Janeiro state coast, showed "favorable (crude) flow" and indicated estimated reserves of 300-500 million barrels, although the company said the commercial viability still had to be confirmed. This is the first significant result by an international operator since the opening of the oil production and exploration sector in Brazil.

December 19

The Federal Energy Regulatory Commission (FERC) outlines its schedule for a proposed rulemaking to set national standards for competitive power markets. FERC also unveils a staff paper with a proposal for a standardized market design in preparation for the formal launching of the rulemaking process in March. FERC also announces that it would delay, possibly for several months, a strict new market power rule that could limit a utility's electricity rates if it controls too much generation in a given area.

December 19

A 30-year, $6 billion deal between Venezuela and China for the export of 6 million tons per year of the Orimulsion boiler fuel is concluded. China's state-owned National Petroleum Corporation (CNPC) and Venezuela's Orimulsion boiler fuel producer Bitumes de Orinoco, or Bitor, will build a 6.5-million-metric ton, $300 million Orimulsion plant in Venezuela. Most of the plant's production will be exported to China. The plant is scheduled to come online by early 2004. CNPC has a 70 per cent stake in the venture and Bitor, which is owned by Petroleos de Venezuela, has the remaining 30 per cent, but Bitor has a "golden share" giving it decision-making power.

December 20

Iraq announces that it will renew its oil export agreement with Jordan. Iraq will supply Jordan with around 110,000 barrels per day of crude oil and petroleum products next year. Jordan's oil purchases from Iraq are exempted from United Nations sanctions. The deal is worth about $700 million.

December 21

The Indian Oil Minister announces that India will build a strategic reserve of crude oil and petroleum products. When India had previously considered building a strategic reserve in 1995, the plans sought to build tanks to stock 12.55 million metric tons of petroleum products, and also proposed facilities to store 4.25 million metric tons of imported crude oil.

December 23

Qatar's state oil company signs a $3.5 billion deal with Dolphin Energy for the construction and development of a natural gas pipeline from Qatar's offshore reserves to the United Arab Emirates (UAE). The Dolphin Gas Project plans to pump 2 billion cubic feet per day of natural gas from the Khuff reserves in Qatar's North Field to the gas gathering and processing plant at Ras Laffan, and then to a 275-mile subsea pipeline to Tawilah in Abu Dhabi and Jebel Ali in Dubai. Dolphin is owned by the UAE's state-owned Offsets Group (70 per cent) and by TotalFinaElf (30 per cent).

December 26

Phillips Petroleum states that the government of East Timor has agreed on tax rates and other conditions, thereby allowing work to proceed on the $1.6 billion Bayu-Undan natural gas project in the Timor Sea off northern Australia. It had been delayed on August 1, 2001 because of taxation uncertainties between Phillips and East Timor. Phillips will make additional on-shore infrastructure and community investments in East Timor.

December 26

Crude oil prices on the New York Mercantile Exchange (NYMEX) record one of their largest one-day jumps of the year as traders become convinced that OPEC will follow through on production cuts. Prices per barrel for February delivery settle at $20.27 per barrel, an increase of $1.65, or 8.4 per cent higher than the December 21 closing price (the last day of trading before the holiday weekend). Also contributing to the price increase was the return of cold weather to the northeastern United States and forecasts that show that the cold weather pattern May continue. Nevertheless, prices are still considerably lower than one year ago.

December 28

OPEC oil ministers meeting in Cairo agree to reduce their crude oil output quotas by a combined 1.5 million barrels per day (about 6.5 per cent) for a six-month period beginning January 1, 2002. OPEC ministers also announce that they will meet again in March. OPEC received commitments for 462,500 barrels per day of the 500,000 barrels per day in cuts that it had requested from non-OPEC exporters, close enough to the target for OPEC to go ahead and implement its concomitant cuts. This month, Russia announced an export cut of 150,000 barrels per day on December 5. Oman announced a cut of 25,000 barrels per day on December 11, and raised it to 40,000 barrels per day on December 20. Angola announced a cut of 22,500 barrels per day on December 14. Norway announced a cut of 150,000 barrels per day on December 17. Mexico had already announced an export cut of 100,000 barrels per day in November.

December 31

Leaders of the member states of the Gulf Cooperation Council (Bahrain, Kuwait, Oman, Qatar, Saudi Arabia and the United Arab Emirates) sign an agreement at the end of a two-day annual summit in Oman to move forward the setting up of a customs union to 2003 from 2005 and to establish a single currency by 2010–part of a planned joint trade zone.

100

Monthly Energy Chronology 2002

The following information contains a listing of the major energy events that occurred in 2002.

January 2002

January 1

The OPEC crude oil production quota cuts of 1.5 million barrels per day, announced on December 28, officially go into effect for six months. Crude oil production or export cuts of 462,500 barrels per day by five non-OPEC oil exporters also go into effect.

January 2

At a rally celebrating new hydrocarbons laws that were announced on November 13, 2001, and went into effect yesterday, Venezuelan President Chavez announces the need for "a new oil strategy." The new hydrocarbons laws give the state more control over the petroleum industry and impose higher royalty rates on companies operating Venezuela's oil fields.

January 2

The U.S. Minerals Management Service approves the use of floating production, storage, and offloading (FPSO) vessels in the Gulf of Mexico on a case-by-case basis. This will potentially make it easier to produce oil and natural gas from deepwater fields that are not near any existing infrastructure.

January 3

Royal Dutch/Shell's Shell Gas & Power agrees to purchase all the available expansion capacity at El Paso's newly reopened liquefied natural gas (LNG) terminal at Elba Island, Georgia. Under the terms of the 30-year deal, Shell will receive 3.3 billion cubic feet of LNG storage and 360 million cubic feet per day of regasified LNG send-out capacity. This is equivalent to 2.5 million metric tons of LNG per year.

January 5

Saudi Arabian Oil Minister Ali Al-Naimi announces that OPEC's price band mechanism is suspended due to imbalances in the world oil market. He states that OPEC's top priority is to stabilize the market.

January 8

Royal Dutch/Shell has reportedly reached an agreement with the Chinese government to help construct a 2,500-mile natural gas pipeline from the Tarim oil and natural gas region in remote western China to Shanghai on the coast. The agreement represents evidence of foreign confidence in the project and lays out the framework for the joint-venture company composed of PetroChina, Gazprom of Russia, and Shell. The pipeline is expected to cost about $18 billion.

January 8

Royal Dutch/Shell and its joint-venture partners Nigeria National Petroleum Company, TotalFinaElf, and Eni's Agip announce a five-year investment of $7.5 billion in the oil and natural gas industry in Nigeria. A large part of the investment will go to the Nigeria Liquefied Natural Gas project, the biggest industrial project in sub-Saharan Africa.

January 9

U.S. Secretary of Energy Spencer Abraham announces that the Partnership for a New Generation of Vehicles program, started in 1993 in an effort to develop mass-produced vehicles that would get 80 miles per gallon of gasoline by 2004, will be replaced by a new program called Freedom Car. The Freedom Car program will emphasize developing fuel-cell vehicles, powered by oxygen and hydrogen, by an unspecified later date.

January 9

Mexican President Vicente Fox calls for reform of the country's electricity sector in order to encourage more private investment in the sector at a ceremony inaugurating six power plants in the state of Sonora that will add 1,519 megawatts to the national grid. Reform of the electricity sector is more likely now that several tax reforms and the 2002 budget have been passed.

January 9

The U.S. Department of Justice confirms that a criminal probe of Enron has been launched. A task force has been formed to investigate whether the former giant energy company defrauded investors by deliberately withholding or falsifying crucial financial information. The U.S. Securities and Exchange Commission has been investigating Enron since October 2001. A number of civil suits have already been filed against Enron, which declared bankruptcy in December 2001.

January 9

Argentine President Duhalde proposes that oil companies operating in Argentina contribute $1.2 billion to a stabilization fund, instead of paying a 20 per cent tax on oil exports. Oil companies operating in Argentina are already expected to lose millions of dollars because of the default on Argentine government debt and the devaluation of the Argentine peso. The most heavily exposed company is Repsol-YPF, which could lose over $2 billion.

January 10

Kuwait creates a new company, Kuwait Gulf Oil Company, to handle the country's interests in the Neutral Zone, a border area shared with Saudi Arabia. The new company will take over when Japan's Arabian Oil Company's 40-year concession in the Neutral Zone expires in January 2003. Oil production from the Neutral Zone (by Saudi Arabia and Kuwait combined) was about 640,000 barrels per day in 2001.

January 10

The U.S. Department of Energy recommends construction of the Yucca Mountain nuclear waste storage facility in Nevada. The project is expected to cost $40 to $50 billion and be able to store 77,000 tons of radioactive waste. The final decision still requires President Bush's approval and, if vetoed by the governor of Nevada, Congressional approval.

January 14

According to preliminary data reported today by the U.S. Energy Information Administration, Mexico was the leading source for U.S. crude oil imports in November 2001 for the second consecutive month, having displaced Saudi Arabia from the top spot in October. Saudi Arabia had been the leading supplier since February 2000.

January 15

The Nigerian government asks oil producing companies in the country to cut output by 6 per cent in order to achieve better compliance with its OPEC production quota cut commitment.

January 15

The U.S. Department of the Interior announces that royalty fees for deepwater fields in the Gulf of Mexico will be reduced as of February 14 for any firm able to demonstrate that it would be a financial burden to drill in the deep waters, and also that a reduction in royalty payments would be needed to make oil projects economically feasible.

January 16

The American Petroleum Institute (API) releases data showing that oil demand dropped in the United States in 2001, the first annual decline in 10 years. According to the API's report, petroleum deliveries declined by 0.6 per cent, to 19.587 million barrels per day in 2001 from 19.701 million barrels per day in 2000. The report also shows that that in 2001, U.S. crude oil output rose for the first time in a decade to 5.85 million barrels per day, a 0.5 per cent increase over 2000.

January 16

New oil reserves are found at the Buzzard field in the British North Sea, representing the largest find in the North Sea in over a decade. The field is expected to yield 400 million barrels, far in excess of initial estimates of 100-200 million barrels.

January 16

Labor unions begin a general strike in Nigeria as rioters roam the streets of the country's major cities. The government had declared on January 15 that a strike by the 29-union Nigeria Labor Congress was illegal. The strike is in protest of the government's decision on January 2 to raise gasoline and diesel prices. The strike is called off on January 17 after a Nigerian court rules that the strike action is illegal.

January 17

Crude oil prices on the New York Mercantile Exchange close below $18 per barrel for the first time since mid-November at $17.97 per barrel for February delivery. Prices have been in the $18-to-$22-per-barrel range for several months. This latest decrease reflects a generally bearish attitude by traders in the face of weak world oil demand.

January 17

Mexico's state oil company PEMEX declares force majeure on 100,000 barrels per day of oil exports in order to comply with the January 1 export cut Mexico agreed to with OPEC in December in order to shore up oil prices.

January 21

Gaz de France signs a 20-year deal with BG Group, Edison International, the Egyptian General Petroleum Corporation, and the Egyptian Natural Gas Holding Company to import 3.6 million metric tons of liquefied natural gas (LNG) per year for 20 years starting in 2005. This is Egypt's second major LNG export deal. As part of the deal, a $900 million liquefaction plant will be constructed at Idku, near Alexandria.

January 22

The U.S. Department of Energy opens the bidding process for oil companies to deliver 22 million barrels of crude oil to the Strategic Petroleum Reserve instead of making cash royalty payments. The royalty-in-kind oil is the first phase of the Bush administration's plan, announced last November, to fill the Strategic Petroleum Reserve to its capacity of 700 million barrels.

January 22

Germany's federal competition office blocks E.On, Germany's second largest electricity company, from taking control of Ruhrgas, Germany's dominant natural gas transmission company, that controls about 60 per cent of natural gas sales. However, the companies still hope to persuade Chancellor Schröder and Economics Minister Müller to override the decision. The proposed deal is important on a European level because if the takeover does go through, it will be seen as a blow to the efforts of European anti-trust regulators and the European Commission to liberalize energy markets.

January 22

The Energy Ministers of Iran and Turkey inaugurate the official opening of a new Iran-Turkey natural gas pipeline. Turkey will import 106 billion cubic feet of natural gas in the first year, with imports rising to 353 billion cubic feet per year in 2007 under a 22-year contract. Iran hopes that natural gas exports to Turkey May eventually be able to transit onward to lucrative European markets.

January 23

An U.S. official states that the United States and Libya have held "positive" negotiations that could lead to the lifting of United Nations sanctions against Libya. The official stated that "If the Libyans comply with the U.N. Security Council resolutions, then the U.N. Security Council will look at lifting the U.N. sanctions on Libya." In August, President Bush signed a five-year extension of the

Iran Libya Sanctions Act (ILSA). ILSA sanctions foreign companies that provide new investments of over $40 million for the development of petroleum resources in Iran or Libya, or that violate existing United Nations prohibitions against trade with Libya.

January 27

PanCanadian Energy agrees to purchase Alberta Energy for over $6 billion in stock. The combined company, to be named EnCana, will be Canada's largest natural gas producer. According to the companies, EnCana would produce 2.7 billion cubic feet per day of natural gas and 255,000 barrels per day of oil and natural gas liquids in 2002.

January 28

Tanker-tracking consultancy Petrologistics announces that OPEC countries, excluding Iraq, cut crude oil production in January by 1.34 million barrels per day. The January estimate puts OPEC production 659,000 b/d over its new production quota of 21.701 million barrels per day, which OPEC confirmed at its December 28 meeting. In contrast, a Dow Jones survey of analysts on January 29 puts OPEC's combined cuts, excluding Iraq, at 884,000 barrels per day.

January 28

The state-run Japan National Oil Corporation (JNOC) establishes an 11-member panel led by the company's president in order to take "full-scale" preparations for dissolving itself. Last month, the cabinet approved the dissolution of the heavily indebted company as part of government structural reforms.

January 28

At a meeting of the United Nations Security Council's committee on Iraq trade sanctions, Britain requests a formal "clarification" from Security Council member Syria on allegations that Iraqi crude oil is flowing through a pipeline to Syria and being exported, or at least substituted in Syrian refineries allowing for more Syrian crude oil exports, in violation of United Nations sanctions. Analysts have placed the amount of crude oil being sent from Iraq to Syria at between 150,000 and 200,000 barrels per day.

January 28

British officials announce that the policy of setting Iraqi oil prices retroactively, enforced by the United States and Britain, has kept $25 million from Saddam Hussein's government. The U.S./ British policy of blocking or holding Iraqi price proposals has meant the United Nations sanctions committee has had to price retroactively, lowering prices for Iraq and leaving less room for oil middlemen to pay illegal surcharges to Iraq.

January 29

Brazil reportedly plans to invest $6.9 billion in 2002 to expand and modernize its power plants in order to ease the possibility of electricity shortages of the type that occurred throughout the country in 2001. The investment will be made through 63 government-controlled power companies.

January 29

U.S. President George Bush delivers his State of the Union address. In his speech he identifies Iraq, Iran, and North Korea as part of an "axis of evil" that supports terrorism. President Bush also states, "The United States of America will not permit the world's most dangerous regimes to threaten us with the world's most destructive weapons."

January 30

Data released by the U.S. Energy Information Administration today shows that U.S. refinery capacity utilization has decreased by 5.4 percentage points in the past two weeks, dropping to just 86.8 per cent in the week ended January 25. This is a result of poor petroleum product demand, low refinery profit margins, and more than adequate product supply.

January 30

The state oil companies of Kuwait and Qatar and ExxonMobil Middle East Gas Marketing sign an agreement for natural gas to be transported by pipeline from Qatar to Kuwait for use in power plants. First shipments are expected to arrive in 2005, and total value of the deal is expected to be about $2 billion. Details of the deal are expected to be finalized by mid-2002.

January 30

Petro-Canada states that it has agreed to buy the international oil and natural gas operations of Veba Oil & Gas from Veba Oel AG and its parent company BP, for about $2.0 billion. Petro-Canada will acquire production and exploration interests mainly in the North Sea, North Africa, and northern Latin America, expanding the company beyond its current interests, which are almost exclusively in Canada and the United States. The deal is expected to close between May and September 2002. The combined firm would have production of 400,000 barrels per day of oil equivalent.

January 31

Russian Prime Minister Mikhail Kasyanov signs a resolution that reduces the export duty on fuel oil in order to ease a glut on Russia's domestic market following Russia's December decision to cooperate with OPEC by trimming exports.

January 31

In response to President Bush's State of the Union address two days earlier, former Iranian President Rafsanjani calls on Muslim oil-exporting countries to boycott the United States. Iran currently does not export oil to the United States because of sanctions.

February 2002

February 1

An explosion and fire occur at an oil-gathering center near Kuwait's northern Rawdatain oil field, killing four workers and injuring seventeen. As a result, about 600,000 barrels per day of crude oil production go offline. However, Kuwait's Oil Minister, Adel al-Subaih, declares that the use of oil from stockpiles and increased production from other fields will mean that exports will be unaffected in the next two weeks. Some production in the area gradually returns throughout the month as repairs proceed. The Kuwait Petroleum Corporation declares force majeure on liquefied petroleum gas (LPG) exports because of damage to the LPG booster station. Oil Minister al-Subaih resigns later in the month, taking responsibility for the incident, though reportedly also in protest of cost cutting and political interference in Kuwait's oil industry.

February 4

PetroChina, China's largest oil and natural gas producer, announces that it has reached an agreement with a consortium led by Royal Dutch/Shell, and including Gazprom and Hong Kong & China Gas, to build a 2,600-mile natural gas pipeline from the natural gas-producing Tarim basin in western China to Shanghai on the coast, with deliveries to start in 2004. The cost of the pipeline is expected to be about $18 billion. Chinese authorities also give Shell the right to discuss investment opportunities in the natural gas pipeline project with ExxonMobil.

February 6

Saudi Arabia's Crown Prince Abdullah states that Saudi Arabia is facing a "suffocating" economic downturn and that government spending must be cut. The Saudi American Bank (SAMBA) forecasts a 2 per cent decline in gross domestic product (GDP) in Saudi Arabia in 2002, due in large part to low world oil prices.

February 6

Nigeria cuts the typical size of its crude oil loadings by 5 per cent in order to better comply with its promised OPEC quota cuts. Nigeria has a new OPEC quota of 1.787 million barrels per day, effective from January 1, but output was 2.094 million barrels per day through January 20, according to Nigeria's Department of Petroleum Resources.

February 7

Turkish government officials state that Turkey aims to import about 80,000 barrels per day of Iraqi Kirkuk crude oil in 2002, an increase over the 50,000 barrels per day imported in 2001. The oil, which is taken into Turkey by truck, is transported by Turkey's foreign trade undersecretariat, for refining by Tupras. The trade is outside of the United Nations oil-for-food program.

February 8

Colombia's state oil company Ecopetrol announces two substantial discoveries of crude oil, each estimated at over 100 million barrels, in the Piedemonte and Upper Magdalena Valley of Colombia. The new fields are expected to begin full-scale production in 2005.

February 8

The U.S. Minerals Management Service announces that it has awarded the first four contracts to deliver royalty-in-kind crude oil to the Strategic Petroleum Reserve. The contracts were awarded to ChevronTexaco, Williams, Equiva, and ExxonMobil for some 60,000 barrels per day of royalty oil from federal properties in the Gulf of Mexico.

February 9

Venezuelan President Hugo Chavez replaces General Guaicaipuro Lameda with central bank Vice President Gaston Parra as the head of the state-owned Petroleos de Venezuela, S.A. (PdVSA), South America's largest oil company. General Lameda's replacement comes amidst public expression of discontent by several military officers with President Chavez's leftist government. On February 20, the Ministry of Energy and Mines announces the replacement of five of the seven members of the board of directors of PdVSA with new board members more closely aligned with President Chavez. This move sparks daily protests by PdVSA employees.

February 11

Oil prices rise to a one-month high as rumors of a very large purchase of Brent crude by Royal Dutch/Shell trading subsidiary Equiva for the U.S. Department of Energy's Strategic Petroleum Reserve (SPR) stoke fears that there could be a supply shortage for Brent crude oil in March. Brent crude oil for March delivery settles $1.63 per barrel higher, at $21.35 per barrel. However, oil prices ease somewhat the next day, with Brent falling 32 cents per barrel. As Equiva has the option to deliver the 18.6 million barrels of sweet crude oil to the SPR between April 1, 2002 and May 1, 2003, it is not clear that the company will meet its commitment by taking most of the Brent-for-March-delivery market, though a Shell executive states that the company sees "potential demand for all the Brents in March."

February 12

A spokesman for Marathon Oil announces that the U.S. State Department has given Marathon, as well as Conoco and Amerada Hess, permission to begin renegotiating dormant oil field contracts with Libya. The approval was issued January 22, 2002. In September 2001, Libya's Foreign Minister announced that U.S. companies would be given one year to resume oil operations in the country before Libya decides whether their licenses should be revoked and given to other firms.

February 13

Venezuelan President Hugo Chavez decides to let the Venezuelan bolivar float against the U.S. dollar. The bolivar loses 19 per cent of its value on its first day of trading under the new system. The decision comes as part of a package of fiscal and monetary reforms that are aimed at closing a projected government budget deficit. Low world oil prices have reduced projected government revenues for 2002. Oil revenues account for about half of government revenues and about one third of Venezuela's gross domestic product (GDP).

February 13

Iraq says that it will not allow United Nations (U.N.) arms inspectors to return to Iraq. Iraqi Vice President Taha Yassin Ramadan states, "There is no need for the spies of the [U.N.] inspection teams to return to Iraq since Iraq is free of weapons of mass destruction." The United States has hinted that actions May be taken against the Iraqi government if U.N. arms inspectors are not allowed to return.

February 13

The government of Argentina announces a 20 per cent tax on energy exports. Argentina exports over 300,000 barrels per day of oil, as well as some natural gas, worth about $2.4 billion and $637 million, respectively, in 2001. The companies most affected include Repsol-YPF, ChevronTexaco, Perez Companc, and U.S. Pan American Energy.

February 13

El Paso Energy announces that it will build a $450 million, 380-mile, 500,000-barrel-per-day oil pipeline from the western Gulf of Mexico to Port Arthur and Texas City, Texas. The pipeline is expected to begin functioning in the third quarter of 2004.

February 13

The Federal Energy Regulatory Commission (FERC) agrees to launch a probe into whether Enron and other energy traders deliberately inflated electricity prices during the California power crisis. FERC has come under pressure from Democratic legislators from the West Coast to undertake such a probe.

February 14

The U.S. Nuclear Regulatory Commission announces that it is increasing security at U.S. nuclear facilities, including 104 reactors. The new measures are the result of a review by the Nuclear Regulatory Commission of nuclear plant security that was undertaken in response to the September 11 terrorist attacks.

February 14

U.S. President George Bush announces a plan that addresses the issue of "greenhouse gas" emissions, linked by many scientists to global warming. The plan proposes spending on research and new technology, plus tax incentives to promote voluntary (as opposed to mandatory) reductions. The plan links reductions in carbon emissions to U.S. economic output, focusing on emissions intensity per unit of GDP rather than on fixed targets.

February 15

U.S. President George Bush authorizes the construction a large, centralized site for the storage of nuclear waste at Yucca Mountain, Nevada, after the Department of Energy recommends its construction. In his letter notifying the U.S. Congress of his decision, Bush states, "Proceeding with the repository program [at Yucca Mountain] is necessary to protect public safety, health, and the nation's security." Nevada Governor Kenny Guinn has indicated that he will oppose the project, so Yucca Mountain will require congressional approval to go forward. The site's selection is also being challenged in court.

February 18

ExxonMobil signs an agreement with the government of Papua New Guinea for the construction of a $3.5 billion sub-sea natural gas pipeline from that country to Queensland, Australia. The other partners in the project include: ChevronTexaco, Oil Search, Orogen Minerals, and Japan PNG Developments, as well as Mineral Resources Development Company, which represents the interests of local landholders. The 1,988-mile pipeline would transport natural gas from the Hides and Kutubu gas fields in the Central Highlands of Papua New Guinea, to a coastal processing facility and then via a section of sub-sea pipeline to northern Australia. The pipeline would supply Australia with around 6 trillion cubic feet of natural gas over a thirty-year period.

February 19

Oman's oil minister states that the non-OPEC oil exporter will continue to abide by its 40,000-barrel-per-day production cut agreement with OPEC until September 2002.

February 20

Iranian Oil Minister Bijan Zanganeh asserts that Iran will proceed with oil and natural gas projects in the Caspian Sea, despite a lack of demarcation of sectors among the five littoral states. Zanganeh adds, "[We] will prevent the activities of others in the parts we consider to be ours."

February 20

Colombian President Andres Pastrana breaks off peace talks with the Revolutionary Armed Forces of Colombia (FARC) guerrillas and orders the Colombian army to re-enter FARC's demilitarized area. President Pastrana takes the action after FARC guerrillas hijack an aircraft carrying 34 people and kidnap Senator Jorge Gechem, president of a Senate peace talks committee. The following day, the Colombian air force bombs FARC targets in the former demilitarized zone, marking an intensification in the conflict in Colombia.

February 21

Moody's Investors Service puts the long-term debt of AES Corporation on review for a possible downgrade as the energy company's shares drop 11 per cent. AES, once the 17th-largest publicly-traded energy company in the world in terms of market value, with 181 electricity generating facilities and 19 million customers, has lost over 90 per cent of its stock value in the past 52 weeks, including 36 per cent on February 15. On February 6, AES announced plans to sell up to $1 billion in assets in order to raise cash. Credit analysts are concerned that AES has insufficient cash flow to pay its debt of $22.3 billion.

February 21

The California Public Utilities Commission reaches an accord with the California Department of Water Resources that allows California to sell up to $11.1 billion in bonds to repay the State's considerable debt from purchasing power during the State's power crisis. The electricity rate agreement assures a revenue flow to the state agency, but requires the agency to attempt to renegotiate long-term contracts that were locked in when rates were higher than they now are. On February 24, State officials announce their intention to ask the Federal Energy Regulatory Commission (FERC) to void over $40 billion in long-term contracts signed when, according to these officials, the market was being manipulated by sellers.

February 22

Angolan rebel leader Jonas Savimbi is killed by government troops, possibly presaging an end to a 30-year conflict that has displaced 4 million people in the country. With most of Angola's 900,000-barrel-per-day oil production located offshore, the fighting had little effect on the petroleum industry, but proceeds from the oil have been diverted to fight the civil war rather than being used for economic development.

February 25

Venezuela declares force majeure on some crude oil loadings in order to comply with the OPEC quota cuts begun on January 1. Venezuela had agreed to reduce its quota by 170,000 barrels per day.

March 2002

March 1

Electricity rationing ends in Brazil after having been in effect since June 2001. Most companies and households were forced to cut electricity usage by 20 per cent after a drought seriously reduced hydropower capacity. Brazil is more than 90 per cent reliant on hydropower for its electricity consumption.

March 4

The U.S. Supreme Court upholds rules set by the Federal Energy Regulatory Commission (FERC), which hold that once electric power moves onto interstate transmission lines, it is no longer under a state's regulatory control. This freer market, according to the Court, should "encourage lower electricity rates" and create "competitive bulk power markets." Nine states, including California and New York, had challenged the FERC's right to act in this area.

March 4

Strikes and demonstrations over Ecuador's 450,000-barrel-per-day heavy crude oil (OCP) pipeline that is currently under construction, largely from environmental and indigenous groups, come to an end as the government and the pipeline builder agree to provide millions of dollars in local development assistance. The area had lost about 40,000 barrels per day of oil production out of a total of 190,000 barrels per day over a 54-day period from already completed projects that were also blocked, and the pipeline builder lost about $2 million because of delays.

March 5

The United Nations announces that Iraqi oil exports for the previous week rose 1 million barrels per day to 2.49 million barrels per day, taking the four-week average to 1.9 million barrels per day. This is the highest level since November 2001.

March 5

The Energy Ministry of the United Kingdom sets new rules for investment in the country's North Sea oil and natural gas sector. The rules, which are intended to slow declining output, set new investment deadlines on existing licenses, make trading licenses between companies more transparent, and set shorter investment deadlines in new concessions.

March 6

At a joint news conference, oil ministers of major non-OPEC oil exporters Mexico and Norway announce that they plan to maintain their respective export and production cuts through the end of the second quarter of 2002. This same day, non-OPEC Persian Gulf exporter Oman announces that it is willing to maintain its relatively small production cut through the end of the year.

March 7

Light, sweet crude oil for April delivery on the NYMEX closes at $23.71, the highest price since September 21, 2001, when oil prices had temporarily spiked because of the September 11 terrorist attack. Oil prices have been on the rise because of OPEC and non-OPEC production cuts, an improving U.S. economy, and concern over U.S. intentions toward Iraq.

March 8

Administrative workers of Venezuela's state oil company PdVSA, who have been holding protests for a week, stage a four-hour work stoppage. The stoppage does not affect exports, but signals rising tensions between PdVSA employees and the government of President Hugo Chavez, who replaced PdVSA's board of directors in February, a move unpopular with employees.

March 8

A team of scientists reports in the journal Science that small-scale nuclear fusion has been achieved using a new technique. However, only a miniscule amount of energy is produced and the technique's prospects as a means or step towards using fusion as a practical energy source are still unclear.

March 11

The U.S. Energy Information Administration (EIA) releases the results of its weekly survey of gasoline prices, showing that the average price nationwide rose 7.9 cents per gallon in the previous week. This is the second-highest one-week increase in the past 10 years. Gasoline prices continue rising for the remainder of the month.

March 11

Employees at Venezuelan state oil company PdVSA stage a work slowdown, including a halt to overtime work. It is unclear how long the work slowdown will continue.

March 12

Shareholders of Conoco and Phillips Petroleum approve a proposed $15.6-billion merger of the two major oil companies. The new company would be the third-largest oil company in the United States and the sixth-largest investor-owned oil company in the world. The company would also be the largest oil refiner in the United States. Joint reserves of the two companies are about 8.7 billion barrels of oil equivalent.

March 13

The U.S. Senate rejects a bill that would have raised the Corporate Average Fuel Economy (CAFE) standards from the current 24 miles per gallon to 36 miles per gallon by 2016. Congress last increased fuel economy standards in 1975.

March 15

OPEC oil ministers meeting in Vienna decide to maintain their quota restrictions, established January 1, 2002, through the end of the second quarter of the year. On January 1, 2002, OPEC cut its crude oil production quotas by an aggregate 1.5 million barrels per day.

March 15

Syrian Oil Minister Ibrahim Haddad confirms that Syria is receiving Iraqi oil through a pipeline between the two countries. However, the minister states that the pipeline is only being tested and is functioning irregularly due to damage. According to Syria, the pipeline is pumping less than 100,000 barrels per day when at maximum. However, data on Syrian exports loadings for March show an increase of about 115,000 barrels per day.

March 15

California Governor Gray Davis issues an executive order delaying the implementation of a statewide ban on methyl tertiary butyl ether (MTBE) in gasoline for one year. MTBE was to have been banned as of January 1, 2003 because of leakage of the chemical, a possible carcinogen, into the groundwater from underground storage tanks. MTBE is added to gasoline in order to make it burn cleaner in areas susceptible to smog. The alternative to MTBE, ethanol, was viewed by the governor as problematic because suppliers would not be able to adequately replace MTBE with ethanol, creating supply problems and leading to price increases.

March 16

A summit of European Union (EU) leaders meeting in Barcelona agrees to open the EU's electricity and natural gas markets to competition by 2004. However, after heavy lobbying by France, which still has a completely closed electricity market, the leaders agreed only to require that the market be open for industrial (large) consumers, and not for households.

March 17

Venezuelan President Hugo Chavez threatens to have the military take over operations of state oil company PdVSA if the company's operations are halted by a strike. Spokesmen for groups representing employees assert that such a military takeover of the company's operations would not be possible. White-collar PdVSA employees do stage a voluntary one-day strike on March 21, but the company's operations are not seriously affected.

March 18

NYMEX crude futures settle above $25 per barrel for the first time since September 26, 2001, with crude oil for April delivery settling at $25.11 per barrel. Crude oil prices have been rising during the month of March and continue to post six-month highs later in the month.

March 19

Bulgaria's government announces that it has reached an agreement with Greece for an equal stake with Greece and Russia in a joint company to be set up to manage a planned oil pipeline from the Black Sea to the Aegean Sea. This resolves what had been a key sticking point in negotiations for over six years. The project envisages carrying 700,000 barrels per day of crude oil from the Russian port of Novorossisk by tanker to the Bulgarian Black Sea port of Bourgas, from where the underground pipeline would transfer the oil to the port of Alexandropoulis in northeastern Greece. This would enable crude oil to reach the Mediterranean without having to pass through the congested Bosporus Strait.

March 20

Central Gulf Lease Sale 182 takes place, with over 70 companies bidding for oil and natural gas leases in the Gulf of Mexico offshore Louisiana, Mississippi, and Alabama. About $363 million in high bids are received by the U.S. Minerals Management Service, which managed the sale. The Gulf of Mexico currently accounts for about one-fourth of U.S. oil and natural gas production.

March 20

Russian Prime Minister Mikhail Kasyanov announces that Russia will extend its voluntary crude oil export cuts of 150,000 barrels per day through the end of the second quarter of 2002. Russia, the biggest non-OPEC oil exporter, had agreed to implement the cuts beginning on January 1, 2002 as a cooperative move with OPEC. Many analysts question whether Russia has complied at all with its pledged cuts, and some data actually points to Russian exports rising since the beginning of January.

March 21

OPEC announces that its members, excluding Iraq, produced 779,000 barrels of crude oil per day above the agreed production quota of 21.7 million barrels per day during the month of February. This is a 65,000-barrel-per-day decline from January's output.

March 21

United Nations (U.N.) Security Council permanent member Russia blocks attempts by other members of the U.N. Sanctions Committee to either set out a stricter pricing mechanism or eliminate middlemen from Iraq's oil trade. The U.N. suspects that many middlemen pay surcharges to the Iraqi government, in violation of U.N. rules. Russian companies have been the largest lifters of Iraqi crude oil since the start of the U.N. "oil-for-food" program.

March 22

The U.S. Senate passes a bill that will require utilities to produce 10 per cent of their power from renewable energy sources such as wind, solar, and geothermal by 2020. On March 14, the U.S. Senate had rejected a bill that would have required 20 per cent of U.S. electricity to be produced from renewable sources by 2020. There are 14 states that already require a percentage of their state's energy to be from renewable sources.

March 22

Germany's largest utility, RWE, announces that British power company Innogy has accepted its $4.42-billion takeover bid. The purchase of Innogy will be RWE's first foothold in a major electricity market outside of Germany.

March 25

Mexican state oil company PEMEX reports that Mexican crude oil exports in February were at 1.575 million barrels per day, well below non-OPEC Mexico's export cap of 1.66 million barrels per day agreed to with OPEC and implemented on January1, 2002. This is also lower than January's crude oil exports, which were 1.591 million barrels per day, according to PEMEX.

March 25

Oil exports resume at the Turkish port of Ceyhan, following a two-week stoppage that resulted from declining Iraqi exports since the beginning of the month. Iraqi oil exports have been increasingly irregular since after-the-fact pricing was implemented in early 2001. This makes the payment of surcharges to Iraq's government more difficult, but many purchasers now wait until the last few days of the month to load their cargos.

March 25

Sudanese rebel leader John Garang states that rebels of the Sudan People's Liberation Movement and Army (SPLM/SPLA) will continue to attack oil installations in the center of the country despite an agreement to protect civilians and civilian targets. Sudan produces about 210,000 barrels of oil per day, and exports about 175,000 barrels per day.

March 26

Angola and Congo (Brazzaville) sign an agreement delineating the countries' maritime border. This allows U.S. major ChevronTexaco to pursue exploration in the formerly disputed border area offshore. The two countries are among Africa's largest oil exporters, with most of that production coming from offshore areas.

March 27

At an Arab summit meeting in Beirut, Iraq pledges "non-interference" in Kuwait's internal affairs and recognition of Kuwait's borders. Iraqi Foreign Minister Naji Sabri states, "We are for the prosperity and independence of the state of Kuwait and also for the normalization of ties, diplomatic, economic, political." On March 28, delegates at the meeting endorse a peace plan for Israel/Palestine put forward by the Saudi delegation.

March 27

Royal Dutch/Shell confirms that it will build a $500 million liquefied natural gas (LNG) regasification terminal in Baja California with a send-out capacity of 1.3 billion cubic feet per day. Shell intends to source LNG for the project from the Asia-Pacific region, and sell it to power plants and other Mexican and Californian consumers. Completion is expected in 2006.

March 28

A U.S. District Court ruling restricts underground coal mining beneath national parks, inhabited residences, and other protected areas. The court rules in favor of the Citizens Coal Council, an environmental advocacy group, in its suit challenging the way the U.S. Department of the Interior has allowed permitting of underground coal mining that May cause ground subsidence in specified protected areas. The National Mining Association and the U.S. Department of the Interior pledge to appeal the ruling.

March 28

The Russian Foreign Ministry reports that it has reached a basic agreement with the United States on revisions of the United Nation's (U.N.) "oil-for-food" program. The draft documents will now be sent forward to the U.N. Security Council for consideration. Of particular importance is that Russia and the United States appear to have reached an agreement concerning a "goods review list" of supplies that cannot be exported to Iraq without approval by the Security Council.

March 29

A U.S. Geological Survey study concludes that opening Alaska's Arctic National Wildlife Refuge to oil drilling could harm caribou and other wildlife. The U.S. Senate will take up debate in April on whether to open the refuge to drilling.

April 2002

April 1

Iraq calls for an oil embargo by Arab oil exporters to punish the United States and other countries for their support of Israel. This, along with intensification of the Israeli-Palestinian conflict, contributes to oil prices reaching new six-month highs.

April 1

India liberalizes its oil and natural gas sector by putting in place a series of market reforms, including: the end of government-fixed prices for gasoline and diesel; the end of subsidized cooking gas and kerosene prices; market competition for state-run downstream companies; and assigning the Oil Ministry the role of energy watchdog.

April 2

Royal Dutch/Shell agrees to buy Enterprise Oil for $5 billion in cash. This will increase Royal Dutch/Shell's production in the North Sea by 30 per cent and overall production by 6 per cent, according to the company. The acquisition will also add 1.5 billion barrels of oil to Royal Dutch/Shell's reserves. The company is also assuming $1.15 billion in Enterprise's debt.

April 3

Venezuela sends out its first commercial shipment of 550,000 barrels of synthetic crude to a U.S. Gulf Coast refinery. Venezuela's Sincor heavy crude upgrade plant, which was inaugurated last month, refines ultra-heavy crude oil into 32 degree API syncrude.

April 4

The Angolan army signs a ceasefire accord with rebels of the National Union for the Total Independence of Angola (Unita). The agreement includes amnesty for former Unita soldiers and their demobilization and reintegration into society. The civil war, which began in 1975, has killed thousands of Angolans and taken much of the government's revenues from Angola's substantial oil production and exports.

April 5

Thousands of workers at Venezuelan state oil company PdVSA stay home, close gates of facilities, and engage in protests. This is the largest disruption of PdVSA's operations in 2002, though it is not a full-blown strike by all PdVSA workers. Oil production and refining slows, and two of Venezuela's five main oil export terminals are unable to operate. The government of President Hugo Chavez threatens to militarize PdVSA's operations.

April 5

Nigeria's Supreme Court rules that the federal government, not state or regional authorities, has control over offshore oil and natural gas. Six southern states had threatened to take control of natural resources in their territories, which could have meant rules changes for operators in the areas and a concentration in hydrocarbon revenues in those states.

April 5

Iran's supreme leader, Ayatollah Khamenei, joins Iraq in calling for an oil embargo against countries supporting Israel. However, other Iranian government officials indicate that Iran will not go ahead with any embargo without support from all Muslim major oil exporters. On April 7, the Libyan government also calls for an embargo, but with the provision that such an embargo would have to include all Muslim oil exporters. Saudi Arabia and Kuwait reject these proposals.

April 6

A BP pipeline near Little Lake, Louisiana ruptures, spilling 90,000 gallons of oil into marshland and coastal waters. Clean-up crews are sent to the area and some of the oil is recovered the following day.

April 7

Venezuelan President Hugo Chavez announces that he has fired seven executives of state oil company Petroleos de Venezuela (PdVSA) and sent 12 others into early retirement. This action comes in response to the executives leading protests and work stoppages that have intensified since April 5. These protests began after President Chavez installed new PdVSA board members on February 20, 2002.

April 8

Nevada Governor Kenny Guinn vetoes the use of the Yucca Mountain facility as a repository for high-level nuclear waste. Under a 1982 federal law on nuclear waste policy, a governor can file a "notice of disapproval" to reject a president's selection of a site in his state. This gives the U.S. Congress 90 legislative days to debate and vote on the issue. Simple majorities in both chambers can override the veto.

April 8

Iraq announces that it will halt its "oil-for-food" exports for 30 days as a "gesture of support" for the Palestinians' struggle with Israel. Iraq also requests that other OPEC countries do not raise production to make up for lost Iraqi exports. Iraqi "oil-for-food" exports had averaged about 1.7

million barrels per day to date in 2002. Major Arab OPEC exporters Saudi Arabia, Kuwait, and Qatar have expressed unwillingness to join in any embargo.

April 9

A general strike begins in Venezuela, shutting down many stores and factories and nearly halting oil production, refining, and export terminals. Strike organizers decide to extend the strike through April 10.

April 9

California Attorney General Bill Lockyer files suit against four power companies, charging them with profiteering during the power crisis of 2000-2001. The attorney general claims that companies could face penalties of as much as $1 billion.

April 12

Venezuelan President Hugo Chavez is ousted by the country's military after three consecutive days of general strikes during which oil production, refining, and exports-the mainstays of the Venezuelan economy-were seriously affected. After at least a dozen protesters are killed, high-ranking military officers, including the army commander general, request President Chavez to step down.

April 12

Pedro Carmona is named interim President of Venezuela by the military high command. He dissolves the National Assembly and Supreme Court. PdVSA operations that had been halted start up again, but rioting begins again the following day.

April 14

Interim President Pedro Carmona announces that he has resigned following large, and sometimes violent, pro-Chavez protests and a lack of support among many military officers. Carmona also loses support of Venezuela's largest labor group. Several hours later, Hugo Chavez returns to power in Caracas and claims that he never resigned the presidency. Chavez announces that PdVSA's Board of Directors had submitted its resignation on April 11.

April 15

Venezuelan President Hugo Chavez attempts some rapprochement with the employees of PdVSA as oil exports return to normal. PdVSA executives support continued normalization of the company's operations. Minister of Energy and Mines Alvaro Silva asserts that Venezuela will continue to abide by its OPEC quotas, but the government also asserts that it will be a reliable oil supplier to the United States and will not participate in any oil embargo.

April 18

El Paso Energy agrees to build, install, own, and operate a production hub platform for the Falcon natural gas field in the U.S. Gulf of Mexico. The field is owned by Pioneer Natural Resources and Mariner Energy. The platform will have an initial capacity of 300 million cubic feet per day, and as a hub eventually will service an additional nine blocks in the area. A 20-mile natural gas pipeline will link the development to an offshore natural gas gathering system.

April 19

A Venezuelan government spokesman announces that OPEC Secretary General Ali Rodriguez will quit this position in order to become the new president of PdVSA. It is not clear who will succeed Mr. Rodriguez as OPEC Secretary General at this time.

April 22

National Grid Group, which runs Britain's power transmission network, announces that it plans to merge with Lattice Group in a stock deal worth $9.7 billion. Lattice is a natural gas pipeline operator that was spun off from British Gas. National Grid will also assume $10 billion of Lattice's debt under the plan. This creates one firm with significant power over Britain's natural gas and electricity networks. The new company will be the UK's largest utility.

April 22

California State officials announce that California has renegotiated $15 billion in power contracts with five of 12 companies with which the State has contracts. As a result of the restructuring, the State will lower costs by about $3.5 billion over the decade. These long-term contracts were signed during the height of the power crisis of 2000-2001, locking in what many State officials now consider higher-than-market rates for electricity. The companies that have renegotiated extricate themselves from possible legal action on the part of California in the future.

April 22

The government of Venezuela and executives of PdVSA agree on a new board of directors for the state oil company. Recent protests and strikes by PdVSA managers and employees that led up to President Chavez's temporary ouster on April 12 began in response to a board of directors appointed by President Chavez on February 20.

April 22

Russian Finance Minister Alexei Kudrin announces that Russia will not extend the 150,000-barrel-per-day crude oil export cut it has promised OPEC beyond the end of June 2002. Data from several sources, including Russia itself, appear to cast doubt on whether Russian exports were ever reduced since the cuts were supposed to become effective January 1, 2002.

April 22

Russia's largest oil producer, OAO Lukoil, unveils a restructuring plan that aims to double the company's output by 2010. In addition to investments in new fields, the company also plans to increase output by outsourcing its drilling and well-management to oil services companies. The company plans to increase production 20 per cent this year alone, to 1.4 million barrels per day.

April 24

A summit of the leaders of the five littoral states of the Caspian Sea ends without an agreement on how to divide the Caspian's resources between the five countries.

April 25

The U.S. Senate passes an energy bill that differs substantially from the version passed by the U.S. House of Representatives in August 2001. The Senate version, for instance, has no provision for oil exploration in the Arctic National Wildlife Refuge. Another provision of the bill would require that companies gradually increase the percentage of electricity that they generate from renewable sources. The bill now goes to conference.

April 25

President George W. Bush and Saudi Crown Prince Abdullah meet in Crawford, Texas. The main topic of discussion is how to arrive at a resolution of the Israeli-Palestinian conflict. Also today, Saudi officials deny fresh reports that they will reduce oil exports to express displeasure with U.S. policy toward the Middle East. Oil prices rise substantially on the rumor, but fall after denials by U.S. and Saudi officials.

April 29

An investigating panel of the Senate Governmental Affairs Committee issues a report on gasoline price and the degree of market concentration of refiners and retailers. The report states that, "In a number of instances, refiners have sought to increase prices by reducing supply." The American Petroleum Institute and several oil refiners and retailers have contested the report's conclusions.

May 2002

May 1

The Canadian province of Ontario opens its electricity market to competition, including suppliers from the United States. Under the new rules, suppliers other than the current monopoly supplier, Ontario Power Generation, are able to sell electricity under long-term contracts or on the spot market. Initially, large industrial consumers, local utilities, and new electricity retailers will be able to choose amongst suppliers. The U.S. Federal Energy Regulatory Commission (FERC) has licensed Ontario Power Generation to sell electricity in the United States.

May 1

Oman approves plans to add a third production train at its liquefied natural gas (LNG) plant at Qalhat. The new train will add 3.3 million metric tons per year to the plant's current 6.6-million-metric-tons-per-year capacity in two existing trains. In January, Oman signed a financial agreement to borrow $1.3 billion from a consortium of 12 international banks in order to be able to finance the expansion.

May 2

Hawaii becomes the first state to enact legislation that would regulate retail gasoline prices. Prices are to be capped starting in July 2004. The state's Public Utilities Commission would set a maximum price based on an average of prices in U.S. West Coast markets and would allow a maximum profit of 16 cents per gallon for retailers.

May 3

The Environmental Protection Agency approves rules that will allow coal-mining companies to dump waste and dirt from mountaintop mining into rivers and streams. The mining industry greets the decision with approval, while several environmental groups are critical of the decision.

May 3

A two-day G8 (the world's seven largest economies plus Russia) energy summit in Detroit concludes. The energy ministers focus their discussions on ensuring adequate, safe, supplies as demand increases, as well as on new technologies that have efficiency and environmental benefits. U.S. Secretary of Energy Spencer Abraham states that, "We [energy ministers] agreed on the importance of providing adequate protection for our energy facilities as well as the importance of having multiple links between suppliers and consumers to reduce our vulnerability to disruption of energy supplies."

May 3

Iraq and the United Nations (U.N.) end three days of talks on disarmament programs and weapons inspectors without any plan for the return of weapons inspectors to Iraq. U.N. weapons inspectors were expelled from Iraq in 1998.

May 3

A U.S. Appeals Court upholds the Environmental Protection Agency's (EPA) diesel fuel sulfur content regulation in ruling on a lawsuit filed by the National Petrochemical and Refiner's Association that challenged the regulations. The rule requires that refiners reduce the sulfur content in diesel from the current 500 parts-per-million requirement to 15 parts per million in 2007.

May 6

The Federal Energy Regulatory Commission (FERC) releases documents showing that Enron used a number of deceptive trading strategies to manipulate California's power market during the 2000-2001 state energy crisis. On May 8, the FERC orders all power sellers operating in California to preserve their records from 2000 and 2001, particularly in regard to trading strategies employed. On May 21, the FERC orders firms to report any "round-trip" electricity trading in sworn statements. Several firms alleged to have participated in these schemes lose over 50 per cent of their market values in May. Such trades are thought to have inflated companies' revenue statements. This is all part of a FERC probe into the causes of electricity price volatility in California in 2000-2001.

May 7

A $125 million study commissioned by ExxonMobil, BP, and Phillips, that examined two proposed routes for a natural gas pipeline from Alaska's North Slope to the lower 48 states, finds that neither route is economically feasible. The pipeline would transport 4.5 billion cubic feet per day of natural gas. Company spokespeople indicate that the project will be delayed until market conditions support the project.

May 8

Iraq starts pumping crude oil to its export terminals, following the country's announcement on May 5 that it would end its oil export embargo after one month, ***i.e.***, May 8. Iraq also submits price proposals for May crude oil loadings to the United Nations for approval.

May 8

The United Arab Emirates (UAE) announces that it has selected Occidental Petroleum to take over Enron's 24.5 per cent stake in the $3.5 billion Dolphin natural gas project. Enron had sold back its stake to UAE state investment company UOG before Enron's bankruptcy in December 2001.

May 9

The U.S. House of Representatives votes to approve the use of the Yucca Mountain facility as a repository for high-level nuclear waste, the first step in overriding the Nevada governor's veto of the site's selection by the U.S. Department of Energy. The U.S. Senate must also approve the site's use for the governor's veto to be overridden and for the project to proceed.

May 10

Scientists at the University of Munich's Nanoscience Center in Germany succeed in harnessing the power of light to produce mechanical energy for the first time in history. In the experiment, the scientists shine a selected frequency of light on a single molecule of the plastic azobenzene, causing the molecule to crimp and pull down a board to which it is attached.

May 10

Thailand announces that it will go ahead with the construction of a natural gas pipeline from the Thai-Malaysian Joint Development Area (JDA) natural gas project in the Gulf of Thailand to the southern Thai province of Songkhla, and then on to Kedah state in northern Malaysia. The pipeline's approval had been delayed over environmental concerns, and the pipeline route will be adjusted slightly because of these concerns. Malaysia had previously stated that it might re-route its share of the natural gas from the Gulf of Thailand if the Thai government failed to make a decision on the pipeline.

May 13

Russia and Kazakhstan sign an accord defining ownership rights in the northern part of the hydrocarbon-rich Caspian Sea. A summit last month of the five littoral nations of the Caspian Sea failed to arrive at a comprehensive agreement between all five countries.

May 14

Near-month WTI crude oil on the New York Mercantile Exchange (NYMEX) settles above $29 per barrel for the first time since the temporary price spike on September 14, 2001 when the NYMEX first opened after being closed in the wake of the events of September 11. Tensions in the Middle East, the effect of OPEC production cuts and the lingering effects of Iraq's one-month embargo, as well as low U.S. crude oil stock levels, are seen as factors influencing oil prices.

May 14

The United Nations (U.N.) Security Council approves an overhaul of the "oil-for-food" program for Iraq that makes use of an extensive list of "dual-use" goods (goods that could have a military as well as civilian use). Iraq will be able to use its oil revenues, which go into a U.N. escrow account out of which suppliers exporting products to Baghdad are paid, in order to purchase items not on the list. The resolution renews the U.N. program until November 25, 2002. On May 16, official Iraqi news agency INA announces that it will comply with the new six-month tranche of the "oil-for-food" program voted by the U.N. Security Council on May 14, despite condemning the Security Council resolution in the same statement. Iraq officially accepts the U.N. proposal on May 29.

May 15

Private Iranian energy company Petro Iran finalizes a $585-million buy-back deal to renovate the offshore Foroozan-Esfyandar oilfields. Petro Iran is expected to acquire a foreign partner in order to give the company the financial resources necessary to fund the project. Commenting on the deal, Iranian Oil Minister Bijan Zanganeh states, "With the investments already made and those about to be secured.Iran's crude oil production capacity and its quota in OPEC will not decrease."

May 15

Russian customs data released today show that Russia's crude oil exports rose to 3.49 million barrels per day in the first quarter of 2002, an all-time high, despite Russia's agreement with OPEC to reduce crude oil exports by 150,000 barrels per day during the same period.

May 16

The U.S. Geological Survey (USGS) announces that the mean estimate of technically recoverable resources in Alaska's National Petroleum Reserve is 9.3 billion barrels. The USGS pegs natural gas reserves in the National Petroleum Reserve at between 39.1 and 83.2 trillion cubic feet. The Reserve's eastern border is about 100 miles west of the Arctic National Wildlife Refuge (ANWR). The USGS also declares that at current prices, more oil could be pumped from ANWR than from the National Petroleum Reserve. The Reserve is still undeveloped, and the recovery of natural gas would necessitate the construction of a pipeline.

May 17

Russian Prime Minister Mikhail Kasyanov announces that Russia will not extend its 150,000-barrel-per-day crude oil export cut, agreed to with OPEC, into the third quarter of 2002 and furthermore, that Russia will gradually phase out the export cut in the remainder of the second quarter of 2002. Russia is the world's second-largest oil exporter.

May 17

The Norwegian parliament approves a $5.24 billion investment in the Snoehvit liquefied natural gas (LNG) project on the Barents Sea. The project is expected to produce 5.7 billion cubic meters of LNG per year after it comes on line in 2006.

May 17

The comments of the Environmental Protection Agency (EPA) on the Bureau of Land Management's (BLM) Draft Environmental Impact Statement (DEIS) on coalbed methane production in Wyoming's Powder River Basin give the BLM's environmental plan the EPA's lowest rating. This will make it difficult for increased production from the Powder River Basin to take place without the EPA's concerns being addressed. The DEIS assesses the impact of development of over 60,000 coalbed methane wells in Wyoming and Montana.

May 20

Representatives of East Timor and Australia sign an accord giving East Timor 90 per cent of the revenues from the Bayu-Undan natural gas and condensate field and 20 per cent of the revenues from the Greater Sunrise field. The division of the remaining 80 per cent of the Greater Sunrise's revenues is still under discussion. The Greater Sunrise field, about twice the size of the Bayu-Undan field, contains an estimated 9.5 trillion cubic feet of natural gas.

May 23

The U.S. Department of Energy changes a rule concerning increased efficiency of air conditioners and heat pumps, such that manufacturers will be required to make them 20 per cent more energy efficient by 2004 instead of 30 per cent, as had been proposed by the Department under the previous administration.

May 23

Chinese officials claim to have discovered the country's largest natural gas field. The field, in the Ih Ju Meng area of the region of Inner Mongolia, is estimated to contain 21 trillion cubic feet of natural gas.

May 24

U.S. President George Bush and Russian President Vladimir Putin agree to a major new energy partnership that will entail more investment from the United States in Russia's oil and natural gas sector. The leaders also agree to joint efforts to improve ports, pipelines, and refineries in order to expedite export flow. This could mean more Russian hydrocarbon exports to North America.

May 25

Alvaro Uribe Velez is elected president of Colombia. He has promised to try to end Colombia's conflict with two guerrilla groups, and to provide greater security for energy infrastructure so as to attract more foreign investment. He also has indicated his support for a bill that would replace Colombia's fixed 20 per cent royalty payment on all upstream ventures with a sliding scale of royalty payments.

May 28

Canadian oil company Sheer Energy signs a contract that makes the company the first Western company to commit finances to Iran since U.S. President George Bush's assertion that Iran forms part of an "axis of evil."

May 28

A law firm representing the government of Libya announces that Libya will pay $2.7 billion to the families of those who lost their lives in the Pan Am flight 103 bombing if U.S. and United Nations sanctions are first lifted. Some Libyan government officials deny that any deal has been proposed, but the lawyers employed by the Libyan government continue to affirm that a deal is possible. On May 29, U.S. Secretary of State Colin Powell comments, "Just reading press accounts of what has been said about the offer, it certainly is a step in the right direction, but I don't think it resolves all the outstanding issues that have to be dealt with, with respect to Libya and Pan Am 103."

May 28

India announces that it has a 60-day reserve of all petroleum products ready should current tensions with Pakistan erupt into war. Pakistan responds on May 30 by stating that it has a 45-day reserve of all petroleum products ready to use if war occurs. Pakistan and India dispute areas of Kashmir, and several militant groups operate in the area.

May 28

The U.S. government decides to buy back leases for oil and natural gas drilling on the Florida coast and in the Everglades for $235 million because of environmental concerns. Skip Horvath, president of the Natural Gas Supply Association states, "The natural gas fields taken off the table today represent enough natural gas to supply a million U.S. families for over 30 years." However, Secretary of the Interior Gale Norton has asserted that there are only 40 million barrels of oil equivalent in the area to be protected, about two days' worth of U.S. consumption.

May 30

The Federal Energy Regulatory Commission (FERC) rejects the State of California's demand to expand an existing lawsuit seeking refunds from companies for alleged over-charges from October 2000 to June 2001, to instead start in May 2000. This would have had the potential to increase refunds by $2.8 billion.

May 30

Elections are held in Algeria, but turnout is very low. The ruling National Liberation Front (FLN) appears to have won the election.

May 31

Japan releases statistics that show that crude oil imports for April 2002 were down 9.4 per cent year-on-year with April 2001, at 4.69 million barrels per day. Japan's gross domestic product (GDP) fell by 0.4 per cent in 2001, and it is projected to fall by 1.1 per cent in 2002. Japan is the world's second-largest oil importer and consumer.

June 2002

June 2

Occidental Petroleum acquires the 24.5 per cent stake in Dolphin Energy previously held by Enron for $310 million. The Dolphin project will transfer natural gas from Qatar to Abu Dhabi and Dubai, with deliveries expected to begin in 2005.

June 3

The U.S. Environmental Protection Agency sends the "U.S. Climate Action Report" to the United Nations. The report asserts that human activities, in particular the burning of fossil fuels, are the principal cause of global warming.

June 3

The U.S. Bureau of Land Management conducts an oil and natural gas lease sale for Alaska's National Petroleum Reserve. Companies submit high bids of $63.76 million for 60 tracts. The lease sale, covering an area of 3 million acres, offers tracts that had not received bids during the last sale in 1999.

June 4

The European Court of Justice rules that France's current golden share in oil major TotalFinaElf is illegal, as France's shareholding goes "..beyond what is necessary in order to obtain the objective indicated," *i.e.*, ensuring "supplies of petroleum in times of crisis." A golden share is usually defined as a situation in which a government owns less than 50 per cent of the shares of the corporation but retains a veto right on decisions in respect to the management of the enterprise.

June 6

Papua New Guinea signs an agreement on a major natural gas pipeline from the producing area in its southern highlands to the Australian State of Queensland on the other side of the Torres Strait. It is estimated that there are six trillion cubic feet (Tcf) of natural gas in the southern highlands area of Papua New Guinea.

June 7

Russia and Kazakhstan sign an agreement that permits annual shipments of no less than 127 million barrels of Kazakh oil through Russia per year over the next 15 years. This agreement covers Kazakhstan crude shipments through two pipelines connected to Russian state pipeline monopoly Transneft's system. The agreement also sets up a joint venture between Gazprom of Russia and

KazMunaiGaz of Kazakhstan that will transport 124 billion cubic feet of natural gas per year through Russian pipelines in the first few years of the agreement. (Reuters, DJ)

June 7

U.S. Secretary of the Interior Gail Norton rejects California Governor Gray Davis' request that the State of California be able to buy back offshore drilling rights. Secretary Norton writes in her response to the governor that, "[A] major difference between Florida and California is that Florida opposes coastal drilling and California does not." Currently, 36 undeveloped oil and/or natural gas drilling tracts remain in place off the coast of California.

June 10

The U.S. Department of the Interior's Minerals Management Service issues a report showing crude oil production in the U.S. Gulf of Mexico growing by an additional 500,000 barrels per day, to 970,000 barrels per day, by 2006. Natural gas production is forecast to rise to 16.39 billion cubic feet (Bcf) per day in 2006 in the high scenario from about 13.9 Bcf per day in 2001, but the low scenario shows a decline to 10.97 billion Bcf per day.

June 17

Violent protests occur in Peru's second-largest city, Arequipa, because of the privatization of two state-owned electricity companies. Hundreds of troops are sent in to quell the riot, and constitutional guarantees are suspended for 30 days. The two companies, Egasa and Egesur, are being sold to a Belgian company for $167 million.

June 18

A report released by the General Accounting Office (GAO) states that the Federal Energy Regulatory Commission (FERC) "..is not adequately performing the oversight that is needed to ensure that the prices produced by these [electricity] markets are just and reasonable, and therefore, it is not fulfilling its regulatory mandate." FERC Chairman Patrick H. Wood III responds that, "The commission [FERC] had not previously focused its efforts clearly enough to succeed..that has now changed."

June 18

American Electric Power (AEP) announces that it will sell its United Kingdom Seeboard unit to London Electricity Group for $975 million in cash. London Electricity will also assume $1.05 billion of Seeboard's debt. Seeboard sells electricity and natural gas to 1.9 million customers in the United Kingdom.

June 20

OPEC releases a report showing that its ten members subject to production quotas (*i.e.* excluding Iraq) produced 23.1 million barrels per day in May, 1.4 million barrels per day above the ten members' combined quota of 21.7 million barrels per day. This is an increase over April's production of 220,000 barrels per day.

June 20

Norway's Oil and Energy Ministry states that, "The Norwegian government has decided not to extend the restriction on oil production into the second half of 2002." Norway had agreed with OPEC to reduce its crude oil production by 150,000 barrels per day for the first two quarters of 2002.

June 21

Russia's largest natural gas company, Gazprom, and Ukraine's state oil and natural gas monopoly, Naftogaz, sign an agreement to transport natural gas across Ukraine, resolving a decade-long dispute on natural gas exports. Russia now has the right to transit 3.9 trillion cubic feet of natural gas through Ukraine every year from 2003 until 2013, ending Kiev's denial of foreign access to its pipeline system. On June 10, the leaders of Germany, Russia and Ukraine signed a statement on cooperation in the use of the Ukrainian gas pipeline network, as well as the creation of a consortium to manage the system.

June 24

ExxonMobil signs an agreement with state oil and natural gas company Qatar Petroleum for the development of two liquefied natural gas (LNG) trains to be used for export to the United Kingdom. ExxonMobil states that the two trains will be "the largest ever built" by the industry.

June 24

Centrica of the United Kingdom signs the second of two ten-year natural gas contracts, this time with Gasunie of the Netherlands, that will provide 10 per cent of total natural gas demand in the United Kingdom. On June 9, Centrica had signed a ten-year deal with Norway's Statoil. The combined deals involve about 459 billion cubic feet per year.

June 25

Russia formally announces that it will raise its crude oil exports by 150,000 barrels per day in the third quarter of 2002 and thereby, end its agreement with OPEC to limit crude oil exports by 150,000 barrels per day for the first and second quarter of 2002. Many analysts believe that Russia has already been exporting near capacity for some months.

June 26

OPEC ministers meeting in Vienna decide to leave their combined output quota, excluding Iraq, unchanged at 21.7 million barrels per day for the third quarter of 2002. It is estimated that OPEC-10 countries (*i.e.* excluding Iraq) are producing between 1 million and 1.5 million barrels per day above the quota agreement. OPEC members also agree to appoint Venezuelan Oil Minister Alvaro Silva as the cartel's new secretary general, replacing Ali Rodriguez, who will now head Venezuelan state oil company PdVSA. At the meeting, Algeria requests a larger share of OPEC's total quota, but the issue will not be taken up until the OPEC Board of Governors meeting in August.

June 26

BP announces that the cost of developing the Shah Deniz oil and natural gas field in Azerbaijan will be an estimated $3.2 billion. BP expects the field to eventually account for 7 per cent of the company's total hydrocarbon production.

June 27

Mexico announces that it will continue its agreement with OPEC to limit crude oil exports to 1.66 million barrels per day into the third quarter of 2002. A statement from the Energy Ministry said that the decision was "based on national interests and conditioned upon the future conduct of the world oil market." Mexico is among the five largest oil exporters to the United States.

June 28

Angola fines U.S. major ChevronTexaco $2 million for environmental damage. This is the first time an African country has levied a fine against a foreign oil company operating in its waters. Angola held ChevronTexaco liable for failing to replace obsolete pipes, which caused several oil spills.

June 29

An official at Oman's Oil and Gas Ministry announces that the non-OPEC country will continue its 40,000-barrel-per-day production cut into the third quarter of 2002. Oman had agreed with OPEC to cut production 40,000 barrels in the first and second quarters of 2002.

June 29

Egypt's Electricity and Energy Minister Hassan Ahmad Younes confirms media reports that Egypt will construct a nuclear power plant, the country's first, on the northwest coast of the country near Alexandria.

July 2002

July 1

The California State Legislature passes a bill that limits vehicle emissions of carbon dioxide, the first such bill to pass a state legislature. The specific regulations, to be developed by 2005, would take effect on 2009 model-year vehicles. The limits, enacted because of carbon dioxide's putative effect on global climate change, are likely to have significant repercussions beyond California because the State represents some 10 per cent of the U.S. automobile market. Governor Gray Davis signs the bill into law on July 22.

July 2

The World Bank suspends consideration of $370 million in loans and guarantees for AES's $550 million Bujagali Dam hydroelectric power project on the Nile River in Uganda. The decision, taken because of allegations that AES's main construction contractor bribed a Ugandan government official, effectively halts progress on the dam, the largest single private investment in East Africa's history.

July 3

The supertanker Astro Lupus arrives offshore of the Port of Houston, carrying the first direct shipment of Russian crude oil to the United States. The oil, about 2 million barrels of Urals Blend, was exported by Yukos, Russia's second-largest producer and destined for two ExxonMobil refineries in Texas. Yukos hopes to make six such shipments this year.

July 3

It is revealed publicly that several large energy companies were fined a total of $122 million by the California Independent Systems Operator, the state agency that control's California's power grid for failing to deliver electricity during emergencies in the midst of California's power crisis. The largest fines were assessed against Dynegy ($44.8 million), the Williams Companies ($25.5 million), and Reliant Energy ($25.1 million). On July 26, Williams reaches a global settlement with the State of California and other power partners, including refunds for Williams' activities during the California power crisis.

July 4

Eight large energy companies conclude a framework joint-venture agreement for the development of a natural gas field in western China and a 2,500-mile "West-to-East" pipeline to transport the natural gas to eastern China. The pipeline will be designed to carry about $2 billion worth of natural gas per year. PetroChina, with a 50 per cent stake in the project, put the pipeline and initial field development cost at $8.5 billion. Royal Dutch/Shell, ExxonMobil, and Gazprom each have a 15 per cent stake in the venture.

July 5

Energy merchant Reliant Resources, in a filing with the U.S. Securities and Exchange Commission, admits that the company overstated its profits for 1999, 2000, and 2001 by a cumulative $7.8 billion, of which $6.4 billion came from phantom energy trades. Phantom trading entails selling and then immediately buying back power, thereby increasing revenues, but without actual power changing hands.

July 5

The German Economics Ministry approves the takeover of Ruhrgas, Germany's largest natural gas distributor, by E. On, the world's largest non-oil energy company, also based in Germany. The approval contains a list of additional conditions that must be satisfied in order to go through with the takeover. This conditional approval overrides a previous ruling by the German cartel office, which had disallowed the takeover. The cartel office had feared that a takeover of Ruhrgas by E. On would undermine attempts to liberalize the German energy market, as the takeover will give E. On control of 60 per cent of Germany's natural gas imports.

July 9

An undersea power cable with a capacity of 500 megawatts in each direction begins operation between Italy and Greece. The cable is 75 per cent owned by Enel and 25 per cent owned by the Greek Public Power Corporation.

July 10

Pricing agency Platts begins to include two additional North Sea crudes, Oseberg and Forties, in its valuation of Brent. This is an attempt to increase volumes, and hence liquidity in order to avoid market squeezes that can occur when large amounts of crude are bought by a single trading company in the context of a small overall volume. BP supports the move, but Shell opposes it. Brent serves as a benchmark price for global crude oil trade of about $150 billion per year.

July 17

Indonesian President Megawati Sukarnoputri approves the formation of a new "Implementing Body" to monitor activities of oil companies operating in Indonesia. This body, whose effective date is still unclear, will replace in this role state oil company Pertamina, which had its monopoly revoked last year. Production sharing contracts will be between the new body and foreign companies, instead of between Pertamina and the foreign company.

July 17

The Federal Energy Regulatory Commission (FERC) decides to raise the electricity price cap in California from $91.87 per megawatthour to $250 per megawatthour, effective October 1, 2002. The

FERC proposes a computer system that will analyze bids above $91.87 per megawatthour to determine whether price gouging is occurring. Also today, the FERC announces that the current board of the California Independent System Operator (ISO), appointed by Governor Gray Davis, is not truly independent, and that a new board appointed in consultation with a number of entities involved in the energy industry must be put in place by January 2003.

July 17

The U.S. House of Representatives votes to cut off federal funds necessary to permit new oil drilling off the coast of California. The funds would have gone toward helping oil companies act on outstanding drilling leases in federal waters.

July 17

Venezuelan President Hugo Chavez appoints Rafael Ramirez as the country's new oil minister, replacing Alvaro Silva, who was chosen in June to be OPEC's new Secretary General. Ramirez is currently head of Venezuela's state natural gas regulator, Enagas.

July 19

Japan's Parliament passes a bill that mandates the liquidation of the heavily indebted Japan National Oil Company (JNOC), beginning in March 2004.

July 19

British Energy Minister Brian Wilson awards 25 new offshore oil and natural gas licenses to companies under new terms that require companies to develop fields quickly or relinquish their claims. The British government is attempting to slow production declines in the British North Sea by encouraging the use of new technologies and the involvement of small companies for whom the development of smaller fields is economic.

July 19

British Gas, Edison International, and Shell are given final permission by the Egyptian government to develop the Rosetta and West Delta Deep Marine offshore gas concessions for domestic consumption and export. The exported natural gas will be liquefied at a new plant under construction at Idku, near Alexandria. The liquefied natural gas (LNG) will be exported to France, where it is expected to eventually supply 10 per cent of consumption.

July 22

Brazilian state oil company Petrobras announces that it has a preliminary accord to take over 58.6 per cent of Perez Companc, Argentina's second-largest oil company, for $754.6 million in cash and $370.5 million in bonds. The deal would expand Petrobras' total production capacity by about 12 per cent.

July 23

President George Bush signs a bill into law that makes the Yucca Mountain site in Nevada the United States' central depository for nuclear waste. Use of the Yucca Mountain site required federal legislative approval because Nevada Governor Kenny Guinn had objected to the site. The U.S. House of Representatives voted for use of the site on May 9, and the U.S. Senate voted for use of the site on July 9.

July 24

A report from the Saudi American Bank projects that Saudi Arabia's budget deficit will be about $5.1 billion for 2002, a reduction of the kingdom's earlier official forecast of a $12 billion deficit. The report attributes the lower projection to higher-than-expected oil prices.

July 25

Israel cancels a natural gas pipeline tender after Paz Oil and Africa Israel Investments run out of time to build a viable consortium. Israel intends to build a significant natural gas production/importation capacity, but this will be impossible without a natural gas pipeline network. The failure to attract a major international partner after Tractebel of Belgium pulled out in April 2002 and BG pulled out of any deal earlier this month is seen as the reason for the cancellation.

July 26

Russia releases plans to build five more nuclear reactors in Iran as part of a ten-year plan for enhancing political, economic, and scientific ties with Iran. Russia is currently constructing a 1,000-megawatt light water reactor for Iran at Bushehr.

July 26

The U.S. Department of Energy announces that it intends to increase the rate at which the Strategic Petroleum Reserve (SPR) is filled by increasing the "royalty-in-kind" exchange program by 40,000 barrels per day. Under the "royalty-in-kind" program, oil companies deposit oil that is produced on federal leases in the SPR as a form of payment for those leases.

July 29

Dynegy agrees to sell its Northern Natural Gas pipeline unit to Berkshire Hathaway's MidAmerican Energy Holdings for $928 million. In addition, MidAmerican will assume $950 million in outstanding debt. Dynegy is raising its cash reserves in order to carry out a restructuring plan.

July 29

President George Bush announces that he has submitted legislation to Congress as part of his "Clear Skies" initiative that he claims will reduce power plant emissions 70 per cent by 2018 using market incentives.

July 29

Russia announces that it has granted state-owned medium-sized oil firms Rosneft and Zarubejneft the right to develop jointly with Kazakhstan a large oil structure known as Kurmangazy in the Kazakh sector of the Caspian Sea. Some analysts believe that Kurmangazy May be as large as Kazakhstan's Kashagan field, which is the largest oil discovery of the past thirty years.

July 29

Norway and the European Commission (EC) resolve their dispute over non-competitive natural gas sales. Norway had previously ended its monopoly gas sales committee (GFU), but the EC had threatened to invalidate long-term contracts negotiated when the GFU was still in place. In return for the closure of EC action against Norway, Norway agrees to sell an additional 530 billion cubic feet of natural gas to new customers in the European Union.

July 30

Estonia closes the energy chapter of its accession negotiations with the European Union. Estonia receives special status for its shale oil industry, which provides about 65 per cent of the country's primary energy supply.

July 31

ChevronTexaco announces the resumption of crude oil exports from Nigeria after protests and a fire caused the company to declare force majeure on its exports for a ten-day period. Between 300,000 and 400,000 barrels per day were temporarily halted. ChevronTexaco has not fully resolved the issues between the company and protestors who disrupt operations. Before the fire, about 110,000 barrels per day were interrupted at times by protestors. Nigeria's army moved in to prevent protestors from damaging equipment, but declined to remove the protestors from the facilities.

July 31

In the most far-reaching regulatory proposal to potentially affect U.S. energy markets since 1996's FERC order that gave new competitors access to utility power lines, the FERC releases a 600-page "Standard Market Design" that outlines "best practices" of wholesale markets regionally and internationally. If the plan were implemented, it would make the new rules uniform throughout the United States, eliminating incompatibility problems.

July 31

China begins using a new $450 million, 780-mile petroleum products pipeline, the longest product pipeline in China. The pipeline can move 5 million metric tons of petroleum products per year (about 108,000 barrels per day) from Gansu province to the southwestern cities of Chengdu and Chongqing.

August 2002

August 1

BP Chief Executive John Browne announces that the company will be investing $15 billion through 2010 in the development and production of deepwater oil and natural gas in the U.S. Gulf of Mexico.

August 1

U.S. President Bush extends U.S. sanctions against Iraq for another year, having notified the U.S. Congress that the government of Saddam Hussein continued to act in ways "hostile to U.S. interests."

August 2

The U.S. Office of Management and Budget approves U.S. Environmental Protection Agency regulations that authorize new penalties for manufacturers of diesel engines that exceed various pollutant levels, to take effect October 1, 2002. The new rules are part of long-term plan, begun in the previous administration, to require diesel trucks and buses to reduce emissions by 90 per cent by 2007.

August 4

Russia announces that cooperation with Iran in the area of nuclear energy will be limited to the construction of one nuclear reactor, already under way, at Bushehr. This would appear to cancel plans released July 26 to build up to five additional reactors in Iran and expand scientific ties.

August 5

Algeria formally requests an increase in its current 693,000-barrel-per-day OPEC crude oil production quota. The request will be examined later this month at a meeting of OPEC's Board of Governors, before being submitted to the full OPEC ministerial conference in September.

August 5

Nigerian Presidential Adviser on Petroleum and Energy Rilwanu Lukman announces that Nigeria's crude oil production capacity is now 2.6 million barrels per day and that its reserves are now 32 billion barrels, up from 26 billion barrels in 1999. Nigeria's current OPEC crude oil production quota is 1.79 million barrels per day.

August 7

Mexican Energy Minister Ernesto Martens announces that Mexico will continue to limit its crude oil exports to 1.66 million barrels per day, in coordination with OPEC, although Mexico is not a member of the cartel. Mexico is the only major non-OPEC exporter cooperating with the cartel, after Norway and Russia ended their cooperation earlier in the year.

August 7

Libyan Foreign Minister Mohammed Abderrahmane Chalgam indicates that Libya may be ready to pay compensation for the 1988 airliner bombing over Lockerbie, Scotland, that killed 270 people and to address U.N. demands that it accept responsibility for the attack. The statement comes during the visit of a high-ranking British official to Libya for the first time since relations between the two countries were restored.

August 9

Brazilian state oil company Petrobras announces that it has discovered a very large oil field off the coast of the state of Espirito Santo. Petrobras said in a statement to the Sao Paulo Stock Exchange that the find has an estimated 600 million barrels of oil reserves.

August 15

Dynegy settles a lawsuit with Enron over their failed $9 billion merger with a payment of $25 million plus $62.9 million already held in escrow by Enron, ending a potential $10 billion lawsuit that was a major uncertainty for current and potential investors in Dynegy. This settlement also clears the way for Dynegy's July sale of the $928 million Northern Natural Gas Pipeline to MidAmerican Energy to go through. Had it not gone through because of claims from an Enron lawsuit, Dynegy would have faced a possible separate lawsuit from MidAmerican Energy.

August 19

An OPEC report has OPEC 10 (*i.e.* excluding Iraq) producing at 1.8 million barrels per day above the quota of 21.7 million barrels per day for the month of July. This production rate is 338,000 barrels per day greater than the revised June figure reported by OPEC.

August 19

Mexican President Vicente Fox puts forth a bill to the Mexican Congress that would modify certain parts of the Mexican constitution in order to allow private companies to sell electricity directly to corporations and other large consumers. This is part of an effort to increase generation capacity through private investment, though passage of the bill by the congress is not assured.

August 20

The NYMEX near-month crude oil futures price closes above $30 per barrel for the first time since February 2001. Concern over possible conflict in Iraq, OPEC quotas, and declining crude oil and product stocks are among the factors leading to a nine-straight-session rise in NYMEX prices.

August 21

Williams of the United States agrees to sell its 33 per cent stake in Lithuania's Mazeikiai Nafta oil company to Yukos of Russia for $85 million. This will give Yukos, which already has a 26.85 per cent share, control of the company. Mazeikiai Nafta owns Lithuania's (and the Baltic republics of Former Soviet Union's) only refinery and accounts for about 10 per cent of the country's GDP.

August 23

The state-owned China National Offshore Oil Corporation (CNOOC) announces that it has reached a preliminary agreement to buy a stake in Australian natural gas fields for $320 million. The 5 per cent stake in the Northwest Shelf gas fields comes on the heels of an agreement by CNOOC on August 8 to buy between $11 billion and $13 billion worth of liquefied natural gas (LNG) from the field over the next 25 years.

August 25

Mexican state oil company Pemex announces that will invest $4.3 billion over the next eight years in the Ku-Maloob-Zaap project in Campeche Sound. Pemex commented that "With these works it is estimated that in the year 2011 the development of the Ku-Maloob-Zaap project will reach its maximum in production of Maya crude of some 800,000 barrels per day, compared with 263,000 barrels per day currently being extracted."

August 28

Norway and the United Kingdom unveil a joint plan to strengthen cooperation on their North Sea oil and natural gas fields. The plan aims to boost output, cut operational costs and promote cross-border cooperation in an effort to save energy companies $2 billion by 2010.

August 29

U.S. Vice President Cheney states that a new round of U.N. weapons inspections in Iraq is likely to be insufficient to guarantee that Iraq has ended its biological, chemical, and nuclear weapons programs. That same day, Iraqi Vice President Ramadan declares that future inspections by the United Nations are a "waste of time," as the U.S. administration has already decided upon "changing the regime by force."

August 30

The Federal Trade Commission approves the merger of Conoco and Phillips Petroleum after the two companies agree to sell certain assets to maintain competition in gasoline refining and marketing and other business sectors. The $15.12 billion merger will create the world's sixth-largest publicly-held oil and natural gas company, to be based in Houston.

September 2002

September 2

The National Iranian Oil Company (NIOC) denies reports that Iran had decided to suspend all new oil development deals with foreign companies for one year. Iran has not signed any new oil agreements with foreign companies for about 14 months. The buy-back deals that Iran uses with foreign companies are not favored by some in the Iranian government.

September 4

Researchers at the University of California-Riverside release findings of a study confirming that two Japanese automobile manufacturers have developed gasoline engines that meet super ultra-low emissions standards, standards heretofore only achieved by gasoline-electric hybrids. The engines require specially formulated gasoline in order to achieve these results. This May pave the way for more stringent California emissions standards to take effect in the future.

September 7

Venezuela resumes shipments of oil to Cuba under a unique preferential sales agreement, some five months after they were suspended during the brief April coup and a failure by Cuba to keep up with its payments. Opponents of President Chavez accuse him of squandering Venezuela's resources by selling oil at less than market rates and/or conditions to Cuba.

September 11

The International Energy Agency's (IEA) monthly oil market report notes that global oil stock levels have fallen to "uncomfortably low" levels that could lead to higher prices and more price volatility in the coming months. According to the IEA, OECD crude oil inventories fell by 790,000 barrels per day in August compared with July.

September 11

The European Union (EU) releases a plan for coordination of member countries' crude oil reserves, including raising the minimum level of national oil stocks to 120 days of consumption from 90 days and putting one third of reserves into a stockpile which could be drawn on in times of crisis. The European Commission would have the power to release oil from the stockpile onto the market if prices rose to a level that, if sustained for a year, would raise the EU's external oil bill by an amount equal to 0.5 per cent of its gross domestic product. Energy Commissioner Loyola de Palacio predicts that the new system will be in place in 2007.

September 12

California Governor Gray Davis signs a bill into law that will require the State's private utilities to increase the share of electricity coming from renewable sources by 1 per cent per year, such that a level of 20 per cent for all utilities in California is reached by 2017. The bill also has a provision for renewable power to receive a partial subsidy.

September 12

U.S. President George Bush addresses the United Nations. President Bush declares in regard to Iraq that "The Security Council resolutions will be enforced–the just demands of peace and security will be met–or action will be unavoidable..and a regime that has lost its legitimacy will also lose its power."

September 13

The World Bank approves lending for a controversial oil pipeline between Chad and Cameroon. The bank is funding $140 million of the $4 billion project to develop the oil fields of Doba in southern Chad and construct a 665-mile pipeline to an offshore oil-loading facility on Cameroon's Atlantic coast.

September 15

Iran awards a $1.6 billion contract to a consortium of Iranian and South Korean companies to develop phases nine and ten of the South Pars natural gas field. The South Korean companies will have a 42 per cent stake, and the contract will be foreign financed, a change from Iran's usual practice of buyback deals. The South Pars natural gas field is estimated to have resources of about 280 trillion cubic feet.

September 16

Iraqi Foreign Minister Naji Sabri sends a letter to U.N. Secretary General Kofi Annan indicating that Iraq will accept the return of U.N. weapons inspectors "without conditions." The following day, Iraqi Deputy Prime Minister Tariq Aziz says, "All the reasons for an attack have been eliminated," though White House spokesman Scott McClellan describes the move as "a tactical step" to avoid strong U.N. Security Council action, going on to say, "As such, it is a tactic that will fail." On September 19, Foreign Minister Sabri asserts that Iraq is free of nuclear, biological, and chemical weapons.

September 17

A report released by the California Public Utilities Commission alleges that five independent power producers-Duke, Dynegy, Mirant, Reliant, and Williams-deliberately withheld output during California's 2000-2001 energy crisis in order to drive up prices, sometimes precipitating blackouts. Independent Energy Producers, a trade group representing the companies, has denied the allegations of the report.

September 18

Work begins on the $2.9 billion Baku-Ceyhan Pipeline, which will transport oil from the landlocked Caspian Sea to Turkey's Mediterranean coast. The BP-led pipeline will be 1,110 miles long when completed in 2005. Work begins on the Turkish section on September 26.

September 18

According to United Nations officials and representatives of the oil industry, Iraq has stopped attempting to impose illegal surcharges on oil it sells through the United Nations' "Oil-for-Food" program. Though the surcharges have provided funds to the regime, Iraq May be attempting to cooperate more closely with U.N. resolutions in the face of increased scrutiny by the United States and Britain.

September 19

OPEC, meeting in Osaka, Japan, decides that its ten members subject to quotas (*i.e.* excluding Iraq) will not raise their current 21.7-million-barrel-per-day production ceiling. However, OPEC's communiqué states that OPEC is committed "to taking any further measures, including convening extraordinary meetings when deemed necessary. .to maintain prices [OPEC basket price] within the range of $22-$28 [per barrel]." Also at the meeting, Qatari Oil Minister Abdullah bin Hamad al-Attiyah is appointed as the new OPEC President, replacing Rilwanu Lukman of Nigeria.

September 19

The German government approves the $10 billion merger of energy conglomerates E. On and Ruhrgas with tougher conditions than the previous conditional approval granted by the Economics Ministry on July 5. A German court had since ruled that July 5 approval unlawful due to certain procedural flaws. E. On and Ruhrgas indicate that they will accept the new conditions.

September 24

Energy trading company Dynegy agrees to pay a $3 million fine to settle a U.S. Securities and Exchange Commission (SEC) fraud investigation. The SEC had specifically investigated Dynegy's use of so-called "special purpose entities" in its accounting and "round trip" energy transactions that can improperly inflate a company's revenue.

September 24

Tropical Storm Isidore halts most of the U.S. Gulf's 1.4-million-barrel-per-day crude oil production, as well as about one-third of the region's natural gas production, for a short time. The storm also temporarily halts oil imports and shuts down the Louisiana Offshore Oil Port (LOOP). Some Gulf coast refineries reduce their output. On September 30, oil and natural gas workers are forced to again evacuate offshore platforms as Hurricane Lili approaches.

September 24

The United Nations confirms that Iraq's oil exports under the U.N. "Oil-for-Food" program jumped to 1.9 million barrels per day last week, more than doubling the 914,285-barrel-per-day rate for the prior week. This is the highest export rate since the current phase of the program began (5/20/02). The increase is largely due to Iraq's informing oil companies earlier in the month that Iraq will no longer demand an illegal surcharge on its export sales.

September 26

Indonesia's state oil firm Pertamina and China's National Offshore Oil Corp (CNOOC) sign a $8.5 billion deal to supply liquefied natural gas (LNG) to a terminal in China's Fujian province. Under the terms of the deal, CNOOC will be supplied with 2.6 million metric tons of LNG per year from the giant Tangguh field in the remote eastern Indonesian province of Papua, 49.66 per cent owned by BP, Britain's BG Group holds a 10.73 per cent stake in Tangguh, whose other shareholders include a string of Japanese firms such as Mitsubishi, Nippon and Kanematsu. On September 27, CNOOC indicates that it will buy from BP an additional 12.5 per cent stake in the project.

September 27

The U.S. Federal Trade Commission gives conditional approval of Shell's $1.8 billion acquisition of Pennzoil-Quaker State, one of the largest producers of motor oil and other lubricants.

September 27

The U.S. Energy Information Administration (EIA) announces that U.S. crude oil reserves increased by 1.8 per cent in 2001. Total discoveries of crude oil in existing and new fields totaled 2.565 billion barrels in 2001, and oil reserve additions exceeded domestic crude production by 21 per cent. Separately, dry natural gas proved reserves jumped by 3.4 per cent or 22.758 trillion cubic feet, as additional natural gas reserves exceeded output by 31 per cent.

September 29

Union workers at Mexican state oil company Pemex accept a lower raise than initially requested, but higher than management's first offer, averting a threatened strike that would have seriously affected production had it gone on for any extended period of time. Mexico is among the top three crude oil suppliers to the United States.

October 2002

October 2

Brazilian state oil company Petrobras announces that it exported more oil than it imported for the first time in the company's history in the month of July. Petrobras President Francisco Gros asserts, "This is the beginning of a trend on a growth basis." Production has increased in Brazil's Campos Basin, where Petrobras recently made a very large find.

October 3

Hurricane Lili makes landfall on the U.S. Gulf coast after passing through offshore hydrocarbon production areas and the Louisiana Offshore Oil Port (LOOP). Nearly all offshore production (about 1.5 million barrels per day of oil production), as well as some onshore refineries, the LOOP and the Capline crude oil pipeline are shut down. Refineries and offshore operations begin to come back on line on October 4, with most operations fully online by the second half of the month. There is little permanent damage.

October 6

A French oil tanker chartered by Malaysian state oil company Petronas is attacked off the coast of Yemen, seriously damaging the ship and killing one crew member. The VLCC, with about 400,000 barrels of oil aboard, catches fire. The tanker does not sink, and is towed to port. Later, investigators determine that a terrorist suicide attack by a small boat is the cause of the explosion. The tanker was on its way to load additional oil in Yemen when attacked.

October 9

American Electric Power (AEP) admits that some of its traders provided inaccurate energy price data to a major publisher of pricing indexes. This follows a disclosure by Dynegy in September that some of its traders also provided inaccurate energy price data to pricing indexes. Regulators are investigating traders for providing incorrect information in order to manipulate energy markets.

October 9

The U.S. Energy Information Administration (EIA) releases data showing that crude oil stocks in the previous week fell to their lowest levels (270.5 million barrels) since the agency began keeping weekly records over 20 years ago. Crude oil stocks have fallen by over 50 million barrels since February of this year and are now 39 million barrels below the year ago level and only 0.5 million above the EIA's "Lower Operational Inventory." While not implying shortages, operational problems, or price increases, the Lower Operational Inventory means that supply flexibility could be constrained.

October 10

The International Court of Justice at The Hague rules that Cameroon is the rightful owner of the Bakassi Peninsula in the Gulf of Guinea, which Nigeria also has claimed. The Bakassi Peninsula is suspected of containing significant oil reserves, particularly offshore. On October 23, Nigeria issues a statement refusing to withdraw its military or officials from the peninsula.

October 10

Construction is completed on a new oil terminal at Georgia's Black Sea port of Poti. Annual capacity of the new terminal is two million metric tons of oil (about 40,000 barrels per day). The second phase of the project, envisaging construction of another four reservoirs aimed at raising the terminal's capacity to four million tons (80,000 barrels per day), is expected to be completed by June 2003. Azerbaijan, Kazakhstan and Turkmenistan transport the oil from their fields by rail over the strip of land between the Caspian Sea and Georgia's Black Sea ports.

October 10

The European Commission announces that the European Union (EU) will dramatically increase its spending on fuel cell research by spending $2.09 billion over the next three years. During the previous three years, the EU had only spent about $118 million, far less than Japan or the United States.

October 11

The U.S. Senate votes to give President George Bush the authority to use force, if necessary, to persuade Iraqi President Saddam Hussein to abandon programs for the development of biological, chemical or nuclear weapons. The U.S. House of Representatives passed a similar measure the previous day.

October 11

The International Energy Agency, in its Monthly Oil Market Report, reduces its forecast of world oil demand growth in 2003 by 100,000 barrels per day, to 1.04 million barrels per day. According to the report, "The cut [in the demand forecast] reflects the slowdown of the U.S. and global economic recovery, the impact of high oil prices on oil consumption and the broader economy."

October 14

U.S.-based TXU announces that it will withdraw financial support for its European units and put them up for sale. TXU's largest European business, in Britain, has 5.6 million retail electricity customers. Analysts point to problems with Britain's new deregulated electricity market, in which wholesale electricity prices have fallen 40 per cent in the past 18 months. Many companies are having difficulties covering their generation costs in the new system.

October 17

The former head of Enron's western energy trading desk pleads guilty to wire fraud in connection with schemes to manipulate California's wholesale electricity market in 2000 and 2001. U.S. Deputy Attorney General Larry Thompson states that the government's investigation into illicit profits made by energy companies during the California power crisis "is active and ongoing." The crisis cost consumers and taxpayers billions of dollars.

October 21

U.S. President George W. Bush signs a law that will impose certain financial penalties on the Sudan if the government there is found not to be negotiating in good faith to end the 19-year civil war in that country. Talks between the government of the Sudan and rebels representing the Christian and Animist south are underway, with a truce to have begun October 17.

October 22

ExxonMobil signs a $3 billion agreement with the government of China to produce refined petroleum and petrochemical products in China under its own brand name. ExxonMobil will build a $3 billion integrated refinery and ethylene complex in southeastern China's Fujian province. ExxonMobil will have a 25 per cent stake in the 160,000-barrel-per-day refinery and the 800,000-

metric ton-per-year ethylene plant. Sinopec's unit Fujian Petrochemical Co. Ltd. will hold a 50 per cent stake and Saudi Arabia Oil Co., or Saudi Aramco, will control the remaining 25 per cent. The project is expected to start operations in 2006.

October 22

A report released by Salomon Smith Barney shows that U.S. refining margins have risen to their highest level since September 2001, at $5.88 per barrel, which is approximately triple the August 2002 low point.

October 22

Venezuela's Minister of Finance forecasts the Venezuelan economy to contract by 4.8 per cent to 5 per cent for 2002, despite relatively high oil prices. Venezuela's government has been dealing with a number of nationwide strikes, as well as high-ranking military officers calling for the end of the government of Hugo Chavez.

October 23

China approves reforms of the country's electric power market, including the break-up of the monopoly State Power Corporation into five generation and two grid corporations, encouraging some competition in the world's second-largest power market as well as better linking of fragmented transmission networks.

October 24

The Nigerian Senate approves a bill put forward by Nigerian President Olusegun Obasanjo and already passed by the Nigerian House of Representatives, which will give state governments a share of revenues from offshore oil and gas. This law reverses a decision by the Nigerian Supreme Court in April 2002 that awarded all offshore oil revenues to the central government. Several state governments protested that decision and warned that instability in the Niger delta region was likely to occur unless that decision were reversed.

October 28

Statoil of Norway signs a $300 million deal with Iran's semi-state Petropars company to become an operator in the Pars offshore Iranian natural gas field (Phases Six, Seven, and Eight), part of a $2.6-billion project. This is part of a concerted Iranian effort to increase foreign investment in its oil sector, and is Statoil's first major investment in Iran.

October 30

White House spokesperson Ari Fleischer denies that the United States has any interest in taking over Iraq's oil fields if there were a military conflict. When asked if the United States would take over Iraq's oil fields if the U.S. attacked Iraq, Fleischer responded: "No. The purpose of any plan the United States has is to make certain that Saddam Hussein complies with all U.N. resolutions..The only interest the United States has in the region is furthering the cause of peace and stability not his [Saddam Hussein's] country's ability to generate oil."

November 2002

November 1

Greece, Bulgaria, and Russia agree to equal stakes in the $699 million Trans-Balkan Pipeline. The 159-mile pipeline will bypass the Bosphorus Strait in order to bring Russian oil from the Bulgarian

Black Sea port of Burgas to the Greek Mediterranean port of Alexandroupolis. The pipeline will be able to carry about 697,000 barrels per day.

November 3

The Trans-Alaska oil pipeline is shut down temporarily after a major earthquake in Alaska. The pipeline transports about 1 million barrels of oil per day from the North Slope oil fields to the port of Valdez. The pipeline restarts November 6.

November 5

A report from Salomon Smith Barney indicates that average U.S. gross refining margins jumped 64 cents per barrel to $6.31 per barrel of crude oil distilled for the week ended November 2. This is the highest level since September 11, 2001, caused in part by "drawdowns in stocks of gasoline and distillate as refineries struggle to increase utilization rates," according to the report.

November 7

KazTransOil, the operator of Kazakhstan's oil pipeline network, announces plans to upgrade the 429-mile crude oil export pipeline from Atyrau to Samara in Russia from a current capacity of about 16.5 million metric tons per year (about 329,000 barrels per day) to a capacity of 25 million metric tons per year (about 498,000 barrels per day), potentially increasing exports. This is one of only two oil export pipelines in Kazakhstan.

November 8

The United Nations (UN) Security Council unanimously adopts Resolution 1441, that Iraq must accept or reject within seven days, giving United Nations inspectors the unconditional right to search anywhere in Iraq for banned weapons. Furthermore, Iraq will have to make an "accurate full and complete" declaration of its nuclear, chemical, biological and ballistic weapons and related materials used in civilian industries within 30 days. The resolution requires violations to be reported back to the Security Council by inspectors before any actions could be taken against Iraq for violating weapons bans.

November 12

Canada's most populous province, Ontario, ends electricity deregulation that had begun on May 1 of this year. Electricity prices have risen by more than 25 per cent since deregulation began and the provincial government will be issuing refunds to consumers to offset some of the extra cost.

November 13

In a letter to United Nations (UN) Secretary General Kofi Annan, Iraq accepts UN Security Council resolution 1441 of November 8, granting UN inspectors the right to conduct unfettered inspections in Iraq, "despite its bad contents." In the letter, Iraq also denies that it possesses any weapons of mass destruction.

November 14

The TengizChevroil consortium, a consortium of companies led by operator ChevronTexaco that is developing the estimated 2.7-billion-barrel Tengiz oil field in Kazakhstan, announces that the consortium has decided to indefinitely suspend investment in the second phase of the project. Production from the first phase was about 12.5 million metric tons in 2001 (about 249,000 barrels per day). The second phase would require about $3 billion of investment in order to boost the project's output by about 3 million metric tons per year (about 60,000 barrels per day).

November 14

U.S. President George W. Bush decides that November will be the last month that the U.S. government finances fuel oil shipments to North Korea. The shipments are part of a 1994 arms control agreement with North Korea, and are being halted after revelations of a covert North Korean nuclear weapons program in violation of that agreement. The decision is later backed by Japan, South Korea, and the European Union, the other three entities that comprise the Korean Peninsula Energy Development Organization, which administers the fuel oil shipments.

November 15

The U.S. Strategic Petroleum Reserve, an emergency crude oil stockpile administered by the U.S. Department of Energy, reaches 592 million barrels, the largest amount in the reserve since it was initiated in 1977.

November 18

The tanker Prestige, loaded with 24 million gallons of Russian fuel oil, splits in two and sinks 155 miles off the coast of northwest Spain. The tanker, flying a Bahamian flag and owned by a Liberian company based in Athens, Greece, spills about 2.5 million gallons of the fuel oil from a crack before sinking, polluting beaches in the region and harming marine life. Fuel oil may continue leaking from the sunken ship.

November 18

BG (formerly British Gas) announces that it will build a $322 million liquefied natural gas (LNG) import terminal in Brindisi, Italy, having finally been granted approval by government authorities to go ahead with the construction, due to be completed in 2006. The terminal will initially be able to handle imports of 3 million metric tons per year, imported from Egypt. This will be Italy's second LNG terminal.

November 19

Philippines President Gloria Macapagal Arroyo calls on Energy Secretary Vincent Perez to keep Meralco (the Manila Electric Supply Company) functioning. Meralco, which supplies 60 per cent of the Philippines' electricity, is facing bankruptcy as the Philippines Supreme Court ordered the company to pay compensation for overcharging going back to 1994. The court later orders the company to pay over $500 million. The government already owns 10 per cent of the company, but the company refuses a proposed government intervention on November 25 that would offer government support in exchange for greater government control.

November 22

The U.S. Environmental Protection Agency (EPA) announces that it will change certain rules regarding industrial air pollution. According to the EPA, these changes will encourage plant improvements to clean the air, though critics describe the changes as a loosening of rules requiring plants to invest in pollution control equipment when modernizing.

November 25

European Union energy ministers reach an agreement allowing consumers to choose their natural gas and electricity suppliers by 2007. An accord reached on March 16, 2002, allowed large companies to be able to choose their suppliers by 2004.

November 25

The International Energy Agency (IEA) announces that its members are prepared to insert large quantities of oil into the world market on short notice, in case of any supply disruption. IEA Executive Director Robert Piddle states: "From public, or government stocks, we could release as much as 12 million barrels a day in the first month."

November 25

The United Nations Security Council extends the "Oil-for-Food" program that permits Iraq to sell oil for a nine-day period, instead of the usual six-month phase, in order to give Security Council members time to reach agreement on which goods would be placed on the so-called "goods review list" which are goods that require approval before Iraq can use its revenues to purchase them.

November 26

Alaska State officials and representatives of six oil companies sign an agreement extending the lease agreement between those companies and the state for the trans-Alaska oil pipeline by 30 years, to 2034. The trans-Alaska pipeline delivers nearly a fifth of domestically produced oil and has shipped over 13.5 billion barrels of crude since it began operations in 1977.

November 26

Murphy Oil of the United States announces the discovery of 400-700 million barrels of oil in the Kikeh field off the coast of Malaysia's Sabah region on the island of Borneo. This is one of the largest discoveries in Southeast Asia in recent years.

November 26

Exxon Mobil and TotalFinaElf sign an agreement with the government of Qatar for the construction of a 146,000-barrel-per-day gas condensate refinery. The two companies will also each hold a 10 per cent stake in the refinery, to be completed in 2006. The remainder will be owned by state-owned Qatargas.

November 27

Officials of four of Russia's largest oil companies, Lukoil, Yukos, Sibneft, and Tyumen, announce a preliminary agreement for a joint project to build a $1.5 billion dollar Arctic oil port near the town of Murmansk. This would enable Russia to expand ocean-going tanker exports.

November 27

Congo (Brazzaville) Oil Minister Jean-Baptiste Tati announces that Congo and Angola have reached an agreement for the development of oil deposits on their maritime border, estimated at about 950 million barrels. The two countries will split royalties 50-50 from a consortium headed by ChevronTexaco to develop the area.

December 2002

December 1

State-owned Venezuelan oil company Petroleos de Venezuela S.A. (PdVSA), Royal Dutch/Shell, and Mitsubishi sign an agreement for the development of the $2.7 billion Mariscal Sucre liquefied natural gas (LNG) project. The project, if completed, would produce 4.7 million metric tons of LNG per year, all for export.

December 2

ChevronTexaco announces that it has submitted an application with the U.S. Department of Transportation (DOT) to build a U.S. Gulf Coast liquefied natural gas (LNG) import terminal. The terminal, to begin operations in 2006, would be an offshore facility with an initial design capacity of about 800 million cubic feet per day of gas.

December 2

Business and labor groups in Venezuela, including employees of state-oil company PdVSA, begin a strike in order to obtain an early referendum on the rule of Venezuelan President Hugo Chávez. The strike has little effect on its first day, but as the strike continues through the end of the month, oil production, refinery runs, and crude oil and refined petroleum product exports fall dramatically. Several refineries in the Caribbean dependent on Venezuelan crude are also adversely affected. This has a serious impact on the Venezuelan economy, but no agreement between President Chávez and the opposition forces leading the strike is reached by the end of the month.

December 4

The United Nations (U.N.) Oil-for-Food program is unanimously renewed by the Security Council for another six months, and shortly thereafter accepted by the Iraqi government. The Oil-for-Food program allows Iraq to sell unlimited quantities of oil, with revenues going into a U.N. account that pays vendors for approved goods that Iraq orders.

December 4

The government of Russia sells off 5.9 per cent of Lukoil for $775 million, further privatizing the oil company and giving the government additional funding to service the foreign debt.

December 9

The government of India announces that it will sell a large stake in Hindustan Petroleum, India's largest oil refining company, as well as shares in Bharat Petroleum, India's largest fuel retailer. The government is currently the majority stakeholder in both companies, but does not detail how much of its stakes will be sold nor a timetable for the sale.

December 10

The Canadian House of Commons (the lower house of Parliament) votes to support government plans to ratify the Kyoto Protocol on cutting greenhouse gases. Although the government is empowered to approve the treaty without parliamentary approval, the vote gives support to the government of Prime Minister Jean Chretien, who states after the vote that he intends to ratify the treaty.

December 12

A federal regulatory judge at the Federal Energy Regulatory Commission (FERC) rules that energy companies overcharged California consumers by $1.8 billion during the 2000-2001 power crisis. This is less than the $8.9 billion that the State of California had sought in refunds. However, the FERC is still considering some allegations of price manipulation by electricity and natural gas suppliers and a final ruling is not expected until March 2003.

December 12

The government of Iraq cancels a $3.8 billion contract with three Russian companies-Lukoil, Zarubezhnest, and Machinoimport-to develop the very large West Qurna oilfield. Although the reasoning for the decision is not made clear by Iraq, it is thought that it is in response to Russian political decisions regarding United Nations inspections and the Oil-for-Food program.

December 12

OPEC oil ministers, meeting in Vienna, decide to raise OPEC-10's (*i.e.* excluding Iraq) total production quota from 21.7 million barrels per day to 23 million barrels per day. OPEC ministers also urge strict compliance with the new quotas in an effort to cut back production, as OPEC-10 production is widely regarded to be exceeding even the new production quota of 23 million barrels per day.

December 12

The government of North Korea announces that it will end a freeze on the use of nuclear power in response to last month's decision by the United States and its allies to end fuel oil shipments to North Korea.

December 13

Statistics are released in Japan indicating that the country's ten electric power utilities increased their oil consumption 164 per cent in November 2002, compared with November 2001. The thermal electric plants had to make up for the temporary closure of several nuclear power plants.

December 13

It is reported that Reliance Industries has discovered recoverable reserves equal to more than five trillion cubic feet of natural gas, one of the world's biggest discoveries this year, in the Krishna Godavari basin, offshore Andhra Pradesh, India. Cairn Energy has also found an additional 800 billion cubic feet of natural gas while drilling in an adjacent block of the sea floor.

December 16

The near-month crude oil futures price on the NYMEX tops $30 per barrel for the first time since October 2, as the general strike in Venezuela impacts the world oil market. Later in the month, on December 27, the near-month crude oil futures price rises to $32.72 per barrel, the highest price since November 2000. (WSJ, AP)

December 16

Croatia, Russia, Belarus, Ukraine, Slovakia and Hungary sign an agreement for the integration of the Druzhba and Adria oil pipelines, which would enable Russia to increase exports of its crude oil to the Mediterranean and other world markets through the Croatian deep-sea port of Omisalj. The project is expected to cost about $300 million and involve a gradual increase of exports from five million metric tons to 15 million metric tons of crude oil per year (100,000 to 300,000 barrels per day). Completion of the first stage is planned for the end of 2003.

December 17

The U.S. Department of Energy allows several oil companies to postpone delivery of an additional 430,000 barrels of crude oil to the Strategic Petroleum Reserve in an attempt to keep more oil in the market during the strike in Venezuela. The oil companies will have to deliver the oil at a later date.

December 18

New estimates are released by the United States Geological Survey showing that oil reserves in the Rocky Mountain region of the United States are about 1.9 billion barrels. This figure is about two-thirds of the 2.76 billion barrels estimated previously, in 1995.

December 19

U.S. Secretary of State Colin Powell declares that Iraq is in "material breach" of United Nations resolutions after reviewing Iraq's weapons of mass destruction declaration released December 7 to the United Nations. States Powell: "Our [U.S.] experts have found it to be anything but currently accurate, full or complete. The Iraqi declaration totally fails to meet the resolution's requirements."

December 28

A tanker with 22 million gallons of gasoline arrives in Venezuela from Brazil, providing crucial supplies to the country, as the strike by employees of state-oil company PdVSA has meant severe reductions in refinery runs in that country. Crude oil production, that was in excess of 3 million barrels per day before the strike, is less than 500,000 barrels per day for many days in December.

December 29

Acting Kuwaiti Oil Minister Sheikh Ahmed Al-Fahad Al Sabah says that Kuwait can keep producing and exporting oil in the event of a military conflict in Iraq. States Al Sabah: "I can't go into details of this plan, but I can guarantee that production will continue, exports will continue.and I believe we can also meet the commitments we have made to our clients abroad."

101

Monthly Energy Chronology 2003

The following information contains a listing of the major energy events that have occurred in 2003.

January 2003

January 3

The near-month crude oil futures price on the New York Mercantile Exchange (NYMEX) reaches $33.08 per barrel, its highest settlement price since November 30, 2000, as the strike in Venezuela enters its second month, forecasters call for colder weather, crude oil stocks remain low, and traders fear possible military action in Iraq.

January 6

Venezuelan Minister of Energy and Mines Rafael Ramírez announces that the Venezuelan government plans to split state oil company Petroleos de Venezuela S.A. (PdVSA) into two separate entities as part of a large-scale restructuring of the company, most of whose 40,000 workers are currently on strike. Such a decentralization could limit the power of Caracas-based executives who have joined in the strike, which began on December 2, 2002.

January 10

The U.S. Department of Energy, PG&E Corp.'s Pacific Gas & Electric, and Trans-Elect agree on a $300 million upgrade of electricity transmission lines on the Path 15 electricity route between

Northern and Southern California. This area can be a bottleneck on the transmission grid, so the additional transmission capacity is expected to save $100 million in a normal year and $300 million in a dry year.

January 11

Venezuelan President Hugo Chavez fires 700 PdVSA workers, the first large group of workers that are fired in an effort to break the strike that began December 2. Previously, about 300 PdVSA managers had been fired by the government of President Chavez.

January 12

The Organization of Petroleum Exporting Countries (OPEC), meeting in Vienna, agrees to raise the aggregate production quota of its members (excluding Iraq) to 24.5 million barrels per day, up from the current 23 million barrels per day, effective February 1. Each member will receive a proportionately higher share of the quota, about a 6.5 per cent increase.

January 13

Two Norwegian North Sea oil fields, Gulfaks and Visund are temporarily shut in due to safety concerns resulting from technical problems with the fields' integrated flare system. Shutting in the two fields causes a 130,000 barrel-per-day disruption for two days.

January 14

Koch Industries cancels its planned $2.2 billion, 190,000-barrel-per-day project to develop oil sands in Alberta, Canada. Company officials cite problems reducing their 70 per cent share in the project, the high cost of production, and problems complying with the Kyoto Protocol as reasons for the cancellation.

January 16

Fourteen U.S. corporations or subsidiaries launch the Chicago Climate Exchange, a trading program wherein companies would be able to earn redeemable credits for exceeding emissions reductions goals of 4 per cent of 1998-2001 average emissions over the next four years. Companies unable to meet the goals would buy the credits. The Exchange intends to create means to verify that actual reductions in emissions have taken place.

January 17

Iraq awards a contract to Russian company Stroitransgaz for a small oil field in western Iraq and sets aside two others for Russian companies. Some analysts interpret these awards as an attempt at rapprochement between Iraq and Russia after Iraq canceled a giant contract with Russia's Lukoil in December 2002.

January 17

The International Energy Agency (IEA) estimates that Venezuela has permanently lost 400,000 barrels per day of production capacity owing to the country's ongoing general strike.

January 21

The near-month crude oil futures price on the NYMEX settles at $34.61 per barrel, the highest price since November 29, 2000. The market is experiencing a variety of higher price pressures, including the strike in Venezuela, fears of a conflict in Iraq, a cold winter in the United States, and low commercial oil stock levels in the United States.

January 21

It is reported that oil tankers pilots in Venezuela have agreed to return to work on Lake Maracaibo, indicating that the general strike in Venezuela has lost some of its support.

January 22

One of Venezuela's key refineries, Refineria Isla, located on the island of Curaçao, resumes crude oil processing at partial capacity after being out of service for about one month due to the ongoing strike in Venezuela. The Curaçao refinery, which has a nominal throughput capacity of about 320,000 barrels per day is reportedly processing 32,000 barrels of crude oil daily. Two other refineries, Puerto La Cruz and Paraguana, are also reportedly processing at partial capacity, bringing the country's total crude runs to approximately 150,000 barrels per day, approximately 12 per cent of the country's pre-strike refining capacity.

January 23

Average U.S residential heating oil prices reach $1.453 per gallon, the highest price in almost two years, as colder weather in many parts of the country increases demand, while rising crude oil prices increase the cost of production.

January 26

The government of India announces that it will sell large shares in two state-owned petroleum companies. The government will sell 34 per cent of Hindustan Petroleum to another company and will list 35 per cent of Bharat Petroleum to the public on domestic and international equity exchanges. A further 5 per cent of each company will be allotted to employees. After the privatizations, the government's share of Hindustan will be 12 per cent and the government's share of Bharat will be 26 per cent.

January 27

Kazakhstan reaches an agreement with an international consortium led by ChevronTexaco over the financing of a $3.5 billion expansion of the Tengiz oil field in the Caspian Sea, already the largest single oil development project in the Caspian region. The expansion would raise production from a current 286,000 barrels per day to 460,000 barrels per day by 2006.

January 28

The U.S. Department of Energy approves oil company requests to delay delivery of March shipments to the Strategic Petroleum Reserve (SPR). The announcement will allow 4.4 million barrels of crude oil designated for storage in the SPR, to be marketed to domestic refineries instead.

January 29

Striking managers at Venezuelan state oil company PdVSA confirm that oil production has surpassed 1 million barrels per day once again, after falling to as low as 200,000 barrels per day during the strike that began on December 2. On January 31, PdVSA President Ali Rodriguez announces that production is at 1.5 million barrels per day and that 5,300 striking workers have been fired. Opposition estimates of production are much lower at around 1.05 million barrels per day.

January 29

During his State of the Union address, President Bush proposes $1.2 billion in funding to support the research and development of hydrogen-powered vehicles.

January 29

The Federal Energy Regulatory Commission (FERC) releases a report stating that "Evidence indicates that price manipulation has occurred in certain natural gas markets and May be continuing." The FERC releases no details on the companies under suspicion, but indicates that investigations will continue.

February 2003

February 3

Indian Petroleum Minister Ram Naik announces that the government of India plans to boost the country's strategic crude oil reserves to 45 days from 15 days at an estimated cost of 43.50 billion rupees ($910 million).

February 4

It is reported that the government of Colombia has admitted that oil royalties from foreign companies operating in Colombia are being illegally channeled to leftist guerillas and other illegal groups. Oscar Muñoz, governor of the oil producing state of Arauca states that "Unfortunately, it does look like some of these oil royalties May be ending up in the hands of the rebels." As a result of the Colombian National Royalties Commission's investigation into these allegations, $40 million dollars in royalties assigned to Arauca, location of the Cano Limon field, are frozen.

February 4

Beijing-based industry newsletter "China OGP" reports that Chinese imports of liquefied petroleum gas (LPG) increased 28 per cent in 2002 over 2001 to a record 6.26 million metric tons, on strong domestic demand.

February 6

Iranian Oil Minister Bijan Zanganeh announces that phases two and three of the South Pars natural gas field are now on-line. These phases represent additional production of about 55 million cubic meters (1.9 billion cubic feet) of natural gas per year, 85,000 barrels per day of condensate, and 1 million metric tons (11.6 million barrels) of liquefied petroleum gas per year. The two phases are officially inaugurated on February 15.

February 9

Iranian President Muhammad Khatami announces that Iran has mined uranium for use in nuclear power plants, though he also states that the uranium would only be for civilian uses. This is the first time that Iran has acknowledged having uranium ore reserves. President Khatami also states that uranium-processing facilities have been set up in Isfahan and Kashan.

February 10

According to a spokesman for the Hovensa Refinery in St. Croix, U.S. Virgin Islands, the refinery has increased its runs to normal for the first time since the general strike in Venezuela began at the beginning of December 2002. The refinery, with a crude oil throughput capacity of 495,000 barrels per day, is the largest single refinery in the United States. The Hovensa Refinery is 50 per cent owned by Amerada Hess and 50 per cent owned by Venezuelan state oil company PdVSA.

February 10

The U.S. Department of Energy awards three contracts, one each to ChevronTexaco Global Trading, Shell Trading (U.S.) Co., and Exxon Mobil Oil Corp., to begin providing a total of 116,000 barrels per day of royalty-in-kind crude oil to the Strategic Petroleum Reserve (SPR) starting in May 2003. Also, beginning in April, the Department of Energy plans to begin receiving approximately 15,000 barrels per day of crude directly from producers offshore Texas that ship through the Hoover Offshore Pipeline System. Current plans call for 44.7 million barrels of crude to be supplied to the SPR by the end of January 2004, bringing the total stored in the reserve at that time to about 644 million barrels, or 7.5 per cent more than the current level.

February 11

BP invests $6.75 billion in Russia by creating a new joint venture company with TNK (Russia's fourth largest oil company) and Sidanco, of which BP already held a 25 per cent stake. BP will have a 50 per cent stake in the new company. TNK's shareholders, investment groups Alfa Group and Access-Renova, will hold the other 50 per cent stake of the new firm, and board control will be balanced equally. The investment by BP is equivalent to almost 10 per cent of Russian foreign exchange reserves and around 1.5 per cent of Russian gross domestic product (GDP).

February 12

Data from the U.S. Energy Information Administration (EIA) show that U.S. commercial crude oil stocks have fallen to 269.8 million barrels for the week ending February 7, 2003. This is the lowest commercial crude oil stock level since 1975, and just slightly below the lower operational inventory level of 270 million barrels. The lower operational inventory level, while not implying shortages, operational problems, or price increases, is indicative of a situation where inventory-related supply flexibility could be constrained or nonexistent.

February 14

The lower house of the Russian Parliament (Duma) approves legislation that will allow the breakup of the Russian electricity monopoly, Unified Energy System. Although it is still unclear what will happen to shares after the breakup, eventually Russia's electricity sector will be reorganized into groups of power producers, distributors, and sales units. The government will retain control of the grid itself. Unified Energy System's grid is the largest in the world.

February 18

Exxon Mobil begins construction of the $3 billion Kizomba B offshore development project in Angolan waters. The project, when completed, is expected to produce 250,000 barrels of crude oil per day, beginning in 2006, with total production over the life of the field estimated to be about 1 billion barrels. Besides Exxon Mobil, which has a 40 per cent stake, the other stakeholders are BP (26.67 per cent), Eni (20 per cent), and Statoil (13.33 per cent). The concessionaire is Angolan state oil company Sonangol.

February 21

Devon Energy agrees to buy Ocean Energy for $3.5 billion in stock and the assumption of $1.8 billion in debt. The combined company will produce 2.4 billion cubic feet of natural gas and 250,000 barrels of oil and natural gas liquids per day, making it the largest independent oil and natural gas producer in the United States.

February 21

A strike by public sector Nigerian oil workers ends after six days. Oil majors said the walkout by the PENGASSAN union and its blue-collar sister union, NUPENG, did not disrupt loadings at terminals. The replacement staff had taken over the duties of the strikers who monitor quality and volume of exports.

February 21

A gasoline barge explodes at an oil terminal on Staten Island, New York. New York City Mayor Michael Bloomberg says there is "..absolutely no evidence and reason to think that this is anything other than a very tragic industrial accident." WTI near-month crude oil prices on the NYMEX rise $1.21 per barrel on the news, and finally settle $0.84 per barrel higher for the day at $35.58 per barrel.

February 21

OPEC Secretary-General Alvaro Silva announces that OPEC will assure that a supply shortage does not occur should hostilities in Iraq prevent Iraq from continuing its oil exports. According to Silva, "The Organization has been dealing with the problem of Iraq for the last 10 years, attending to Iraq's market when it was restricted under the oil-for-food deal. This Iraq situation has been handled before and if it were to happen again the Organization would be able to handle it."

February 21

Dissident workers of Venezuelan state oil company PdVSA acknowledge that Venezuela's oil production has risen just above 1.5 million barrels per day, the highest level they have acknowledged since their strike began in early December. The government of President Hugo Chávez estimates production at over 2 million barrels per day.

February 24

U.K. Foreign Minister Jack Straw indicates that the United States and Great Britain want a United Nations decision on Iraq within about two weeks of submitting a new resolution to the Security Council. "Early this week, we shall be tabling a second resolution in respect of Iraq and that will spell out that we are approaching the end of the final opportunity that Saddam Hussein was given in 1441 on November 8 to completely disarm in substance as well as process," states Straw.

February 24

Venezuela lifts force majeure for exports of some lighter grades of its crude oil as well as for liquefied petroleum gas (LPG), though the government does not specify exactly for which crude grades the force majeure is lifted. Also, management of the 180,000-barrel-per-day capacity extra-heavy crude oil Sincor project announces that the project is in the process of restarting, and oil in storage is beginning to be processed.

February 25

The 108,000-barrel-per-day capacity Cerro Negro extra-heavy crude oil project in Venezuela restarts for the first time since the strike in that country began in the beginning of December. According to operator Exxon Mobil, initial production is 55,000 barrels per day.

February 26

The NYMEX near-month crude oil futures price settles at $37.70 per barrel, the highest price (in nominal terms) since October 1990. High winter heating demand, reduced output from Venezuela, low commercial crude oil stocks, and fears of war with Iraq are among the factors that have driven crude oil prices up.

February 26

It is reported that Turkey has decided to halt imports of Iraqi oil by overland tanker truck. Turkish oil company Tupras had planned to import about 19 million barrels of Iraqi oil by tanker truck over the course of 2003, according to statements made by company officials to the press.

February 28

Russian Foreign Minister Igor Ivanov threatens to veto a U.N. resolution in the Security Council authorizing war in Iraq. According to Ivanov, "Russia does not support any resolution which could directly or indirectly open the way to an armed resolution of the Iraq problem. Naturally, Russia has the right of veto. If the interest of international stability demands it, Russia, of course will exercise its right."

February 28

The NYMEX near-month heating oil futures price settles at an all-time high of 125.59 cents per gallon, as many of the same market forces affecting the crude oil market also have driven up the price of heating oil, especially increased demand from the cold winter. High sulfur distillate fuel inventories (also referred to as heating oil) plunged more than 15 per cent over the most recent four-week period to end the week of February 28, at 35.6 million barrels, 32 per cent below the level for the same period last year.

March 2003

March 3

The U.S. Energy Information Administration (EIA) releases data showing that national average retail diesel fuel prices were at a record high in nominal terms the previous week. EIA began collecting retail diesel price information in 1994. This is the third straight week of record retail diesel prices.

March 5

U.N. Security Council permanent members France and Russia, as well as member Germany, pledge to block any proposal put forth in the Security Council that would authorize the use of force in Iraq because of non-compliance with U.N. Security Council Resolution 1441, which calls for Iraq to give up weapons of mass destruction.

March 5

Some 500,000 bbl/d of Venezuelan production in the eastern region begins to come back on-line. It was shut off at the wellhead for a week because of bottlenecks at export terminals as Venezuelan state oil company PdVSA encountered problems in returning loading at terminals to pre-strike levels. The Venezuelan government claims that oil production is over 2 million barrels per day, while fired PdVSA workers claim production is at 1.1 million barrels per day.

March 5

International Energy Agency (IEA) Executive Director Claude Mandil and Saudi Oil Minister Ali al-Naimi meet in Riyadh. Mandil receives a commitment from OPEC to meet "any further loss of supplies to oil markets in a swift and timely manner." Furthermore, according to the Saudi Press

Agency, "In this situation, IEA agreed with OPEC opinion that the producers should utilize their spare capacity before resorting to the oil available in the strategic reserves by consumers."

March 6

U.S. Secretary of Energy Spencer Abraham and officials from the European Atomic Energy Community (a part of the European Union) sign agreements for cooperation and joint research into nuclear power and hydrogen and fuel cell technology.

March 6

Venezuelan President Hugo Chavez announces that force majeure is henceforth lifted on Venezuelan oil exports. Venezuela had declared force majeure on its oil exports shortly after the national strike began on December 2, 2002. It is later revealed that this lifting does not apply to certain petroleum products. President Chavez also refuses to consider rehiring any of the over 15,000 fired PdVSA workers.

March 7

The New York Mercantile Exchange (NYMEX) puts into effect expanded price limits on its energy contracts and reduces to five minutes the time trading is halted when those limits are reached. Under the revised rules, the initial price limits for light, sweet crude oil futures will be expanded to $10 per barrel in all months from the current $7.50 in the first two months and $3.00 in all other months. The initial Henry Hub natural gas futures limits will expand to $3.00 per million British thermal units (MMBtu) in all months from $1 in all months. The initial limits on heating oil, gasoline and propane futures will increase to 25 cents per gallon in all months from 20 cents in the first two months and 6 cents in all other months.

March 7

CNOOC Ltd., the publicly-traded Hong Kong subsidiary of the state-owned China National Offshore Oil Corporation, announces that it is purchasing a $615 million, 8.33 per cent stake in the North Caspian Sea Project from BG Group of the United Kingdom. The field's reserves are estimated at 13 billion barrels of oil equivalent. Mark Qiu, Chief Financial Officer of CNOOC, comments that "Adding CNOOC as a partner adds the possibility of marketing the production in China, which is thirsty for energy."

March 7

Officials in the U.S. Environmental Protection Agency (EPA) announce that new clean water regulations for smaller sites, to take effect March 10, will not apply to the petroleum and natural gas industries. Rather, these two industries will have a two-year exemption, because, according to the EPA, further study of the effects of these regulations upon these two industries is needed.

March 7

British Foreign Minister Jack Straw announces a U.N. Security Council Resolution, supported by the United States and Spain as well, which authorizes the use of military force if Iraq does not cooperate fully with disarmament demands by March 17. The proposed resolution calls on Iraq to hand over to U.N. inspectors all weapons, delivery systems and support systems and structures banned by the United Nations and also provide information about the prior destruction of such items.

March 10

U.N. Security Council permanent member France pledges to veto the proposed resolution on the use of force in Iraq that was proposed by the United States, Great Britain, and Spain on March 7. States French President Jacques Chirac on the proposed resolution, "Whatever happens, France will vote 'no'."

March 11

The Organization of Petroleum Exporting Countries (OPEC) meets in Vienna and decides to maintain crude oil production quotas for its member countries (excluding Iraq) at 24.5 million barrels per day. Most analysts, including EIA, believe that OPEC-10's (excluding Iraq) actual production is higher than the quota amount.

March 11

At an OPEC meeting in Vienna, swing producer Saudi Arabia's Oil Minister Ali al-Naimi assures the world that "There will be no shortage of oil. The test is, when the need is there, whether we will use the capacity or not and I can assure you we will." However, OPEC Oil Ministers agree to leave output official limits at 24.5 million barrels per day "for the time being" and set another meeting for June 11 in Qatar. Also, OPEC President Abdullah al-Attiyah estimates OPEC spare capacity at between two and four million barrels per day.

March 12

The near-month (April) crude oil futures price at the NYMEX settles at $37.83 per barrel, the highest near-month settlement price (in nominal terms) since October 1990. This comes as EIA reports today that commercial crude oil inventories for the previous week declined by 3.8 million barrels to 269 million barrels. This is below the 270 million barrel lower operational inventory level, which, while not implying shortages, operational problems, or price increases, is indicative of a situation where inventory-related supply flexibility could be constrained or nonexistent. This heightens supply concerns before an impending war in Iraq.

March 15

The president of Venezuelan state oil company PdVSA, Ali Rodriguez, announces that Venezuela's crude oil production has surpassed 3 million barrels a day for the first time since a national strike began to affect Venezuelan oil production in early December 2002. He goes on to state that Venezuela will produce above its OPEC quota of 2.8 million barrels per day to make up for lost production in December and thus far in 2003. Executives fired from the state oil monopoly for participating in the strike put crude oil production at 2.1 million barrels per day.

March 16

Oil production from the Forcados and Escravos fields in Nigeria's Western Niger Delta is partially closed as violence occurs between Nigerian security forces and members of the Ijaw ethnic group. Some Ijaws had given a seven-day ultimatum to the Nigerian government to address their grievances or there would be "mass action." The ultimatum expired today. Exact details about how much production is affected are not available.

March 17

U.N. Secretary General Kofi Annan orders U.N. oil inspectors and other staff for the Oil-for-Food program out of Iraq, as the likelihood of military action in Iraq is imminent. Though this means that oil loadings will cease at the Persian Gulf oil export terminal of Mina al-Bakr, there is the possibility of continued loadings at the program's other approved port, Ceyhan, in Turkey. Oil continues to flow through the pipeline to Ceyhan. Tanker loadings in Kuwait are continuing as normal.

March 17

EIA releases a report showing that the previous week's national price for regular unleaded gasoline averaged $1.728 per gallon, the highest nominal price since EIA began recording the data in August 1990. This comes as crude oil prices have been high in recent weeks and stock levels have been low.

March 17

U.S. President George Bush, in a nationally televised address at 8:00 pm, gives an ultimatum to Iraqi President Saddam Hussein to leave Iraq within 48 hours or face military action to be forcibly disarmed at a time of the U.S. government's choosing. About 275,000 American, British, and Australian military forces are massed near Iraq's border or in the Persian Gulf.

March 18

The NYMEX near-month crude oil futures price drops 9.3 per cent, to settle at $31.67 per barrel, as President Bush's ultimatum is perceived by the market as ending uncertainty about a military conflict in Iraq, and amidst speculation that the conflict will be short in duration.

March 19

Officials of the state-owned Kuwait Petroleum Company indicate that they are now selling 300,000-350,000 barrels per day of additional crude oil to help stabilize prices ahead of a probable war in Iraq.

March 19

Military action in Iraq commences with a bombing raid and missile attack on targets in the Iraqi capital of Baghdad (March 20 Baghdad time) by Coalition forces, given Saddam Hussein and his regime's rejection of U.S. President George Bush's March 17 ultimatum. Iraq launches several conventional missiles at Kuwait, but this has no effect on Kuwaiti oil production. However, the Kuwait Petroleum Company does implement an emergency plan to protect its workers and facilities.

March 20

Several oil wells in southern Iraq are set on fire. The last scheduled loading at Ceyhan of Iraqi oil, about 600,000 barrels, is completed.

March 20

ChevronTexaco declares force majeure on its Escravos crude oil exports from Nigeria after shutting in 140,000 barrels per day of production. Royal Dutch/Shell shuts in 126,000 barrels per day of production, but does not declare force majeure. These shut-ins come as lethal violence continues to occur in the Niger Delta region between government soldiers intervening in conflicts between fighters of the Isekeri and Ijaw ethnic groups.

March 20

Indian Oil Minister Raim Naik announces that India has increased its oil stocks to two months. "We have presently crude oil, petrol, diesel stocks.kerosene and LPG gas which will last at least for two months."

March 21

British and American land forces enter Iraq, taking control of areas bordering Kuwait and parts of the Faw Peninsula. British marines seize some oil export facilities and U.S. armored columns push further into Iraq towards the Rumaila oilfields. About 7 oilfields are set on fire by Iraqi forces. U.S. special forces begin conducting significant operations with Kurdish units in the north of Iraq. A primary objective of Coalition forces is to take control of oil wells before they can be sabotaged by Iraqi forces. Coalition air strikes continue on military targets throughout the country.

March 21

Royal Dutch/Shell declares force majeure on the company's Bonny Light and Forcados exports. Production of these two crude streams is normally 800,000-900,000 barrels per day, though the force majeure does not mean that the two crude streams will be entirely shut in. Shell also shuts in an additional 50,000 barrels per day of production, bringing the combined Shell and ChevronTexaco production shut in to about 300,000 barrels per day.

March 23

Outbreaks of violence between soldiers and militants of various ethnic groups in the Niger Delta region of Nigeria prompt three major oil companies operating in the region–ChevronTexaco, Royal Dutch/Shell, and TotalFinaElf–to shut in operations in the area, totaling about 800,000 barrels per day. This represents about 40 per cent of Nigeria's total production, including about 768,000 barrels per day in the West Niger Delta (all operations there for the three companies) and 50,000 barrels per day of Shell production in the East Niger Delta.

March 24

After Coalition forces have pushed further into Iraq securing most of the southern oilfields over the weekend, Kuwaiti fire fighters are able to enter Iraq and are able to extinguish one of the wellhead fires. Iraq's southern fields represent about 40 per cent of the country's output. Damage is assessed to be relatively minimal. Some pockets or Iraqi resistance in the southern oilfields remain, however. Furthermore, heavy Iraqi resistance in some parts of Iraq gives rise to market speculation that the war could last longer than initially thought. The NYMEX near-month crude oil price rises 6.5 per cent, to settle at $28.66 per barrel, as the war in Iraq as well as the situation in Nigeria have traders concerned.

March 24

Venezuelan state oil company PdVSA Chief Operating Officer Luis Marin announces that PdVSA has made a 1 billion barrel crude oil discovery. He describes the oil as light and of high quality, but does not indicate the location of the find. Marin also puts Venezuela's output at 2.95 million barrels per day, much higher than other estimates by industry sources, striking workers, and EIA.

March 25

The Nigerian National Petroleum Corporation (NNPC) shuts down its 125,000-barrel-per-day capacity Warri refinery because its supplies from ChevronTexaco's production in the West Niger Delta have been halted. Shell begins to increase production of Bonny Light crude oil in unaffected parts of the Niger Delta in order to compensate for lost production.

March 25

China's new regulatory body for the electricity sector, the State Electricity Regulatory Commission, starts operations. China now has a government agency tasked with: researching and drafting regulations for the electricity market; drafting national and regional development plans for the power sector; setting technical and safety standards for the industry; and handling commercial disputes among companies in the sector. This is an important role, as the Chinese energy market continues to experience reforms.

March 25

U.N. data shows that Iraq's crude oil exports for the week ending March 21 were 443,000 barrels per day. This is a 75 per cent drop compared to the previous week. A reduced amount of oil is still flowing to rapidly-filling-to-capacity storage tanks at the Ceyhan export terminal, but no ships are scheduled to load there for the time being.

March 25

U.S. President George Bush sends a $74.7 billion budget request to Congress to pay for the war with Iraq. Included in the budget is a request for $489.3 million to pay for repairs to damaged Iraqi oil facilities.

March 26

The Federal Energy Regulatory Commission (FERC) releases the results of its 13-month investigation into the causes of the 2000-2001 California power crisis. The report finds that about 30 companies attempted to manipulate the electricity and natural gas markets in California during that time period. These companies will have to show why they should not be sanctioned for their alleged market manipulation activities during this time period. These companies May have to refund as much as $1.5 billion. Three companies–Reliant, Enron, and BP Energy–May be prohibited from selling energy in competitive markets anywhere in the United States.

March 26

The U.S. Navy announces that Iraq's Persian Gulf oil export terminal of Mina al-Bakr has escaped sabotage and is ready to resume operations.

March 27

U.S. Brigadier General Vincent Brooks states that "..over 600 oil wells at this point now fall into an area that is under Coalition control." He also states that there are still six oil wells on fire but work is continuing to put those out. Also, the U.K. military reportedly is negotiating with Iraqi oil workers to return to work in the country's North Rumaila oil fields. However, the NYMEX near-month crude oil futures price rises 6 per cent, to $30.37, as market perceptions of the duration of the war in Iraq and the length of time before Iraqi crude oil exports will return grow, and production continues to be shut-in in Nigeria.

March 27

A spokesperson for the Kuwait Oil Company announces that 22 wells at the northern Abdali oilfield near the border with Iraq will be reopened. On March 26, Kuwait reopened 17 wells at the nearby Ratqa oilfield. These wells, which were closed as a precaution prior to the initiation of the military conflict in Iraq, will restore production of about 85,000 barrels per day.

March 27

OMV of Austria announces that the consortium of which it is member, along with Lundin of Sweden, Petronas of Malaysia, and Sudapet of the Sudan, is renewing exploration activities in the

Sudan after they had been suspended in January 2002 because of a deteriorating security situation. According to OMV, "As a result of the positive developments in the peace process and the improved conditions in its concession area (the consortium) has decided to carry out repair work on the existing infrastructure in Block 5A. as a first step towards an eventual recommencement of activities."

March 27

Royal Dutch/Shell declares a secondary force majeure on export loadings from its Forcados terminal in Nigeria. "It will mean additional delays," says a Shell spokesman.

March 27

Exxon Mobil signs a $1.1 billion agreement with the state-owned Qatar Petroleum Company for the first phase development of natural gas block al-Kahleej Gas, or AKG-1. The project will produce 1.75 billion cubic feet a day of gas from Qatar's massive North Field, recovering associated condensates and natural gas liquids as byproducts. The AKG-1 project is expected to come on stream in the fourth quarter of 2005 and will be operated by Qatar's RasGas.

March 28

The U.N. Security Council gives U.N. Secretary General Kofi Annan temporary authority to provide food and medicine to Iraq through a 45-day renewal of the Oil-for-Food program. The program had been suspended just before hostilities commenced in Iraq on March 19.

March 28

MOL of Hungary completes a majority takeover of Slovak refiner Slovnaft, increasing its stake in Slovnaft to 70 per cent for $360 million.

March 28

Nigerian President Olusegun Obasanjo orders Nigerian government security forces to arrest militants fighting in the Niger Delta. The fighting there has caused the deaths of over 100 Nigerians and shut in over 800,000 barrels per day of crude oil production, as well as one refinery.

March 28

A 280-mile crude oil pipeline opens in Kazakhstan linking the Kenkiyak northwestern wells with the two major pipelines from Kazakhstan to Russia, Atyrau-Samara and the Caspian Pipeline Consortium. These transport crude from the Caspian Sea region to Baltic and Black Sea export outlets. The new pipeline will initially transport 120,000 barrels per day. It is owned and operated by Kazakhstan's state oil and gas firm KazMunaiGas and China's national oil company CNPC.

March 31

International oil traders report that Syria has cut its April export program by about 10 per cent. Some officials in Syrian state oil company Sytrol attribute this to increased domestic demand, but most traders attribute this to the expectation of decreased crude oil imports from Iraq in April.

April 2003

April 2

U.S. ground forces arrive within 30 miles of Baghdad, beyond the Tigris River and the so-called 'red-line,' *i.e.* the area within artillery and missile range of Republican Guard units defending Baghdad. This increases optimism that the war is going well for Coalition forces and puts downward pressure on oil prices.

April 2

The U.S. Energy Information Administration (EIA) releases data showing that U.S. crude oil stocks (excluding the Strategic Petroleum Reserve) rose by 6.8 million barrels, or 2.8 per cent, in the previous week, to 280.7 million barrels. The EIA also reports that crude oil imports for the previous week increased to about 10.4 million barrels per day, the highest weekly average on record.

April 4

ChevronTexaco announces that "Operations staff and support workers are returning to the Escravos terminal in the Western Niger Delta area of Nigeria and a gradual return to production is beginning." ChevronTexaco shut in about 440,000 barrels per day of production on March 23 because violence between two tribal groups (with government forces intervening) made the Western Niger Delta area unsafe to operate in. The latest decision affects about 320,000 barrels per day of production.

April 4

Coalition forces continue to make progress against the regime of Saddam Hussein in Iraq, with the U.S. military capturing Baghdad's main international airport. Also, according to the U.S. military, 80 per cent -90 per cent of Iraq's southern oilfield production is under coalition control, as well as all related export facilities, as of this date.

April 4

Royal Dutch/Shell restarts production and development work at the Soroosh and Nowrooz fields offshore southwestern Iran, after shutting down work at the two fields on March 19 because of fears that staff could be vulnerable to intentional or accidental attack, given the fields' proximity to the border with Iraq. Soroosh produces about 60,000 barrels per day, and the shut down has delayed the coming on line of the Nowrooz field, scheduled for later this year.

April 6

Venezuelan state oil company PdVSA announces the discovery of a light crude oil field in the eastern state of Monagas with estimated reserves of 350 million barrels, as well as 870 billion cubic feet of associated natural gas.

April 8

Syrian state oil company Sytrol informs customers that it will cut crude oil term export volumes by around 40 per cent (about 150,000 barrels per day) as a result of the halt in Iraqi imports through the Iraq-Syria-Lebanon pipeline that is reported to have been shut down. Sytrol suggests that the reduction will continue for the rest of the year.

April 10

Kurdish Pesh Merga guerilla forces seize control of the key northern city of Kirkuk from Iraqi forces loyal to the regime of Saddam Hussein. Kirkuk is the hub for Iraq's northern oilfields. The previous day, April 9, U.S. military forces in Baghdad declared Saddam no longer in control in the Iraqi capital, though fighting in the capital continues. However, scenes of jubilant Iraqi citizens destroying propaganda of the Baathist regime, coupled with an absence of organized police or military units of the regime maintaining order in central Baghdad, signaled for many observers that today marks the effective fall of the regime of Saddam Hussein. Near-month crude oil futures on the NYMEX drop about 5 per cent.

April 10

Royal Dutch/Shell announces that it has restarted 100,000 barrels per day of production in the south of the Niger Delta, in Nigeria. This still leaves about 200,000 barrels per day of the company's production in the area shut-in.

April 10

The Presidents of Russia and Turkmenistan sign a deal under which Turkmenistan will pipe 1.8 trillion cubic meters (63.6 trillion cubic feet) of natural gas to Russia over a 25-year period. Under the terms of the deal, Turkmen state natural gas company Turkmenneftgas will deliver gradually increasing amounts of gas to Russia's Gazprom, at a set price of $44 per 1000 cubic meters (per 35,315 cubic feet), beginning in 2004. The gas will then be exported or swapped out for export. Turkmenistan is reliant on the Russian natural gas pipeline network in order to export its gas.

April 14

Pumping on the oil pipeline from Iraq's Kirkuk oilfields to the Turkish port of Ceyhan is halted as the storage facilities have reached their maximum capacity of about 6.5 million barrels. There has not been a loading of Iraqi crude oil at the port since March 20.

April 14

Tokyo Electric Power Company (TEPCO) shuts down for inspection the last of its 17 nuclear reactors still in operation. The shut downs result from the discovery last year that TEPCO had falsified data regarding reactor inspections, leading to the decision to shut down by Japan's nuclear authorities. Japan's largest power firm said that unless its reactors were started back up, there would be an electricity shortage of up to 9.55 million kilowatts during the summer, when electricity demand hits its peak.

April 15

U.S. Secretary of Defense Donald Rumsfeld announces that the U.S. military has shut off an oil pipeline from Iraq to Syria that is alleged to have been carrying 100,000-150,000 barrels per day. "We have been told that they have shut off a pipeline," Secretary Rumsfeld told a Pentagon briefing. "Whether it's the only one and whether that has completely stopped the flow of oil between Iraq and Syria, I cannot tell you. I cannot assure you that all illegal oil flowing from Iraq into Syria is shut off. I just hope it is."

April 17

The Russian government is instructed by Prime Minister Mikhail Kasyanov to begin preparing a feasibility study for the construction of a privately-owned $4.5 billion oil pipeline to Murmansk and an oil terminal there. The planned pipeline would carry about 2.4 million barrels per day of crude oil from the Urals oil production region. This would be the first privately-owned pipeline in Russia.

April 18

Dominion Virginia Power Company settles with U.S. government by agreeing to pay a $5.3 million fine and spend $1.2 billion on pollution abatement at eight power plants. This is the largest settlement ever by an electric power utility under the "new source review" provisions of the Clean Air Act. The U.S. Environmental Protection Agency (EPA) had originally cited Clean Air Act violations at eight of Dominion's power plants.

April 19

OPEC-member Nigeria holds a presidential election. Incumbent Olusegun Obasanjo is declared the winner several days later by a 2:1 margin, amidst allegations of vote rigging and intimidation.

April 21

For the first time since the start of the war with Iraq in March, an Iraqi oil refinery begins to process crude oil. The Daura Refinery, Baghdad's main refinery, begins operating at 40,000 barrels per day, about half of the refinery's capacity, with oil from storage. Also, the previous day, a team of U.S., U.K., and Iraqi engineers began to pump oil through Iraq's internal oil infrastructure, preventing a power plant from going offline.

April 22

Yukos Oil Company and Sibneft, Russia's first and fifth largest oil companies, respectively, in terms of production, announce that they will merge in a deal in which Yukos will pay $13 billion in cash and stock for Sibneft. The new company will be the world's fifth-largest publicly traded oil and gas company, with a production of 2.4 million barrels per day. The new company plans to become a major player outside of Russia as well.

April 22

Turkey indefinitely halts electric power imports from Bulgaria, alleging that Bulgaria had delayed a high profile hydro-power project and the construction of a stretch of highway to the border with Turkey, which are part of a bilateral energy deal. Bulgaria was exporting 3.5-4.0 billion kilowatt hours (kWh) of power to Turkey per year, more than half of its total annual exports, under a 10-year energy deal, signed in 1998.

April 22

A spokesperson for Venezuelan state oil company PdVSA announces that the force majeure on gasoline exports was lifted sometime during the previous week. Force majeure has now been lifted on all Venezuelan crude oil and oil product exports. According to PdVSA, the company is in the process of restarting the flexicoking unit at the giant 940,000 bpd Amuay-Cardon plant, the last major refining unit still shut due to the strike.

April 23

According to the American military officer in charge of restarting Iraq's oil production infrastructure, Iraq's southern fields have begun to produce again. Four southern wells have begun producing a modest amount of crude oil, but according to Brig. Gen. Robert Crear of the Army Corps of Engineers, southern wells should soon be producing about 170,000 barrels a day. Initial production would go toward meeting domestic demand, especially as more refineries come back on line. The country's northern oilfields are still offline.

April 24

OPEC oil ministers, meeting for emergency talks in Vienna, decide to simultaneously reduce crude oil production by 2 million barrels per day, as of June 1, and increase their overall production quota by 900,000 barrels per day to a total quota of 25.4 million barrels per day. This is a tacit admission that OPEC production is well in excess of the previous quota of 24.5 million barrels per day. Iraq does not participate in the meetings and is not subject to the quota regime.

April 24

The U.N. Security Council votes unanimously to extend the United Nations Oil-for-Food program for an extra three weeks, ***i.e.*** until June 3. It had been due to expire on May 12. Oil has not been sold through the Oil-for-Food program since the start of the war in Iraq on March 19.

April 24

ChevronTexaco lifts the force majeure on crude oil exports from its Escravos Terminal in Nigeria. Production at fields that export their production through Escravos began again on April 4, and is now at 310,000 barrels per day. The terminal's normal export throughput is 440,000 barrels per day.

April 28

Russian oil company Lukoil announces that its proven oil and natural gas reserves have risen 15 per cent over the past year through acquisitions and discoveries and that its reserves are now the second-highest held by a non-state-owned company in the world. The company, as of January 2003, had proven reserves of 19.3 billion barrel of oil equivalent compared with 16.8 billion as of January 2002, including 15.3 billion barrels of crude oil and 24.2 trillion cubic feet of natural gas.

April 29

Iraq's North Oil Company (NOC) announces that oil output resumed in the Kirkuk fields in northern Iraq on April 24 at 30,000-40,000 barrels per day. Because of war damage to machinery, looting and a lack of workers, it will likely be some time before pre-war output of 850,000 barrels per day is achieved. NOC is using extra stored oil to send 100,000 barrels per day down a pipeline to the Daura and Baiji refineries to feed Iraq's domestic fuel consumption.

April 29

Royal Dutch/Shell makes an announcement asserting that about 200,000 barrels per day of crude oil production is being stolen in Nigeria, 100,000 barrels of which is Shell's production.

April 29

Company and union officials reveal that 97 foreign oil workers have been held hostage by striking Nigerian workers aboard four offshore drilling rigs owned by Houston-based Transocean Inc. since April 19.

April 29

The United States Department of Defense announces that it is ending military operations in Saudi Arabia and removing virtually all of its forces from the kingdom by mutual agreement following the Iraq war. U.S. military forces had doubled to 10,000 personnel during the war in Iraq, but have already begun to pull out, according to the Defense Department. The Combined Air Operations Center (CAOC) at Prince Sultan Airbase will be closed.

April 29

Qataris overwhelmingly approve a new constitution in a referendum that will lead to a more democratic government in the oil-and-gas-rich Persian Gulf emirate, including a mostly-elected parliament in 2004.

April 29

Brazilian state oil company Petrobras announces the largest-ever natural gas discovery in Brazil. The discovery, located about 85 miles off the coast of the state of Sao Paulo, is a field containing an estimated 2.47 trillion cubic feet of natural gas. This field raises Brazil's natural gas reserves by about 30 per cent, according to some estimates.

April 30

Libyan Foreign Minister Mohammed Abderrahmane Chalgam states, in regard to the Lockerbie case, "We have taken on the responsibility for this case on the basis of the international law which states that the state takes on responsibility for what its employees do…The provisioning of that [compensation] fund with the decided amount has started." A Pan Am airliner exploded over the Scottish town of Lockerbie in December 1988, killing 270 people. Libyan intelligence agent Abdel Basset al-Magrahi was convicted for the crime in 2001 by a court sitting in the Netherlands. Total compensation is expected to be about $2.7 billion.

April 30

In Saudi Arabia's third cabinet reshuffle in nearly 30 years, Ali al-Naimi is retained as oil minister, contrary to expectations that he would be replaced with a minister more receptive to foreign investment in the kingdom.

April 30

U.S. and Iraqi oil officials hold their first meeting to discuss rebuilding the country's oil industry. Few details of the meeting are released, though Gary Vogler, an advisor at the U.S. Office of Reconstruction and Humanitarian Affairs (ORHA), states that the two parties had "agreed upon how to reconstruct and rebuild this (oil) sector so that it performs its obligations towards the Iraqi people and towards the international market."

May 2003

May 1

Engineers restart Basra's al-Zubayr oil refinery in southern Iraq. The refinery, which processes crude oil produced at Iraq's southern fields, is running well below its capacity of 140,000 barrels per day.

May 1

U.S. President George W. Bush declares that "major combat operations in Iraq have ended," and that "The battle of Iraq is one victory in a war on terror that began on September the 11th, 2001, and still goes on."

May 3

Expatriate and Nigerian hostages are evacuated from offshore oilrigs where they had been held by striking Nigerian oil workers since April 19. Company officials and the striking workers' representatives negotiated the hostages' release May 2, after which the first captive was soon freed.

May 4

The U.S.-led civilian administration for Iraq selects Thamir Ghadban to run Iraq's oil ministry, with Phillip Carrol, former head of Royal Dutch/Shell in the United States, as head of an advisory board for the ministry. Thamir Ghadban was formerly a top official in Iraq's Southern Oil Company.

May 5

TotalFinaElf receives permission from Angolan state oil company Sonangol to begin development of the Dalia offshore oilfield. Estimated development cost of the Dalia field, due to come online in

2006, is $3.4 billion. Infrastructure will include 34 production wells and an FPSO (Floating Production, Storage, and Offloading vessel) capable of processing 240,000 barrels of oil per day.

May 6

A spokesman for the El Palito Refinery in western Venezuela announces that the refinery recently sent out its first shipment of unleaded gasoline to the United States since the national strike that began in Venezuela in the first days of December 2002.

May 12

Tokyo Gas Company signs a 24-year supply deal with Sakhalin Energy of Russia for a maximum of 1.1 million metric tons per year of liquefied natural gas (LNG), starting in 2007. This is the first long-term LNG supply deal between Japan and Russia.

May 12

A suicide-bomb and car-bomb attack on three residential compounds for foreigners in Saudi Arabia kills several dozen, including 10 Americans and several Saudis. The attack is blamed on Al-Qaeda.

May 14

The Chinese government's official news agency reports that the State Council has approved the construction of 13 new power plants with a combined capacity of 11,880 megawatts and a cost of 51.2 billion yuan.

May 14

Nigeria Liquefied Natural Gas reaches an agreement with BG LNG Services to provide 3 billion cubic meters of liquefied natural gas (LNG) per year, starting in 2005. LNG from two trains under construction in Nigeria will be delivered to the Lake Charles, Louisiana import terminal, where BG has 100 per cent of the capacity rights after 2005.

May 14

The Nigerian government announces that it will privatize 51 per cent the country's four petroleum refineries as well as deregulate unspecified parts of the downstream oil sector. This follows the recommendations of a committee set up to solve the problem of recurring fuel shortages in Nigeria, the world's seventh-largest oil exporter in 2002. Nigeria's refineries have a combined capacity of 445,000 barrels a day, but have been unable to meet local demand because of frequent breakdowns.

May 19

Bulgaria launches electricity exports to neighboring countries Greece and the Former Yugoslav Republic of Macedonia (FYROM), increasing total Bulgarian exports from 200 megawatts (MW) this month to 680 MW in June and 800 MW in July. The exports take the place of electricity exports to Turkey that were halted April 22 because of a dispute between the two countries.

May 22

The United Nations Security Council approves the immediate end of 13 years of economic sanctions on Iraq, dating from the time of Iraq's invasion of Kuwait in 1990. Resolution 1483 effectively grants the United States-led coalition forces control of Iraq until a new Iraqi government can be put in place. The end of the sanctions also makes it easier for Iraqi oil exports to resume without the auspices of the United Nations.

May 23

The Central Bank of Venezuela announces that the Venezuelan economy contracted by 29 per cent in the first quarter of 2003, a record quarterly economic collapse for the world's fourth largest net oil exporter (2002). Much of the contraction is associated with a debilitating national strike that affected Venezuela's crucial petroleum sector and the imposition of tight currency controls.

May 25

The Saudi Arabian Ministry of Industry and Trade announces that it has created a new publicly-held energy services company, Industrialization and Energy Services, to provide operations and services to the petroleum, petrochemical, electric power, and desalination sectors. The new company's shares will not be traded until the company posts two years of profits. The start-up of this company has prompted some analysts to speculate that it May presage the privatization of some of Saudi Aramco's non-core activities.

May 27

The U.S. Department of the Treasury lifts most remaining sanctions on Iraq, thereby implementing U.N. Security Council Resolution 1483. Secretary of the Treasury John W. Snow states that "It is no longer a crime for U.S. companies and individuals to do business with Iraq."

May 27

It is announced that the Iranian National Oil Company (NIOC) has concluded a 50-50 joint venture with BP to develop the Rhum natural gas field in the North Sea in a deal worth $525 million. This is the Iranian state oil company's first entry into the North Sea. The Rhum field has estimated reserves of 812 billion cubic feet and production is expected to commence in late 2005, at 300 million cubic feet per day.

May 28

Iran and Norway agree to closer cooperation in the oil and natural gas sector. States Norwegian Oil and Energy Minister Einar Steensnaes "We signed a memorandum of understanding to increase cooperation in the sector of oil and gas...I understood from my contacts–not least with the NIOC–that Statoil will be invited for further engagement in the Persian Gulf." The Norwegian companies Statoil and Norsk Hydro are already active in Iran.

May 28

Olusegun Obasanjo is inaugurated to a second four-year term in office as president of Nigeria, following his reelection in April. OPEC member Nigeria ranked as the world's seventh-largest net oil exporter in 2002.

May 28

Yukos of Russia signs a $150 billion agreement with China National Petroleum Company (CNPC), wherein CNPC agrees to purchase 5.13 billion barrels of oil between 2005 and 2030 via a $2.5 billion pipeline from Russia's Western Siberia fields to China's Daqing field.

May 29

The Venezuelan government and representatives of opposition groups sign an accord that recommends holding a referendum after August 19, 2003 as the best way of ending political crisis in the world's No. 4 oil exporter. Organization of American States Secretary General Cesar Gaviria also signed the agreement, as the organization had organized the talks leading up to the agreement.

June 2003

June 2

Royal Dutch/Shell signs a $2 billion contract with an alliance of Japanese and Russian companies for the construction of Russia's first natural gas liquefaction plant in Sakhalin. This comes after Tokyo Electric Power Company (TEPCO) and Tokyo Gas agreed two weeks earlier to purchase about one-quarter of the liquefaction plant's planned capacity of 9.6 million metric tons per year. Shell owns 55 per cent of the production rights for the natural gas supplying the planned plant.

June 5

Saudi Arabia cancels the proposed $15 billion Core Venture 1 natural gas project, led by Exxon Mobil, after the Kingdom and the oil companies involved fail to agree upon financial terms of the project, particularly the returns on investment. This means that two other $10 billion natural gas ventures with foreign oil companies are also likely to be cancelled or significantly changed, according to industry analysts.

June 5

ChevronTexaco announces that it has restored 320,000-350,000 barrels per day of crude oil production that was shut in because of dangerous circumstances in Nigeria's Niger Delta in March. This is about 70 per cent of the total Niger Delta output that was shut in.

June 6

Phillip Carroll, Chairman of the Advisory Board to Iraq's Oil Ministry, indicates that there is an organized campaign of sabotage against Iraq's oil industry. Carroll states that "It is very difficult for me to identify who they [the saboteurs] are and what their motives are. I can only say their techniques appear to be very professional and aim at causing harm to significant and important installations." A number of oil and natural gas pipelines in Iraq are bombed throughout the month of June.

June 10

Federal Reserve Chairman Alan Greenspan notes that rising natural gas prices in the United States could have a negative impact on the economy in the months ahead if prices remain at high levels. States Greenspan, "I have no doubt that.if we stay at these very elevated prices we're going to see some erosion in a number of macroeconomic variables which are not evident at this stage. A very significant amount of natural gas using infrastructure in the American economy was based on $2 gas. That means a lot of noncompetitive structures are sitting out there."

June 11

Oil Ministers of the Organization of Petroleum Exporting Countries (OPEC) meeting in Qatar decide to keep OPEC crude oil production quotas unchanged for the ten members (*i.e.* not including Iraq) participating in the quota regime. The combined output quota for the ten members is 25.4 million barrels of crude oil per day. OPEC President Abdullah bin Hamad Al Attiyah, also Qatar's Minister of Energy and Industry, says, "We don't want to cut for the sake of it. We should justify it."

June 12

Iraq's State Oil Marketing Organization (SOMO) announces the results of its first export tender since the start of the war in Iraq that ended the regime of Saddam Hussein. SOMO awards 7.5 million barrels of crude oil to five companies. Terms of the tender are not released.

June 12

Two explosions damage the Kirkuk-Ceyhan oil pipeline, in what is later determined to be an act of sabotage. Several other Iraqi pipelines are damaged in acts of sabotage throughout the month, including a natural gas pipeline in the western desert on June 21, an oil pipeline west of Baghdad on June 22, and the now-stalled Iraq-Syria pipeline on June 23.

June 13

Ecuador declares force majeure on oil exports. A strike by oil workers at state oil company Petroecuador, declared on June 9, has affected not only production at the state-owned company, but also the private operators that use Petroecuador's pipelines and need to mix lighter Petroecuador crudes with their heavier grades. Ecuador normally exports about 240,000 barrels of crude oil per day. The important SOTE pipeline shuts down June 15. The strike ends on June 17 and force majeure is lifted on June 19.

June 14

ConocoPhillips announces that the company will proceed with its $1.5 billion liquefied natural gas (LNG) development project at the Bayu-Undan fields after government officials of Australia and East Timor approved the project in the Timor Sea Joint Petroleum Development Area. Natural gas from the field will be piped to an LNG liquefaction plant in Australia's Northern Territory.

June 16

The United States and the European Union agree to a plan for the two parties to collaborate on hydrogen fuel cell research. Other countries are expected to be invited to collaborate in the future as well.

June 16

The parliament of Hungary passes legislation that will partially liberalize the country's natural gas market at the start of 2004. About 40 per cent of the market, *i.e.*, that which represents consumption by large industrial and electricity generating companies, will be opened to competition.

June 17

Unocal announces that it is planning to build a new deepwater oil port about 100 miles off the coast of Texas. The port, to be called Bulk Oil Offshore Transfer System (BOOTS), could be completed by as early as 2007, according to the company, and would have the capacity to deliver 1.2 million barrels per day to on-shore refineries.

June 17

The head of Iraq's North Oil Company, Adil al-Qazzaz, states that Iraq's main north-south crude oil pipeline, the so-called Strategic Pipeline, will not be operable for some time, especially because the K-3 pumping station was badly damaged during the recent war. Al-Qazzaz goes on to state that because the pipeline is not working, "[W]e don't have export flexibility, and that will have an impact."

June 19

The International Atomic Energy Agency (IAEA) calls on Iran to allow stricter inspections of nuclear sites in the country. Iran is called upon to "promptly and unconditionally conclude and implement an additional protocol to its Safeguards Agreement, in order to enhance the agency's ability to provide credible assurances regarding the peaceful nature of Iran's nuclear activities." Iran initially rejects this request. However, on June 21, Gholamreza Aghazadeh, head of Iran's atomic energy program says, "We will definitely try to cooperate more than before with the IAEA and give them the necessary assurances about Iran's activities."

June 20

Nigeria announces that funding for further investment in improving the output of its four refineries has been exhausted and that gasoline prices will increase by 65 per cent. Despite Nigeria being one of the world's major exporters of crude oil, the country has had to import gasoline, which was subsidized by about $2 billion in government spending. About $700 million was spent repairing and modernizing the refineries, but capacity utilization has only reached about 53 per cent. Furthermore, two of the refineries are shut-in because of a ruptured supply pipeline.

June 22

Iraq exports oil for the first time since March 20, the first day of the war that eventually toppled the regime of Saddam Hussein. The crude oil, 1 million barrels, was part of the June 12 tender and will be sold to Turkish refiners from oil in storage at the Turkish port of Ceyhan. Loading of the oil onto a tanker begins today.

June 26

Saudi Oil Minister Ali al-Naimi, meeting with the Norwegian Oil Minister in Norway, announces the re-launching of Saudi Arabia's natural gas initiative. However, instead of three unified ventures, the projects are now being split along upstream gas development and downstream industry lines as stand-alone ventures. The largest of the previous unified ventures was canceled on June 5.

June 26

It is announced that Colombian state oil company Ecopetrol will be split into three companies. The newly created National Hydrocarbon Agency would handle exploration and oil production. Ecopetrol S.A. will deal with refinery operations, while a third company will handle "nonstrategic" investments. All three will remain publicly owned. According to Ecopetrol president Isaac Yanovich, "We think this is a very positive sign for both foreign and national capital that want to come or continue working in our country's exploration areas." Colombia's oil production has been declining in the last few years.

June 30

A general strike over rising fuel prices begins in Nigeria, though oil production and exports are not immediately affected. Police fire live ammunition and tear gas at crowds of protestors in Lagos and Abuja.

June 30

The Syrian government signs a deal with Royal Dutch/Shell and Petro-Canada to increase Syria's oil output by 30,000 barrels per day as part of the country's plans to raise overall output by 100,000 barrels per day by 2006.

June 30

The 2,000-megawatt capacity Bureya dam on the Amur river in Russia's Far East begins producing power. The $800 million dam was developed without project financing by Russian electricity monopoly Unified Energy System. This is one of the largest hydroelectric dams to come on line in the past 15 years.

July 2003

July 1

The Venezuelan ambassador to the United States announces that U.S. major ChevronTexaco plans to invest $2 billion to develop an offshore natural gas field in that country. ChevronTexaco signed agreements on Feb. 14 to explore and produce from Block 2 in the waters in the northeastern Deltana region, bordering fields in gas-rich Trinidad and Tobago. Company sources confirm the planned investment. The natural gas could feed into a possible natural gas liquefaction facility.

July 2

Thieves cut into one of Iraq's main oil pipelines in the south, near Najaf, causing a fire and an oil spill that close the pipeline. The underground pipeline is used for both transporting oil for export and for supplying Iraqi refineries and power plants.

July 2

The European Parliament votes to cap European industry's carbon dioxide output and let firms trade the right to pollute. As of January 2005, many plants in the oil refining, smelting, steel, cement, ceramics, glass and paper sectors will need special permits to emit carbon dioxide (CO2). "It means that the largest emissions trading scheme in the world to date will be a reality from 2005, and that the architecture foreseen under the Kyoto Protocol is coming to life," according to European Union Environment Commissioner Margot Wallstrom.

July 3

Main Hungarian oil and gas company MOL and Russian oil giant Yukos sign a deal for 7.2 million metric tons (52.3 million barrels) of crude oil per year for MOL's Hungarian and Slovakian refineries. At crude oil prices of $20 per barrel, this is worth at least $7.5 billion per year, as MOL has the option to purchase even more oil. The deal is to take effect January 1, 2004, and will supply 55 per cent to 60 per cent of MOL's refinery capacity.

July 3

A group of 20 Polish companies, led by state-owned Polska Nafta, sign a joint venture agreement with Halliburton subsidiary Kellogg, Brown, and Root, one of the main U.S. contractors rebuilding the Iraqi oil industry. On July 2, Poland sent some 250 soldiers to Iraq to set up a Polish-commanded postwar occupation zone expected to involve more than 9,000 troops from at least 15 countries.

July 3

An offshore exploration and possible production agreement between Pakistan and French major Total is signed by Pakistani President Pervez Musharraf. This represents the first foreign investment in offshore oil and gas exploration in Pakistan. Analysts describe this as a major turning point in the Pakistan's effort to become more self-reliant in oil and natural gas, as well as for the future of foreign investment in the country.

July 3

Sakhalin Energy Investment Company announces that it has loaded the first-ever cargo of crude oil from the Russian Far East for delivery to the West Coast of the United States at its Vityaz production complex offshore Sakhalin Island. Most of the crude oil will go to a Tesoro refinery in Washington State.

July 7

ChevronTexaco announces that two of its flowstations abandoned during ethnic unrest during March in the Niger Delta area of Nigeria have been vandalized and are badly damaged. The need to repair the flowstations, Olero and Dibi, will likely further delay the restoration of some 140,000 barrels per day of ChevronTexaco's Nigerian production still shut-in.

July 7

According to OPEC's news agency, Opecna, Russia's Energy Minister Igor Yusufov and Iranian official Mohammad Forouzandeh sign a protocol in Moscow to cooperate more closely in the energy sector. The agreement is expected to result in construction of power plants, as well as oil and gas exploration and production.

July 8

Leaders of the umbrella Nigeria Labour Congress announce that they are ending a general strike begun on June 30 to protest rising fuel prices in Nigeria. The Labour Congress states that a compromise deal has been struck with the government to increase the fuel price from the previous level of 26 naira ($0.20) per liter to 34 naira, rather than the 40 naira originally announced. Analysts and companies operating in Nigeria had believed that the strike would begin to affect Nigeria's oil production and exports if it went on much longer.

July 8

Iraq's main oil export pipeline to Ceyhan, Turkey, is damaged over the weekend by an explosion (most likely sabotage), according to officials of the de facto Iraqi Oil Ministry. It is estimated that repairs will take two weeks, further delaying the restoration of Iraqi oil exports.

July 9

European Union (EU) energy regulators agree to eliminate the 0.50 per megawatt charge on cross-border electricity sales within the bloc from 2004. The decision is a major breakthrough in the EU's attempts to open its 250 billion energy markets to competition. EU Energy Commissioner Loyola de Palacio said the agreement demonstrated the commitment of EU countries to an "integrated electricity market without "artificial barriers" to trade."

July 9

The government of Chad announces that it has begun its first-ever crude oil production, as wells began pumping on July 1. It will still take weeks before crude is shipped from the $3.5 billion project through a 650-mile pipeline to the Atlantic coast in neighboring Cameroon. The government does not announce the initial flow rate, but eventual production is expected to reach 225,000 barrels per day. Oil begins flowing through the pipeline on July 15.

July 10

The U.S. Army invites bids for two contracts to rebuild Iraq's oil industry, one for the North and one for the South. The contracts, worth up to $1 billion, will replace a no-competition contract awarded in March to Kellogg Brown & Root, a subsidiary of Halliburton. Non-U.S. companies can bid on the contracts, for which the bidding deadline is August 14. The new contracts include extinguishing oil fires, environmental assessments and clean-up at oil sites, the distribution of oil products, maintaining oil fields and the marketing and sale of Iraqi crude oil and refined products.

July 10

Data released by the U.S. Energy Information Administration shows a 111 billion cubic foot build in U.S. natural gas stocks, the sixth consecutive week showing a build. Near-month NYMEX natural gas futures fall 26.2 cents, about 5 per cent, to settle at $5.258 per million Btu. Total U.S. natural gas stocks of 1.773 trillion cubic feet are still 580 billion cubic feet below last year and 317 billion cubic feet below the five-year average.

July 10

The Paris Club of sovereign creditors announces that debts owed by Iraq to its member states amounted to $21.018 billion through August 2, 1990, as well as a similar amount in interest arrears. The largest creditor is Japan, at $4.1 billion dollars. The United States is owed $2.19 billion. The Paris Club numbers do not take account of several key classes of debt, mostly to Gulf States, which amount to some $35 billion, depending on whether they are treated as loans or aid, nor does it take into account commercial debt.

July 11

The Nigerian National Petroleum Company (NNPC) signs a $2.5 billion agreement with ChevronTexaco for the development of the Agbami oilfield. A deepwater offshore field located on Oil Prospecting Lease 216, Agbami is expected to commence production in 2006 at around 200,000-250,000 barrels per day.

July 11

ConocoPhillips announces that it is taking a major stake in one train of the planned Qatargas-3 liquefied natural gas (LNG) project. Total cost of the project is expected to be $5 billion, and will include a liquefaction facility in Qatar, tankers, and a regasification terminal in the United States. It would have two 7.5-million-metric-ton per year trains, together capable of delivering more than 2 billion cubic feet of regasified LNG per day, possibly starting in 2008 or 2009.

July 12

Sakhalin Oil Development Corporation, the Japanese partner in an international consortium in the Sakhalin-1 project, announces that oil drilling offshore has begun. The project, which May eventually see $12 billion invested in oil and natural gas development, is potentially the largest direct foreign investment in Russia. Total recoverable reserves at the Sakhalin-1 area are estimated to be 2.3 billion barrels of oil and 17.1 trillion cubic feet of natural gas.

July 13

The new Iraqi National Council is introduced in a televised ceremony in Baghdad. The Council's makeup was formed from negotiations involving U.S. officials and Iraqis who had opposed former Iraqi President Saddam Hussein. The 25-member governing council brings together Iraqi Shiites, Sunnis, Kurds, Christians and ethnic Turks. It has the power to name ministers and approve the 2004 budget, but L. Paul Bremer, civilian administrator for Iraq, has the last word on Iraq's affairs.

July 14

According to Abolhasan Khamoushi, general director of Iran's state-owned Oil Development and Engineering Company, there has been a major new oil find in Iran. The three new neighboring fields, near the southern city of Bushehr, contain estimated reserves of 38 billion barrels of oil. However, the crude has a low API gravity, and estimated recoverable reserves have not yet been determined.

July 15

The operator of Israel's Eilat-Ashkelon pipeline, a bi-directional pipeline linking the Mediterranean and the Red Sea, announces that the pipeline is operational. The pipeline, with a current capacity of 400,000 barrels per day, but a design capacity of 1.2 million barrels per day, provides an alternative to the Suez Canal, as both Israeli ports can handle VLCCs, whereas Suez cannot. Perhaps even more importantly, with the new southerly flow, Russian crude on small tankers from the Bosporus will be able to eventually load onto VLCCs bound for East Asia.

July 15

Hurricane Claudette hits the Texas coast about 80 miles southwest of Houston. According the U.S. Minerals Management Service, an estimated 2.5 billion cubic feet per day of natural gas had been shut in by Claudette, or about 18 per cent of the Gulf's total gas output. Also, about 330,000 barrels per day of oil, or some 21 per cent of the Gulf of Mexico's daily oil production, has been shut down. Production is quickly restored in the next few days.

July 15

Major energy company Mirant files for bankruptcy in Fort Worth, Texas, in what is expected to be the largest bankruptcy in the United States this year. Atlanta-based Mirant has $20.6 billion in assets and $11.4 billion in debt.

July 16

Italian oil and gas major Eni announces that it has begun exporting oil production from the giant Karachaganak field in Kazakhstan to the Novorossiysk terminal on the Black Sea. In addition, Eni said that it and its partners had completed pipelines and treatment facilities so that output from the oil field could grow by the end of the year to 380,000 barrels of oil equivalent per day from the current 220,000 barrels of oil equivalent per day.

July 16

Royal Dutch/Shell and Total successfully conclude the first deal with Saudi Arabia giving Western companies access to its hydrocarbon reserves since the nationalization of its petroleum industry. The agreement, which is smaller than the original Core Venture Three being negotiated with ConocoPhillips (which has dropped out of the current project), is a natural gas exploration and development deal covering 77,000 square miles in the Kingdom's Empty Quarter.

July 18

United Nations inspectors detect traces of enriched uranium at the Iranian nuclear facility near Natanz, south of Tehran. The Iranian government is constructing a massive uranium processing plant at this location. The inspectors, part of the International Atomic Energy Agency, are preparing to collect additional samples, as the confirmation of the discovery of such materials could indicate a breach by Iran of its agreements under the international nuclear Non-Proliferation Treaty.

July 21

The Brazilian government unveils a new plan for its power sector to reduce risks of shortages and to expand investment. It will involve setting up two different trading systems, one of which would pool old and new electricity sources wherein distributors design mixed contracts to reduce costs for final consumers. The other trading system would set up bilateral contracts between generators and large industrial consumers. Brazilian Minister of Energy and Mines Dilma Rousseff comments, "We want to protect consumers from abrupt price increases but we also want to increase free competition [in the first system]...The independent buyer [in the second system] will be able to negotiate power contracts at its own risk, because that won't affect electricity prices for final consumers."

July 22

Iraq's State Oil Marketing Company (SOMO) announces that it is in negotiations for the first post-war term contracts for its oil exports. Since the end of the war, all Iraqi exports have been spot tenders, but SOMO plans term contract sales of Basra crude to refiners from its southern fields. SOMO also says that it will announce monthly prices for its Basra term sales in the first 10 days of each preceding month. On July 27, SOMO signs 10 term contracts to supply over 20 million barrels of Basra crude a month to global oil majors and refineries and expects to sign several more deals in the next few days, according to SOMO head Mohammed Al-Jiboury.

July 23

The president of São Tomé and Príncipe, Fradique de Menezes, returns to the West African island nation after signing a deal with coup leaders that ends a week-long military takeover. It is believed that the islands May contain large hydrocarbon reserves.

July 23

Exxon Mobil announces that condensate and natural gas liquid (NGL) production from its Oso terminal in Nigeria restarted sometime in the middle of July after being shut-in since a fire in April. Company officials noted that a "significant amount" of the unit's 95,000 barrels of crude oil equivalent per day of condensates and 45,000 barrels of crude oil equivalent per day of NGLs was back in production.

July 25

The first delivery of liquefied natural gas (LNG) since 1980 is made to the reactivated Cove Point LNG regasification plant in Maryland, as a tanker from Trinidad arrives carrying 22 million gallons of LNG. According to Dominion, owner of the facility, the plant will be able to supply 1 billion cubic feet of natural gas per day, and will be the largest LNG regasification facility in the United States.

July 25

It is reported that China National Petroleum Corporation (CNPC) has discovered 2 billion barrels of oil in the Sudan, of which 600 million are estimated as recoverable. The discovery was made in two blocks that CNPC shares with Qatari Gulf Petroleum Company and two Sudanese companies. Production is expected to commence in 2005, with annual production capacity near 60 million barrels (about 164,000 barrels per day).

July 31

The Republican-majority U.S. Senate passes an energy bill that was originally drafted by Democrats last year, an unusual occurrence after there were numerous difficulties passing a Republican-drafted measure. The bill will now go to conference, as Senate and House leaders reconcile their versions of an energy bill.

July 31

Oil Ministers of the Organization of Petroleum Exporting Countries (OPEC), meeting in Vienna, decide to keep their crude oil production quotas unchanged until their next meeting, on September 24. The combined quota for the ten members participating in the quota regime (*i.e.* excluding Iraq) is 24.5 million barrels per day.

July 31

The governor of Nigeria's Delta state announces that theft of petroleum in Nigeria is estimated at about 300,000 barrels per day, which is costing Nigeria $3.6 billion per year in lost revenues.

August 2003

August 4

ChevronTexaco tentatively agrees to purchase at least 2 million metric tons per year of liquefied natural gas (LNG) from Australia's Gorgon project. The LNG is to be shipped to the U.S. West Coast for 20 years beginning in 2008.

August 5

Turkey and Russia reportedly reach a preliminary agreement on natural gas shipments through the Blue Stream pipeline from Russia under the Black Sea. Russia had initiated legal action in July in response to Turkey's move to stop gas imports through the line and to demand a lower price than previously agreed to.

August 7

The United States estimates that restoring Iraq's oil sector to its pre-war status will cost at least $1.1 billion and take nine months to complete. Prior to the war, Iraq was producing around 2.5-2.6 million barrels per day and exporting around 2.0-2.1 million barrels per day. Current production is closer to 1 million barrels per day, with exports of about 600,000-700,000 barrels per day.

August 7

Saudi Oil Minister Ali Naimi says that Saudi Arabia expects $20 billion worth of foreign investment in natural gas projects over the next 6 years. Following the apparent collapse of the giant Saudi Gas Initiative back in June 2003, the country signed an upstream natural gas deal with Shell and Total, and in general appears to be continuing its efforts at attracting foreign investment to its gas sector.

August 8

An oil pipeline near Tucson, Arizona, which ruptured on July 30, is shut down completely, halting the flow of gasoline into Phoenix from refineries in the El Paso area. In following weeks, gasoline prices in the area skyrocket.

August 11

DaimlerChrysler and General Motors agree to drop lawsuits against California regarding clean air regulations requiring production of millions of low-emission vehicles. This does not resolve another issue however, namely California's decision in 2002 to regulate greenhouse gas emissions from automobiles.

August 12

Trinidad and Tobago reaches agreement with Venezuela on a Memorandum of Understanding regarding development of natural gas reserves in their maritime border area. This agreement could facilitate development of Venezuela's offshore Deltana Platform, thought to hold as much as 38 trillion cubic feet of natural gas on Venezuela's side and a slightly lesser amount on Trinidad's side.

August 14

Libya reportedly has agreed to compensate families of the 1988 Lockerbie airplane bombing with $2.7 billion total. The money is to be released in three tranches, the first following a lifting of United Nations sanctions, the second after possible lifting of U.S. sanctions, and the third after Libya is removed from the U.S. State Department's state sponsors of terrorism list.

August 14

A huge electric power blackout hits large parts of the northeastern United States, the Midwest, and southern Canada late in the afternoon. Power is out for at least several hours in major cities like New York, Detroit, Cleveland, and Toronto. President Bush calls the blackout a "wake-up call" and promises, "we will figure out what went wrong and we will address it." Some analysts blame the country's aging power transmission grid.

August 14

Russia approves a $13 billion merger between Yukos and Sibneft, creating "YukosSibneft," Russia's first "supermajor" and one of the world's largest publicly traded oil companies.

August 14

Japan and Russia reportedly have agreed to study the possibility of constructing an oil pipeline connecting Angarsk in eastern Siberia to Russia's Pacific Coast port of Nakhodka, from where the oil could be exported by tanker to Japan. The pipeline is expected to cost $6 billion.

August 15

Iran reportedly aims to have 7,000 megawatts of nuclear power generating capacity by 2020, and has given the go-ahead for construction of a second nuclear power plant at the Bushehr complex. The United States opposes Iran's nuclear program, which Iran claims is for peaceful purposes. Iran's power demand is growing around 7 per cent per year, but the country is also rich in natural gas and oil reserves, calling into question the need for nuclear power.

August 15

Iraq's crucial northern oil pipeline from Kirkuk to the Turkish port of Ceyhan is attacked, stopping flows on the line just two days after it reopened for the first time since the war. The pipeline had a pre-war capacity of 1.1 million barrels per day, but sustained significant damage during hostilities and had started pumping at only around 200,000 barrels per day. Repairs to the line from the latest attack May take weeks, while full restoration of the pipeline's pre-war capacity could take months.

August 15

Oil output from the huge offshore Kashagan oilfield in Kazakhstan is likely to be delayed until 2007, or two years later than originally planned. This could have implications for oil flows through the Baku-Tbilisi-Ceyhan (BTC) pipeline, which is scheduled to open in early 2005.

August 19

U.S. Energy Secretary Spencer Abraham says that his agency will head up a sweeping investigation into what caused the huge power blackout of August 14. Abraham adds, "The electric transmission grid is quite possibly the most vital piece of infrastructure we have."

August 19

Two of China's largest energy companies–CNOOC and Sinopec–agree to cooperate in what could be China's largest offshore natural gas project ever. The area to be developed is located around 250 miles southeast of Shanghai in the Xihu trough. China is attempting to increase its usage of natural gas, which currently makes up only around 3 per cent of the country's energy consumption, and this project would serve the booming Shanghai area.

August 25

Russia's government had approved a joint venture merging BP's assets in Russia and Ukraine with Russian oil major Tyumen Oil (TNK). TNK-BP will be Russia's largest multinational corporation, with a value of around $19 billion.

August 25

U.S. gasoline prices reach a national average of around $1.75 per gallon, following a sharp four-week spike. Gasoline prices are now at their highest ever in nominal terms, but when adjusted for inflation, they remain far below their highs reached in the early 1980s. In 1981, for instance, gasoline prices reached $2.88 per gallon in today's dollars, adjusted for inflation.

August 27

According to the United Nations, Iran has admitted receiving significant foreign assistance in building a secret nuclear enrichment facility south of Tehran.

August 29

The United States, North Korea, and four other countries (China, South Korea, Japan, and Russia) end three days of talks on resolving the North Korean nuclear issue, with no resolution but a decision to meet again.

September 2003

September 1

Ibrahim Bahr al-Uloum, a former Iraqi exile, is appointed Iraq's first post-war oil minister by the country's Governing Council. Uloum replaces Thamir Ghadban, who had been the acting oil minister since early May.

September 2

Russia and Saudi Arabia sign a pact pledging to cooperate to maintain stability in world oil markets during a visit by Crown Prince Abdullah to Moscow, marking the first official visit of a Saudi leader to Russia since the 1930s. Saudi Arabia and Russia are the world's first and second largest oil exporters, respectively.

September 3

The House of Representative's Energy and Commerce Committee launches an investigation into last month's massive power blackout in the U.S. Northeast and Midwest.

September 5

Brazilian state oil company Petrobras significantly revises reserve estimates from its April 2003 hydrocarbon discovery off the coast of São Paulo. The revision, from 2.5 trillion cubic feet to 14.8 trillion cubic feet of natural gas could triple Brazil's total natural gas reserves if confirmed.

September 10

The Inter-American Development Bank approves financing for Peru's Camisea natural gas project. The Camisea fields were discovered by Shell in 1986 and are estimated to hold 13 trillion cubic feet of natural gas and 660 million barrels of condensate, possibly transforming Peru into a net energy exporter.

September 11

The Federal Energy Regulatory Commission approves a plan for the new Cameron liquefied natural gas (LNG) import terminal in Hackberry, Louisiana. Cameron represents the first such project in the United States in over 20 years.

September 11

An accident forces Shell to shut in all four of its platforms located in the North Sea Brent field, affecting approximately 50,000 barrels per day of production according to press reports. Shell restarts one of the platforms on September 15, while the U.K. Health and Safety Executive investigates the other three.

September 12

The United Nations (U.N.) Security Council lifts 11-year-old sanctions against Libya. Development of Libya's sizeable oil resources has been hindered by the sanctions, which were imposed in 1992 in an effort to extradite two Libyans indicted for the 1998 bombing of an American plane over Scotland.

September 12

The International Atomic Energy Agency (IAEA) imposes an October 31 deadline on Iran to answer all of the agency's questions about its nuclear program. Iran is suspected to be developing nuclear weapons facilities, indicating non-compliance with the 1970 Nuclear Non-Proliferation Treaty, which could lead to United Nations sanctions.

September 16

El Paso Corporation confirms that it has reduced production runs at its refinery at Eagle Point, New Jersey, by roughly 50% due to delays in crude oil deliveries associated with Hurricane Isabel, which hit the East Coast on September 18. Motiva also announced that it will moderately cut throughput at its 175,000 barrel-per-day refinery in Delaware City, Delaware, as a precautionary measure.

September 19

Iranian Oil Minister Bijan Zanganeh announces that the deal which granted a Japanese consortium preferential rights to develop Iran's Azadegan oil field has expired. The consortium was granted the rights in late 2000, but had yet to negotiate and sign a contract. The Azadegan field is estimated to hold some 26 billion barrels of oil.

September 21

Iraq's leadership council announces free-market reforms for the country, allowing for, "up to 100% foreign ownership in all sectors except natural resources." In the past, the Iraqi economy was predominantly state controlled.

September 22

Eni (Italy) declares force majeure on loadings at the Brass River oil terminal in Nigeria after a pipeline connecting to the terminal's onshore field ruptured. According to Platts, production in the region is reduced from 200,000 barrels per day to 40,000 barrels per day.

September 23

The second U.S.-Russia Energy Summit ends in St. Petersburg, Russia, with both sides agreeing to enhance cooperation and encourage Russian oil and gas exports to the United Sates. Russian oil production has been growing rapidly in recent years, reaching 7.65 million bbl/d in 2002–a 26% increase over the 1998 level.

September 23

The State of Connecticut appeals a decision by the U.S. Department of Energy to keep the Cross Sound Cable, which runs between Connecticut and New York across the Long Island Sound, in operation. The previously inactive cable was powered up on orders of the Department of Energy on August 15 owing to the northeast blackout and has remained in operation on the orders of Secretary Abraham.

September 23

U.S. authorities in Iraq appoint Robert E. McKee as senior U.S. advisor to the Iraqi Oil Ministry, succeeding Philip Carroll who had served in the position since May.

September 24

OPEC members agree to cut the output ceiling for the ten member countries, excluding Iraq, by 900,000 barrel per day to 24.5 million barrels per day, effective November 1. Iraq attends the OPEC meeting for the first time since 1990. OPEC cited concerns that the world oil market will be oversupplied in 2004 leading to lower prices.

September 25

Under the direction of the Secretary of Energy, the National Petroleum Council, which is made up of representatives from the oil and gas industries, releases a report on the U.S. natural gas industry calling for increasing domestic gas exploration and boosting U.S. capacity for importing liquefied natural gas (LNG).

September 29

Indonesia's oil minister, Purnomo Yusgiantoro, says that OPEC hopes that non-member oil producers will cut production by a combined 500,000 barrels per day following the cartel's decision last week to cut by 900,000 barrels per day.

September 30

The Chicago Climate Exchange announces its first auction of emission allowances. Although emissions cuts are still voluntary, the exchange is considered an important prototype.

October 2003

October 1

Federal regulators at the Commodity Futures Trading Commission (CFTC) charge American Electric Power Inc. with attempting to manipulate the energy markets by falsely reporting the prices of thousands of natural gas trades from 2000-2002.

October 2

British Energy, Britain's biggest power generator, reaches an agreement with its creditors on a $2.5 billion government bailout and debt-for-equity swap. The agreement will help the company avoid filing for bankruptcy protection.

October 2

North Korea announces that it has already processed 8,000 spent fuel rods that could be used in the production of nuclear bombs. If true, the material in the rods would give North Korea access to enough plutonium to make five or six devices.

October 3

Chad's President Idriss Deby announced that the new Chad-Cameroon oil pipeline is officially "onstream." Chad began pumping oil into the pipeline in July 2003 from the Doba field. The $3.7 billion Chad-Cameroon oil pipeline represents the World Bank's single largest investment ever in sub-Saharan Africa.

October 4

The Russian oil companies Yukos and Sibneft complete a $45 billion merger, creating YukosSibneft, the world's fourth-largest private oil producer. The news is accompanied by rumors that major American firms are interested in making a deal with YukosSibneft in order to gain access to the Russian energy market.

October 7

The State of New Hampshire sues 22 major oil companies for their use of the gasoline additive MTBE, which has been linked to water pollution. The suit alleges that the oil companies have been using increasing amounts of the additive despite knowing that it would contaminate water supplies.

October 7

The European Union reports that it has reached agreement with the Italian oil major ENI and the Russian gas giant Gazprom regarding a prolonged dispute over sales restrictions in natural gas contracts. Under the agreement, ENI will be allowed to resell gas it has purchased from Gazprom, while Gazprom will be free to sell gas to other Italian customers without having first received ENI's permission.

October 7

Environmentalists and Native American groups reach an agreement with PPL, a power company, to close two dams and decommission a third on the Penobscot River in Maine. The coalition hopes to resuscitate the river and to revive its Atlantic salmon population. In exchange for the dams' closure, the coalition will pay PPL about $25 million.

October 9

Inspectors at a New Hampshire nuclear plant find a leak in the containment dome similar to one found at the David-Besse plant in Ohio in March 2002. Spokespeople for both the New Hampshire plant and the Nuclear Regulatory Commission say that the leak poses no danger and was detected early enough to prevent any serious damage.

October 13

Abdullah al-Attiyah, president of OPEC, warns that crude oil prices could collapse in 2003-2004 because of the increased flow of Iraqi oil. In al-Attiyah's opinion, this collapse could be avoided if Russia, the second largest exporter, and other large non-OPEC-members cut production.

October 14

Bowing to protests, Bolivian President Gonzalo Sanchez de Lozada announces he will not pursue a plan to export more than one billion cubic feet per day of liquefied natural gas (LNG) to the United States through Chile. The proposal had led to massive popular protests in Bolivia, resulting in the deaths of at least 16 people.

October 16

Spokesmen for both ExxonMobil and Royal Dutch/Shell (RDS) announce that they are separately investing in natural gas projects in Qatar. RDS plans to invest between $4 and $5 billion dollars in an offshore development project and an onshore gas-to-liquids plant. ExxonMobil's deal could supply more than 15 million tons of LNG to the US market, and involves an investment of at least $10 billion in the expansion of Qatar's Ras Laffan Liquefied Gas Co. plant.

October 22

Iran agrees to suspend its uranium enrichment program and accepts a tighter inspection program. The decision comes after months of resistance by Iran to demands from the International Atomic Energy Agency (IAEA) and others regarding its nuclear program.

October 22

Congressional Democrats allege that Halliburton subsidiary KBR is paying significantly more than local buyers to bring fuel into Iraq where it has a controversial infrastructure contract.

October 23

ChevronTexaco goes on trial in Ecuador, accused by 30,000 alleged victims of systematically dumping hazardous waste during its 20 years of operations in the country.

October 25

The owners of the Gorgon plant in Western Australia agreed to supply around $21 billion of LNG over 25 years to China's state-owned China National Offshore Oil Corporation (CNOOC). Gorgon is owned by ChevronTexaco (57 per cent), Royal Dutch/Shell (29 per cent), and ExxonMobil (14 per cent). CNOOC plans to buy a stake in Gorgon, according to officials there.

October 26

Washington State authorities and the U.S. Department of Energy say they have come to an agreement regarding cleanup of the Hanford nuclear complex, the most contaminated nuclear site in the United States.

October 26

Russian authorities arrest Mikhail B. Khodorkovsky, head of the Russian oil giant Yukos, jailing him on charges of fraud and income tax evasion. The arrest caps a months-long governmental probe into the oil company, and prompts some industry members to question the desirability of investing in the Russian energy sector.

October 27

British Petroleum (BP) and the state-owned Algerian oil company Sonatrach sign an agreement with the British transmission system operator, National Grid Transco, to transport 500 million cubic feet per day of liquefied natural gas (LNG) into the United Kingdom starting in 2005.

October 28

A coalition of twelve states and several cities that oppose changes to the Clean Air Act sue to block the revisions that would make it easier to upgrade power plants, factories and other facilities without also installing pollution controls.

October 30

The Russian government seizes 44.1% of oil firm Yukos' stock. Government prosecutors indicate that the shares are being held to be used as payment in the event that a court rules against the company's jailed president Mikhail Khodorkovsky.

October 31

A consortium of international energy firms agrees to build Nigeria's second natural gas liquefication facility in conjunction with the state owned Nigeria National Petroleum Corporation (NNPC). Eni/Agip, ChevronTexaco and ConocoPhillips each will take a 17% stake, with the NNPC taking the remaining 49%. The facility will be built on the Brass River in the central Niger delta.

November 2003

November 4

Mikhail Khodorkovsky, Chief Executive Officer of Russian oil major Yukos, resigns after being arrested on October 25 on charges of fraud and tax evasion. Yukos is Russia's largest oil company in terms of production and has been feuding with the Russian government since July, when Platon Lebedev, an associate of Mr. Khordokovsky was also arrested.

November 4

The International Finance Corporation, the private lending division of the World Bank, approves a $250 million loan for the Baku-Tbilisi-Ceyhan pipeline. The 1-million-barrel-per-day pipeline will enable crude oil exports from the land-locked Caspian Sea region to reach world markets through the Turkish Mediterranean port of Ceyhan.

November 6

The Michigan Public Service Commission releases a 107-page investigative report on the causes of the massive electricity blackout which beset parts of the United States and Canada on August 14, 2003. The report criticizes a lack of communication between several regional utility companies as well as the Midwest Independent System Operator (MISO), an electricity grid reliability coordinating organization based in Carmel, Indiana.

November 7

Iran declares that it will stop manufacturing the Shahab-4 missile, which is thought to be capable of threatening both Europe and Israel. The move is seen as an effort to assuage mounting concerns over Iran's nuclear program.

November 11

The European Bank for Reconstruction and Development approves its $250 million loan for the Baku-Tbilisi-Ceyhan oil pipeline following a similar decision by the International Finance Corporation on November 4. Securing financing from these two international donor organizations could facilitate more commercial loans which will allow the project to proceed according to schedule.

November 12

Iran's representative to the International Atomic Energy Agency (IAEA), Ali Akbar Salehi, acknowledges that his country had failed to comply with international nuclear non-proliferation rules as suggested in a confidential IAEA report. He adds, however, "The failures that Iran has been reproached for are minor, and are only on the order of the gram or milligram." The IAEA investigation reportedly indicates that Iran manufactured small quantities of enriched uranium and plutonium as part of a secret nuclear program.

November 15

Saudi Arabia signs a 40-year natural gas exploration agreement with international oil companies Royal Dutch/Shell and Total. Previous efforts to open up Saudi Arabia's upstream natural gas sector fell apart in June due to conflicts with foreign investors over financial terms. The two companies will be the first international companies to enter Saudi Arabia's upstream energy sector in over two decades.

November 18

ChevronTexaco reports that it has received final approval form the Federal Energy Regulatory Commission (FERC) to build the world's first-ever deepwater liquefied natural gas (LNG) import terminal at Port Pelican in the U.S. Gulf of Mexico. The plant will have a capacity of 1.6 billion cubic feet per day, with construction to begin in 2004 and to be completed in 2007.

November 19

The U.S.-Canada Power System Outage Task Force, led by U.S. Secretary of Energy Spencer Abraham and Canadian Natural Resource Minister Herbert Dhaliwal, releases a 124-page report on the causes of the August 14 electricity blackout. The report concludes that the blackout was "largely preventable" and cites several failures by regional utility companies and regulators. Hearings will be held in the United States and Canada and a final version of the report with recommendations to avoid further blackouts is expected by the end of the year.

November 19

Following several months of negotiations, Turkey and Russia reach an agreement on the price and volume of Russian natural gas exports to Turkey via the Blue Stream pipeline. In March 2003, Turkey halted deliveries through the Blue Stream pipeline, invoking a clause in the contract allowing either party to stop deliveries for six months. The Blue Stream pipeline became operational in December 2001, and represents a significant new export route for Russia, the world's largest natural gas producer.

November 21

The United Nations hands over the "oil-for-food" program in Iraq to the U.S.-led administration in Baghdad. The "oil-for-food" program was established by the United Nations in 1995, and used proceeds from the sale of Iraqi oil to buy food and medicine for Iraqis as well as to finance infrastructure and humanitarian projects. Iraqi oil exports reportedly have reached around 1.5 million barrels per day.

November 21

The Electric Power Research Institute (EPRI) releases its 40-page investigation into the factors causing the August 14, 2003, electricity blackout. The report was commissioned by First Energy, one of the firms implicated by previous investigative reports issued by the Michigan Public Service Commission and the U.S.-Canada joint task force. Unlike previous investigations, the EPRI document reportedly focuses on the issue of insufficient reactive power in the electricity grid, which is essential to maintain stability within the grid.

November 23

Georgian President Eduard Shevardnadze resigns suddenly following widespread protests and a takeover of the country's Parliament by members of the political opposition. New parliamentary elections in the country are scheduled to be held within 45 days, but a date for new presidential elections is not set. Georgia is an important energy transit center for Caspian Sea oil. The United States has supported the principle of multiple export options for Caspian exporters, and three of the largest projects to these ends cross through Georgia (Baku-Tbilisi-Ceyhan, Baku-Tbilisi-Erzurum, and Baku-Supsa, a.k.a the "Western Early Oil Route").

November 24

The U.S. Congress abandons plans to pass an energy bill before the end of the legislative session. The bill was approved in the U.S. House of Representatives on November 18, but then blocked in the Senate as its proponents were unable to close debate on the issue and call for a vote. The legislation has been under construction for three years and represents Congress's first attempt at a comprehensive energy bill since 1992. The bill's proponents intend to revisit the issue in 2004.

November 26

The International Atomic Energy Agency (IAEA) approves a resolution on Iran's nuclear program. The resolution represents a compromise agreement between IAEA members by reprimanding Iran for past violations without referring the issue to the United Nations for possible sanctions.

November 28

Russian oil company Sibneft makes a surprise announcement suspending its merger with Russian oil major Yukos citing technical difficulties. The $13 billion merger was announced in April 2003, and would create the world's fifth-largest publicly traded oil company.

December 2003

December 2

President George W. Bush signs a $27.3 billion energy and water bill that includes funding for a nuclear waste repository at Yucca Mountain, Nevada. The repository remains a source of controversy between state and federal officials.

December 4

OPEC holds its 128th meeting to review oil markets in Vienna, Austria, leaving OPEC 10 output quotas unchanged.

December 8

The United States reaches an agreement with Japan and South Korea on a carefully worded set of principles for ending North Korea's nuclear program. The agreement entails a system of steps to provide North Korea with security guarantees as it begins to verifiably dismantle its nuclear facilities.

December 9

ConocoPhillips agrees to proceed with a $5 billion gas-to-liquids project in Ras Laffan, Qatar to convert natural gas into diesel and other liquid fuels. Sales are expected to be to Asia and possibly Europe.

December 10

A U.S. Department of Defense document limits contracts for 26 Iraqi reconstruction projects to companies from countries that supported the war. Companies from France, Russia, and Germany will still be able to compete for sub-contractor roles.

December 12

An investigation by the Defense Contract Audit Agency of the U.S. Defense Department finds evidence indicating that the Halliburton Company subsidiary Kellogg, Brown & Root overcharged the government as much as $61 million for fuel delivered to Iraq.

December 15

Oil prices fall 4% on the news that U.S. military forces capture Saddam Hussein near his hometown of Tikrit, Iraq.

December 15

The China National Offshore Oil Corporation (CNOOC) reaches agreement with Guangdong provincial authorities to build a natural gas receiving terminal in Zhuhai. The $880 million project should go online in 2006, and is intended to commercialize increased production at the Panyu offshore oil field.

December 16

The U.S. Environmental Protection Agency formally proposes new rules to reduce the total amount of mercury emissions from coal-fired power plants by almost 70% over the next 15 years. The proposal calls for the adoption of a mandatory cap and trade program.

December 17

Apache Corporation, a U.S. energy firm, signs a reported $6 billion Memorandum of Understanding to supply Egypt with a minimum total amount of 2.7 trillion cubic feet of natural gas from its West Mediterranean Concession. The 25-year deal is scheduled to begin by 2007.

December 18

The U.S.-led Coalition Provisional Authority (CPA) in Iraq announces that it will provide the Iraqi Oil Ministry with $600 million for infrastructure improvement. The director-general of the state-owned Iraqi Drilling Company says that he expects that the investment could raise daily production, now at about 1.9 million barrels per day, by as much as 1.05 million barrels per day over an 18 month period.

December 18

Iran signs a protocol to the Nuclear Nonproliferation Treaty that will allow the International Atomic Energy Agency (IAEA) to have more comprehensive access to sites in the country. It is not known when Iran will officially ratify the protocol.

December 18

BP signs a 20-year deal to sell 500 million cubic feet per day of liquefied natural gas (LNG) from its Tangguh facility in Indonesia to the U.S. energy company Sempra Energy. The LNG will be shipped to Sempra's proposed import and regasification terminal in Baja California, Mexico before being distributed to buyers in the United States.

December 19

A consortium of energy companies led by French nuclear engineering firm Areva SA signs a $3 billion contract to build a new nuclear reactor in Finland. It is Western Europe's first new reactor order in 12 years.

December 22

Libya announces that it will abandon its weapons of mass destruction programs and comply with the Nuclear Non-Proliferation Treaty. The United States welcomes the move, but says that it will maintain economic sanctions until it sees evidence of compliance.

December 22

The Sudanese government and the Sudanese People's Liberation Army, the main rebel faction in the southern part of the country, reach an agreement to split oil revenues equally between themselves. The agreement is believed to increase the possibility of a lasting peace settlement in the war-torn country.

December 23

In a move to ensure future energy supplies, the Chinese government announces that it will form as many as 10 large, state-run companies to control 60% of coal production. The move reflects growing concerns that insufficient energy supplies will slow economic growth.

December 25

A federal appeals court blocks a Bush administration rule attempt to relax existing environmental regulations. The change would have allowed aging electric power and industrial plants to make upgrades without installing modern pollution controls.

December 26

A consortium of the Russian state-controlled gas monopoly OAO Gazprom, the state oil firm Rosneft Oil, and the private oil firm Surgut is set up to jointly buy and develop untapped energy reserves in Eastern Siberia.

December 31

The U.S. Department of Defense says that the task of importing gasoline for civilian use in Iraq will be handled by the Defense Energy Support Center, a Department of Defense agency that supplies fuel to the military, instead of a subsidiary of Halliburton Company. Halliburton has been criticized for allegedly overcharging the government under its no-bid contract.

102

Monthly Energy Chronology 2004

January 2004

January 6

A previously undisclosed ruling by Lt. General Robert Flowers, head of the U.S. Army Corps of Engineers, exonerates Halliburton Company and its subsidiary Kellogg Brown & Root of any wrongdoing in its fuel-delivery arrangements with a Kuwaiti supplier. The decision comes after Pentagon auditors asserted that the company had overcharged the U.S. government. A full audit of the company's fuel contracts in Iraq is pending.

January 6

The United States renews economic sanctions against Libya despite Libya's announcement last month that it will abandon its weapons of mass destruction programs and comply with the Nuclear Non-Proliferation Treaty. U.S. officials state that sanctions will be maintained until evidence of compliance is seen.

January 7

The government of India approves plans to build a strategic crude oil reserve. The government plans to build a 37.5 million barrel stockpile, roughly equivalent to two weeks of demand. The project would make India one of the few countries in the world to hold such strategic reserves.

January 7

The government of Sudan and rebel leaders sign an agreement on sharing revenue from the country's oil exports. The two sides agree to equally divide net oil export revenue from southern Sudan for the duration of a 6-year "transition period," which began in July 2002. With the completion of a major oil export pipeline in July 1999, Sudanese crude oil exports have risen rapidly over the past three years, totaling roughly 195,000 bbl/d in 2002. The Sudanese government has been embroiled in conflict with rebel movements for more than 20 years.

January 8

PetroChina cancels deliveries of crude oil to Japan. As both sides wrangle over pricing negotiations and China's domestic supply shortage grows, PetroChina cuts off the flow of Daqing crude oil to Japan, which in previous years had averaged between 60,000 and 100,000 barrels per day.

January 12

Mexican state oil company, Pemex, announces that natural gas drilling has begun on the first four blocks awarded under its first round of "multiple service contract" tenders. The company expects the blocks to increase Mexico's natural gas production by 400 million cubic feet per day. Specially designed "multiple-service contracts" were created to cede more management control to foreign investors, while maintaining deference to Mexico's Constitution, which requires that Pemex retain exclusive rights to oil and gas exploration in Mexico.

January 12

The first of two new gasoline producing units comes online at Kuwait's Mina al-Ahmadi refinery. Once completed, the facilities are expected to turn Kuwait into a net gasoline exporter, having been a net importer in the past. The units will replace facilities which were destroyed during an explosion and fire in June 2000.

January 13

The U.S. Court of Appeals for the Second Circuit in New York rules that the U.S. Department of Energy violated federal energy and environmental regulations by amending a rule requiring manufacturers to increase the efficiency of residential air conditioners by 30% by 2006. In April 2001, the Department of Energy altered the rule to require a 20% increase in efficiency by 2006. Plaintiffs in the case–the Natural Resources Defense Council and Attorneys General from ten states–claimed that as originally intended the rule will eventually reduce electricity peak demand by 14,500 megawatts per year.

January14

Natural gas production begins at the Bayu Undan field, located in the Timor Sea and held jointly by Australia and East Timor. Development of Bayu Undan marks the first stage of development of the Joint Development Area agreed to between Australia and East Timor under the Timor Gap Agreement of March 2003. East Timor was internationally recognized as an independent state in May 2002 after decades of subjugation and civil strife.

January 15

The government of India repeals caps on foreign investment in private oil exploration and marketing companies, allowing overseas companies to develop wholly owned oil and gas units. Previously, such investments were capped at 51%.

January 18

Saudi Aramco formally inaugurates its new Haradh oil and natural gas facility. The Haradh plant is expected to boost Saudi natural gas production capacity by roughly 25%, most of which is slated for the domestic market. The Haradh facility also includes a gas-oil separation plant capable of processing 300,000 bbl/d, as well as infrastructure for delivering up to 170,000 bbl/d of condensates to the Kingdom's Abqaiq processing facility. Developing the country's relatively untapped natural gas potential could allow more oil to be allocated for export in the future.

January 19

A boiler explodes at Algeria's Skikda liquefied natural gas (LNG) export terminal killing at least 27 people. The explosion causes heavy damage to three of the facility's six liquefaction trains and shuts down operations at several adjacent facilities, including a refinery and crude oil and petroleum product loading terminals. Algeria is the world's second-largest LNG exporter, and the Skikda export terminal handles about one-quarter of the country's total LNG exports.

January 20

The prompt month crude oil futures price on the New York Mercantile Exchange (NYMEX) settles at $36.20 per barrel, surpassing $36 per barrel for the first time in 10 months. High prices are attributed to cold weather in the Northeast, which has boosted the demand for heating oil, as well as low commercial crude oil inventories.

January 22

U.S. Interior Secretary Gale Norton approves a plan to open parts of Alaska's North Slope to oil exploration and drilling. Nine million acres of Alaska's National Petroleum Reserve will be opened to long-term production. The site lies adjacent to the Arctic National Wildlife Refuge, which remains closed to oil and gas drilling. A lease sale is expected in June.

January 23

The Alberta Energy and Utilities Board (AEUB) in Canada orders 938 natural gas wells in northeastern Alberta to stop production in an effort to protect the subsurface pressures surrounding the province's sizeable oil sands reserves. The decision is made following a 6-year study by provincial authorities into the effects of natural gas production on oil sands reservoirs. A group of natural gas producers is later granted an appeal. With the inclusion of oil sands, which the AEUB estimates to total 174.4 billion barrels, Canada holds the second largest proven crude oil reserves in the world after Saudi Arabia according to some estimates.

January 26

Shipping of petroleum products resumes from Algeria's Skikda port, one week after a devastating explosion. Repairs continue at the port's oil refinery and LNG facilities.

January 27

Russia's LUKoil announces a new joint venture with Saudi Arabia's state oil company, Aramco, to explore for natural gas near the Kingdom's Ghawar field. Later in the week, similar deals are signed with China's Sinopec, Italy's Eni and Spain's Repsol YPF. The LUKoil project marks the first such deal since July 2003, when Royal Dutch/Shell and Total became the first Western companies to gain access to Saudi Arabia's hydrocarbon reserves since the nationalization of the Kingdom's petroleum industry in the 1970s.

January 28

A federal judge in Alaska imposes punitive damages of $4.5 billion ($6.75 billion with interest) on Exxon Mobil for the March 1989 oil spill by the supertanker, Exxon Valdez. Payment is to go to fishermen and residents of the Prince William Sound area, where the oil tanker ran aground and spilled an estimated 11 million gallons of oil. The decision marks the third effort to award such punitive damages after the previous two were struck down by a federal appeals court in San Francisco. Exxon Mobil plans to appeal.

February 2004

February 2

Front-month crude oil futures on the New York Mercantile Exchange settle at $34.98 per barrel, leaping by $1.93 per barrel in one day. This was the largest one-day jump in crude oil prices since October 12, 2000, when prices reacted to news of an attack on the Navy's USS Cole with a $2.81 per barrel increase. Traders attribute upward price pressure to low oil inventory levels in the United States.

February 3

Core shareholders for Russian oil majors Yukos and Sibneft sign an initial protocol on reversing their planned $13 billion merger. Combined into one entity, YukosSibneft, the company would have become Russia's largest oil firm, with an estimated production of roughly 2.4 million barrels per day. The merger was announced publicly by the two firms in April 2003, but in November 2003, Sibneft backed out of the deal following the arrest of Yukos president Mikhail Khodorkovsky.

February 4

Andriy Klyuyev, Ukraine's Deputy Prime Minister, announces that his country will not reverse the Odessa-Brody pipeline, which extends from Ukraine's Black Sea port of Odessa northward to the city of Brody. The pipeline was designed to carry oil from the Caspian Sea region to Europe, but has been mostly dormant since its completion in 2000. Kiev had previously considered a Russian proposal to reverse the pipeline's flow, which would have provided Russia with a much-needed export outlet.

February 5

The Alberta Energy and Utilities Board grants approval to Shell's Jackpine oil sands project. The Jackpine project is expected produce 200,000 barrels per day of bitumen, which will in turn be processed into synthetic crude oil and possibly exported to the United States. No timetable has been set yet for the project, but some analysts suggest that production is unlikely to begin before 2011.

February 11

OPEC delegates meeting in Algiers agree to lower the cartel's output ceiling by 1 million barrels per day, to 23.5 million barrels per day, effective April 1. OPEC members also urged immediate compliance with the existing OPEC ceiling, as overproduction has been estimated at roughly 1.5 million barrels per day. Assuming full quota compliance, the decision could remove a total of 2.5 million barrels per day from the world market in April.

February 18

The U.S. Department of Transportation announces its decision to extend a rule giving credits to automakers for building vehicles which can run on ethanol. These credits can be used by automakers to meet federally mandated fuel economy targets. The credit, which was originally created under a law passed in 1998, was designed to encourage the use of alternative fuels, but has come under criticism as opponents charge that the duel-use vehicles which earn the credits almost always run exclusively on gasoline. In 2002, the Departments of Transportation and Energy and the U.S. Environmental Protection Agency reported that extending the credit could increase U.S. gasoline consumption by up to 14 billion gallons (total) between 2004 and 2008.

February 19

The Japanese government signs a deal in Tehran giving a consortium of Japanese companies the right to help develop Iran's Azadegan field. Tokyo's negotiations in Iran began in 2000, but heretofore had been slowed due to disagreements over terms with the Iranian side and diplomatic pressure from the United States against the deal. The deal stipulates that production begin in approximately two years and eventually reach a rate of 300,000 barrels per day.

February 19

The Royal Dutch/Shell group announces that the Securities and Exchange Commission (SEC) has begun a formal investigation into the company's restatement of its oil and gas reserves. On January 9, 2004, Royal Dutch/Shell announced that it had overstated its proven oil and gas reserves by 3.9 billion barrels, or 20% due to overly optimistic assumptions about plans for developing its fields around the world.

February 23

Algeria's Ambassador to the United States, Idriss Jazairy, says that preliminary results from an investigation into an explosion at the country's Skikda liquefied natural gas (LNG) terminal on January 19, 2004, indicate that the incident was caused by a leaking gas pipeline, not a faulty steam boiler as has been previously suspected. There are over 30 pending proposals for new LNG import terminals in the United States, and opponents of LNG have pointed to the Skikda catastrophe as evidence that the facilities are unsafe. A formal report from the Algerian government is pending.

February 25

Total (France) and Petronas (Malaysia) sign an estimated $2 billion agreement with the National Iranian Oil Company to build Iran's first liquefied natural gas (LNG) export facility. The two-train facility will have a capacity of 390 billion cubic feet per year, with natural gas to come from Iran's South Pars field. Production of LNG is expected to begin in 2009. Iran holds the world's second largest natural gas reserves-after Russia-and development of LNG facilities would allow the country to export gas around the world.

February 25

Authorities in Kazakhstan announce that construction has begun on phase two of the planned Kazakhstan-China oil pipeline. Phase two entails an 800-mile connection between Atasu in Central Kazakhstan and Alashanku on the Chinese border. Phase one was launched last year, and the entire pipeline is schedule for completion in 2008. China's oil demand has been growing rapidly in recent years, and this growth is expected to continue. The Kazakhstan-China pipeline will allow oil from Kazakhstan's prolific Caspian Sea region fields to be piped directly to Chinese refineries.

February 25

An international consortium of oil companies, known as Agip KCO and led by Italy's ENI, reach an agreement with the government of Kazakhstan on development of Kazakhstan's Kashagan oil field. Kashagan was discovered in 2000 and is estimated to hold proven oil reserves of 9-13 billion barrels of oil. The consortium members and the government had previously disagreed over the project's start-date, but today settled on initial production beginning in 2008, three years later than originally planned. According to the agreement, the field is expected to reach peak production of 1.2 million barrels per day by 2016.

February 26

Veteran energy analyst Matt Simmons outlines the results of his forthcoming analysis of Saudi Arabia's oil reserves at the Center for Strategic and International Studies in Washington, D.C.. Simmons suggests that Saudi Arabia, which is the world's largest oil producer and holds most of the world's spare oil production capacity, could suffer production declines in the next 5-10 years due to rapid depletion of the Kingdom's recoverable reserves. Representatives from Saudi Aramco refute Mr. Simmons's assertion, claiming that current production levels-or even higher-are sustainable for decades to come.

February 26

The United States rescinds a ban on travel to Libya and authorizes U.S. oil companies with pre-sanctions holdings in Libya to negotiate on their return to the country if and when the United States lifts economic sanctions. The United States first imposed sanctions on Libya in 1986 following terrorist attacks in Rome and Vienna. Several U.S. oil companies were forced to abandon their assets in Libya when sanctions were imposed in 1986, including the "Oasis Group" (Marathon Oil, ConocoPhillips, Amerada Hess) and Occidental Petroleum.

February 27

U.S. officials in Iraq announce that the Khor al-Amaya oil export terminal on the Persian Gulf has partially reopened, after the successful completion of repairs. Two of the four loading berths are operational, with a total capacity of around 250,000-300,000 barrels per day, reportedly helping to increase country's export capacity.

March 2004

March 1

Venezuelan President Hugo Chavez states publicly that if Washington were to impose economic sanctions or invade Venezuela, his country would discontinue oil supplies to the United States. "Mr. Bush must know that if he gets the mad idea of trying to blockade Venezuela, or, even worse, of invading Venezuela-if that happened, the people of the United States should know that not a drop more would reach them from Venezuela," said President Chavez. Venezuela was the fourth largest foreign supplier of oil to the United States in 2003.

March 1

Iraq's northern oil export pipeline, Kirkuk-Ceyhan, resumes pumping oil after having been sabotaged several times. Oil flows through the pipeline are only at partial capacity and are being conducted under a "news blackout". to avoid further sabotage efforts. Re-opening of the pipeline, which runs from Iraq, northward to the Turkish port of Ceyhan, gives Iraq another outlet for its lucrative crude oil exports.

March 4

China's State Council approves a plan to build a national strategic petroleum reserve to be utilized in the event of a global oil supply disruption. The plan will make China one of the few countries in the world to hold strategic reserves. China was the world's third largest consumer of petroleum products in 2002, after the United States and Japan.

March 4

Newly appointed Chairman of Royal Dutch/Shell, Jeroen van der Veer, retracts the claims of ousted Chairman Sir Phillip Watts that the company's January 9, 2004, revision of oil reserves was made "in good faith." On January 9, 2004, Royal Dutch/Shell announced that it had overstated its proven oil and gas reserves by 3.9 billion barrels, or 20%, claiming overly optimistic assumptions about plans for developing its fields around the world. Later, on February 19, 2004, it was announced that the U.S. Securities and Exchange Commission had launched a formal investigation and on March 3, 2004, Mr. Watts and another official resigned over the controversy. Other investigations have since been launched by the U.S. Department of Justice and European regulators.

March 5

The supply of liquefied natural gas (LNG) to the United States from Trinidad and Tobago is abruptly halted due to a strike at the country's Atlantic LNG facility. Striking tugboat workers force the closure of two of the facility's three LNG trains. The striking workers reportedly return to work on March 10 and deliveries to the U.S. are resumed. Trinidad and Tobago is the largest LNG exporter to the United States and supplies are expected to increase over the next few years as more production trains are added.

March 12

A landslide ruptures the SOTE, one of Ecuador's two crude oil pipelines, halting exports. The roughly 400,000-barrel-per-day pipeline extends about 300 miles from the Lago Agrio area in the Oriente to the Balao terminal near the port city of Esmeraldas. Ecuador's state oil company is compelled to declare force majeure on petroleum exports. Repairs are completed on March 24, 2004, and in the interim, smaller quantities of crude oil are diverted through Ecuador's other oil pipeline, Oleoducto de Crudos Pesados (OCP).

March 13

Iran bars nuclear inspectors from entering the country for an indefinite period of time after the International Atomic Energy Agency (IAEA) passes a resolution rebuking Iran for failure to fully disclose the details of its past nuclear activity. Iran's top nuclear official, Hassan Rouhani says, "Today, IAEA inspectors were expected to arrive in Iran. We will not allow them to come until Iran sets a new date for their visit. This is a protest by Iran in reaction to the passage of the resolution."

March 16

The Energy Information Administration, the independent statistical and analytical agency within the U.S. Department of Energy, releases its report, Analysis of oil and Gas Production in the arctic National Wildlife Refuge (ANWR). The 20-page report forecasts the impact that development plans for ANWR could have on domestic oil production, U.S. dependence on foreign oil, and oil prices.

March 17

The flow of oil through Iraq's Kirkuk-Ceyhan oil export pipeline is halted after a partial resumption at the beginning of this month. Iraqi authorities reportedly uncover corrosion along the line, forcing them to stop operations until technicians can make repairs. According to press reports, the pipeline remains idle for the rest of March.

March 17

Crude oil for April delivery on the New York Mercantile Exchange (NYMEX) settles at $38.18 per barrel, marking a thirteen-year high (in nominal dollars). The day's settlement price is the highest since October 16, 1990. Recent upward pressure on oil prices is attributed to several factors, including OPEC's pending output cut scheduled for April 1, relatively low U.S. oil inventories, and growing oil demand worldwide.

March 26

Royal Dutch/Shell signs an agreement with Libya's state oil company, National Oil Corporation, establishing a landmark partnership to develop Libya's long-neglected oil and natural gas industries. Since Libya abandoned its weapons of mass destruction program last December, oil and gas companies from around the world have been hoping to sign deals in the country. This includes several American firms which were forced to abandon their assets in Libya when sanctions were imposed in 1986. On February 26, 2004, Washington rescinded a ban on travel to Libya and authorized U.S. oil companies with pre-sanctions holdings in Libya to negotiate on their return to the country if and when the United States lifts economic sanctions.

March 29

The U.S. average retail price for regular gasoline hits a nominal high point, unadjusted for inflation, of 175.8 cents per gallon, which is 1.1 cents per gallon higher than the previous all-time high set on August 25, 2003. The average price of $1.758 per gallon as of March 29 is still a far cry from the inflation-adjusted March 1981 average price, which would be equivalent to $2.99 per gallon in today's dollars.

March 31

OPEC members unanimously agree to implement the cartel's oil production cuts effective April 1, as agreed to in February. Relatively high prices for oil and petroleum products had prompted several consuming countries, including the United States, to suggest that OPEC members vote to postpone the cuts and put downward pressure on oil prices. According to the cartel's official communiqué following the meeting, "Notwithstanding prevailing high prices, the Conference observed that the crude oil market remains more than well supplied as the world moves into the traditionally lower seasonal demand period."

April 2004

April 1

Argentina begins rationing natural gas exports to Chile in an effort to stave off a domestic energy crisis. With Argentine natural gas demand rapidly outpacing supply, the Argentine government announced in late March its intentions to begin limiting natural gas exports to Chilean utilities, effective April 1, 2004. Despite depending on Argentina for roughly 80% of its natural gas supply in 2002, Chile is not compelled to begin rationing electricity. Andean relations are further complicated later in the month when Bolivia pledges natural gas to Argentina on the condition that none of it is passed on to Chile, its historical rival.

April 5

The U.S.-Canada Power System Outage Task Force, led by U.S. Secretary of Energy Spencer Abraham and Canadian Natural Resource Minister Herbert Dhaliwal, releases its Final Report on the August 14th Blackout in the United States and Canada. The report groups the causes of the blackout in to four categories: inadequate system understanding, inadequate situational awareness, inadequate tree trimming and inadequate diagnostic support. The report also includes 46 recommendations for preventing future blackouts.

April 14

The U.S. Securities and Exchange Commission (SEC) announces a change in policy concerning assessments of hydrocarbon reserves in the Gulf of Mexico. The SEC will now accept confirmation of proved reserves in the Gulf of Mexico by several "standard methods," which include recent technological advancements. Before the announcement, the SEC rules stipulated that only a flow test could confirm proved reserves in the Gulf of Mexico.

April 19

Venezuelan President Hugo Chavez reiterates the threats he made last month to stop selling oil to the United States. President Chavez accuses the United States of "intervening in Venezuela's domestic affairs," and suggests that such intervention could, "spark a conflict here, and it would be absurd to continue selling oil to them [the United States]." The United States denies Chavez's charges. Venezuela was the fourth largest foreign supplier of oil to the United States in 2003.

April 21

A car bomb explodes outside a police building in Riyadh, Saudi Arabia, marking the first major attack by militants on governmental targets in the Kingdom. Four people are killed and 148 are wounded. The country's major export facilities are not harmed, but port authorities maintain a "heightened sense of security." Saudi Arabia is the world's largest oil producer and America's second largest foreign supplier of crude oil and petroleum products after Canada.

April 21

The United Nations (UN) Security Council unanimously endorses an independent investigation into charges of corruption in the UN-administered "oil-for-food" program in Iraq. The "oil-for-food" program was established by the United Nations in 1995, and used proceeds from the sale of Iraqi oil to buy food and medicine for Iraqis as well as to finance infrastructure and humanitarian projects. The program was officially handed over to the U.S.-led administration in Baghdad in November, 2003. Recently, allegations have arisen in the Iraqi press that foreign administrators of the "oil-for-food" program were receiving inappropriate proceeds from oil sales before the war.

April 23

The United States eases economic sanctions maintained against Libya since 1986. A written statement from the White House Press Secretary states, " U.S. companies will be able to buy or invest in Libyan oil and products. U.S. commercial banks and other financial service providers will be able to participate in and support these transactions."

April 23

Libya's state-owned National Oil Corporation (NOC) announces its first shipment of oil to the United States in over 20 years. The spokesman did not name the company with whom NOC has contracted to sell the oil, but states that the cargo will be loaded in May and should reach the United States by late May or early June. The United States first imposed sanctions on Libya in 1986 following terrorist attacks in Rome and Vienna.

April 24

Suicide bombers attack Iraq's Basra maritime oil terminal, killing three U.S. Navy sailors and damaging one tanker berth. Roughly 90% of Iraq's crude oil exports are currently loaded on to tankers at the Basra terminal, and the incident marks the first known attack on these maritime facilities since the onset of war in March 2003. Previously, efforts by Iraqi insurgents to dismantle the country's oil industry had focused on facilities on the ground, including pipelines and pumping stations. Iraqi Oil Minister Mohammad Bahr al-Uloum announces the next day (4/25) that "damage was limited and exports are flowing back at the same rates."

April 24

Five people are killed in Nigeria's Delta region, including two Americans and three Nigerians, when gunmen attack a boat carrying oil workers. The five oil workers were reportedly investigating ChevronTexaco facilities which had been abandoned in March 2003 but had since been considered for re-opening. Following the attack ChevronTexaco withdraws all personnel from the Western Delta region and suspends activities in the area pending an assessment of the security situation.

April 27

U.S. Secretary of Energy Spencer Abraham awards $350 million to researchers investigating hydrogen fuel technology. This represents the first tranche of the $1.2 billion in funding for the development of hydrogen powered vehicles promised during President Bush's January 2003 State of the Union address. The money will be divided amongst more than 130 research institutions and auto manufacturers and will address several issues, including the storage, cost and distribution of hydrogen fuel. The funding is intended to help industry meet the Bush Administration's goal of having hydrogen-powered vehicles on the road by 2015.

Glossary

ACBM: Acronym for "asbestos-containing building material."

Account classification: The way in which suppliers of electricity, natural gas, or fuel oil classify and bill their customers. Commonly used account classifications are "Residential," "Commercial," "Industrial," and "Other." Suppliers' definitions of these terms vary from supplier to supplier. In addition, the same customer may be classified differently by each of its energy suppliers.

Account of others (natural gas): Natural gas deliveries for the account of others are deliveries to customers by transporters that do not own the natural gas but deliver it for others for a fee. Included are quantities covered by long-term contracts and quantities involved in short-term or spot market sales.

Accounting system: A method of recording accounting data for a utility or company or a method of supplying accounting information for controlling, evaluating, planning and decision making.

Acid mine drainage: This refers to water pollution that results when sulfur-bearing minerals associated with coal are exposed to air and water and form sulfuric acid and ferrous sulfate. The ferrous sulfate can further react to form ferric hydroxide, or yellow boy, a yellow-orange iron precipitate found in streams and rivers polluted by acid mine drainage.

Acid rain: Also called acid precipitation or acid deposition, acid rain is precipitation containing harmful amounts of nitric and sulfuric acids formed primarily by sulfur dioxide and nitrogen oxides released into the atmosphere when fossil fuels are burned. It can be wet precipitation (rain, snow, or fog) or dry precipitation (absorbed gaseous and particulate matter, aerosol particles or dust). Acid rain has a pH below 5.6. Normal rain has a pH of about 5.6, which is slightly acidic. The term pH is a measure of acidity or alkalinity and ranges from 0 to 14. A pH measurement of 7 is regarded as neutral. Measurements below 7 indicate increased acidity, while those above indicate increased alkalinity.

Acquisition (foreign crude oil): All transfers of ownership of foreign crude oil to a firm, irrespective of the terms of that transfer. Acquisitions thus include all purchases and exchange receipts as well as any and all foreign crude acquired under reciprocal buy-sell agreements or acquired as a result of a buy-back or other preferential agreement with a host government.

Acquisition (minerals): The procurement of the legal right to explore for and produce discovered minerals, if any, within a specific area; that legal right may be obtained by mineral lease, concession, or purchase of land and mineral rights or of mineral rights alone.

Acquisition costs, mineral rights: Direct and indirect costs incurred to acquire legal rights to extract natural resources. Direct costs include costs incurred to obtain options to lease or purchase mineral rights and costs incurred for the actual leasing (*e.g.*, lease bonuses) or purchasing of the rights. Indirect costs include such costs as brokers' commissions and expenses; abstract and recording fees; filing and patenting fees; and costs for legal examination of title and documents.

Acre-foot: The volume of water that will cover an area of 1 acre to a depth of 1 foot.

Acreage: An area, measured in acres, that is subject to ownership or control by those holding total or fractional shares of working interests. Acreage is considered developed when development has been completed. A distinction may be made between "gross" acreage and "net" acreage:

1. Gross–All acreage covered by any working interest, regardless of the percentage of ownership in the interest.
2. Net–Gross acreage adjusted to reflect the percentage of ownership in the working interest in the acreage.

Active solar: As an energy source, energy from the sun collected and stored using mechanical pumps or fans to circulate heat-laden fluids or air between solar collectors and a building.

Actual peak reduction: The actual reduction in annual peak load (measured in kilowatts) achieved by customers that participate in a utility demand-side management (DSM) program. It reflects the changes in the demand for electricity resulting from a utility DSM program that is in effect at the same time the utility experiences its annual peak load, as opposed to the installed peak load reduction capability (*i.e.*, potential peak reduction). It should account for the regular cycling of energy efficient units during the period of annual peak load.

Adjustable speed drives: Drives that save energy by ensuring the motor's speed is properly matched to the load placed on the motor. Terms used to describe this category include poly phase motors, motor over sizing, and motor rewinding.

Adjusted electricity: A measurement of electricity that includes the approximate amount of energy used to generate electricity. To approximate the adjusted amount of electricity, the site-value of the electricity is multiplied by a factor of 3. This conversion factor of 3 is a rough approximation of the Btu value of raw fuels used to generate electricity in a steam-generation power plant.

Adjustment bid: A bid auction conducted by the independent system operator or power exchange to redirect supply or demand of electricity when congestion is anticipated.

Administrative and general expenses: Expenses of an electric utility relating to the overall directions of its corporate offices and administrative affairs, as contrasted with expenses incurred for specialized functions. Examples include office salaries, office supplies, advertising, and other general expenses.

Advance royalty: A royalty required to be paid in advance of production from a mineral property that may or may not be recoverable from future production.

Advances from municipality: The amount of loans and advances made by the municipality or its other departments to the utility department when such loans and advances are subject to repayment but not subject to current settlement.

Advances to municipality: The amount of loans and advances made by the utility department to the municipality or its other departments when such loans or advances are subject to current settlement.

Adverse water conditions: Reduced stream flow, lack of rain in the drainage basin, or low water supply behind a pondage or reservoir dam resulting in a reduced gross head that limits the production of hydroelectric power or forces restrictions to be placed on multipurpose reservoirs or other water uses.

Affiliate: An entity which is directly or indirectly owned, operated, or controlled by another entity.

Afforestation: Planting of new forests on lands that have not been recently forested.

Aftermarket converted vehicle: A standard conventionally fueled, factory-produced vehicle to which equipment has been added that enables the vehicle to operate on alternative fuel.

Aftermarket vehicle converter: An organization or individual that modifies OEM vehicles after first use or sale to operate on a different fuel (or fuels).

Agglomerating character: Agglomeration describes the caking properties of coal. Agglomerating character is determined by examination and testing of the residue when a small powdered sample is heated to 950 degrees Centigrade under specific conditions. If the sample is "agglomerating," the residue will be coherent, show swelling or cell structure, and be capable of supporting a 500-gram weight without pulverizing.

Aggregate ratio: The ratio of two population aggregates (totals). For example, the aggregate expenditures per household is the ratio of the total expenditures in each category to the total number of households in the category.

Aggregator: Any marketer, broker, public agency, city, county, or special district that combines the loads of multiple end-use customers in negotiating the purchase of electricity, the transmission of electricity, and other related services for these customers.

Agriculture: An energy-consuming sub-sector of the industrial sector that consists of all facilities and equipment engaged in growing crops and raising animals.

Agriculture, mining, and construction (consumer category): Companies engaged in agriculture, mining (other than coal mining), or construction industries.

Air cleaner: A device using filters or electrostatic precipitators to remove indoor-air pollutants such as tobacco smoke, dust, and pollen. Most portable units are 40 watts when operated on low speed and 100 watts on high speed.

Air collector: A medium-temperature collector used predominantly in space heating, utilizing pumped air as the heat-transfer medium.

Air conditioning: Cooling and dehumidifying the air in an enclosed space by use of a refrigeration unit powered by electricity or natural gas.

Air conditioning intensity: The ratio of air-conditioning consumption or expenditures to square footage of cooled floor space and cooling degree-days (base 65 degrees F). This intensity provides a way of comparing different types of housing units and households by controlling for differences in housing unit size and weather conditions. The square footage of cooled floor space is equal to the product of the total square footage times the ratio of the number of rooms that could be cooled to the total number of rooms. If the entire housing unit is cooled, the cooled floor space is the same as the total floor space. The ratio is calculated on a weighted, aggregate basis according to this formula:

Air Conditioning Intensity = Btu for Air Conditioning/(Cooled Square Feet * Cooling Degree-Days)

Air pollution abatement equipment: Equipment used to reduce or eliminate airborne pollutants, including particulate matter (dust, smoke, fly, ash, dirt, etc.), sulfur oxides, nitrogen oxides (NO_x), carbon monoxide, hydrocarbons, odors, and other pollutants. Examples of air pollution abatement structures and equipment include flue-gas particulate collectors, flue-gas desulfurization units and nitrogen oxide control devices.

Alcohol: The family name of a group of organic chemical compounds composed of carbon, hydrogen, and oxygen. The series of molecules vary in chain length and are composed of a hydrocarbon plus a hydroxyl group; $CH(3)\text{-}(CH(2))_n\text{-}OH$ (*e.g.*, methanol, ethanol, and tertiary butyl alcohol).

Alkylate: The product of an alkylation reaction. It usually refers to the high-octane product from alkylation units. This alkylate is used in blending high octane gasoline.

Alkylation: A refining process for chemically combining isobutane with olefin hydrocarbons (*e.g.*, propylene, butylene) through the control of temperature and pressure in the presence of an acid catalyst, usually sulfuric acid or hydrofluoric acid. The product, alkylate, an isoparaffin, has high octane value and is blended with motor and aviation gasoline to improve the antiknock value of the fuel.

All-electric home: A residence in which electricity is used for the main source of energy for space heating, water heating, and cooking. Other fuels may be used for supplementary heating or other purposes.

Alternate energy source for primary heater: The fuel that would be used in place of the usual main heating fuel if the building had to switch fuels.

Alternating current (AC): An electric current that reverses its direction at regularly recurring intervals.

Alternative fuel: Alternative fuels, for transportation applications, include the following:

1. methanol
2. denatured ethanol, and other alcohol
3. fuel mixtures containing 85 percent or more by volume of methanol, denatured ethanol, and other alcohol with gasoline or other fuels–natural gas
4. liquefied petroleum gas (propane)
5. hydrogen
6. coal-derived liquid fuels
7. fuels (other than alcohol) derived from biological materials (biofuels such as soy diesel fuel)
8. electricity (including electricity from solar energy.)

"any other fuel the Secretary determines, by rule, is substantially not petroleum and would yield substantial energy security benefits and substantial environmental benefits." The term "alternative fuel" does not include alcohol or other blended portions of primarily petroleum-based fuels used as oxygenates or extenders, *i.e.* MTBE, ETBE, other ethers, and the 10-percent ethanol portion of gasohol.

Alternative-fuel vehicle (AFV): A vehicle designed to operate on an alternative fuel (*e.g.*, compressed natural gas, methane blend, electricity). The vehicle could be either a dedicated vehicle designed to operate exclusively on alternative fuel or a non-dedicated vehicle designed to operate on alternative fuel and/or a traditional fuel.

Alternative fuel vehicle converter: An organization (including companies, government agencies and utilities), or individual that performs conversions involving alternative alternative fuel vehicles. An AFV converter can convert (1) conventionally fueled vehicles to AFVs, (2) AFVs to conventionally fueled vehicles, or (3) AFVs to use another alternative fuel.

Alternative-rate DSM program assistance: A DSM (demand-side management) program assistance that offers special rate structures or discounts on the consumer's monthly electric bill in exchange for participation in DSM programs aimed at cutting peak demands or changing load shape. These rates are intended to reduce consumer bills and shift hours of operation of equipment from on-peak to off-peak periods through the application of time-differentiated rates. For example, utilities often pay consumers several dollars a month (refund on their monthly electric bill) for participation in a load control program. Large commercial and industrial customers sometimes obtain interruptible rates, which provide a discount in return for the consumer's agreement to cut electric loads upon request from the utility (usually during critical periods, such as summer afternoons when the system demand approaches the utility's generating capability).

Amorphous silicon: An alloy of silica and hydrogen, with a disordered, noncrystalline internal atomic arrangement, that can be deposited in thin-film layers (a few micrometers in thickness) by a number of deposition methods to produce thin-film photovoltaic cells on glass, metal, or plastic substrates.

Amortization: The depreciation, depletion, or charge-off to expense of intangible and tangible assets over a period of time. In the extractive industries, the term is most frequently applied to mean either (1) the periodic charge-off to expense of the costs associated with non-producing mineral properties incurred prior to the time when they are developed and entered into production or (2) the systematic charge-off to expense of those costs of productive mineral properties (including tangible and intangible costs of prospecting, acquisition, exploration, and development) that had been initially capitalized (or deferred) prior to the time the properties entered into production, and thereafter are charged off as minerals are produced.

Ampere: The unit of measurement of electrical current produced in a circuit by 1 volt acting through a resistance of 1 Ohm.

Ancillary services: Services that ensure reliability and support the transmission of electricity from generation sites to customer loads. Such services may include: load regulation, spinning reserve, non-spinning reserve, replacement reserve, and voltage support.

Annual operating factor: The annual fuel consumption divided by the product of design firing rate and hours of operation per year.

Annual requirement: The reporting company's best estimate of the annual requirement for natural gas to make direct sales or sales for resale under certificate authorizations and for company use and unaccounted-for gas during the year next following the current report year.

ANSI assembly identifier: The serial numbering scheme adopted by the American National Standards Institute (ANSI) to ensure uniqueness of an assembly serial number.

Anthracite: The highest rank of coal; used primarily for residential and commercial space heating. It is a hard, brittle, and black lustrous coal, often referred to as hard coal, containing a high percentage of fixed carbon and a low percentage of volatile matter. The moisture content of fresh-mined anthracite generally is less than 15 percent. The heat content of anthracite ranges from 22 to 28 million Btu per ton on a moist, mineral-matter-free basis. The heat content of anthracite coal consumed in the United States averages 25 million Btu per ton, on the as-received basis (*i.e.*, containing both inherent moisture and mineral matter).

Anthropogenic: Made or generated by a human or caused by human activity. The term is used in the context of global climate change to refer to gaseous emissions that are the result of human activities, as well as other potentially climate-altering activities, such as deforestation.

API: The American Petroleum Institute, a trade association.

API gravity: American Petroleum Institute measure of specific gravity of crude oil or condensate in degrees. An arbitrary scale expressing the gravity or density of liquid petroleum products. The measuring scale is calibrated in terms of degrees API; it is calculated as follows:

Degrees API = (141.5 / sp.gr.60 deg.F/60 deg.F)–131.5

Apparent consumption, (coal): Coal production plus imports of coal, coke, and briquets minus exports of coal, coke, and briquets plus or minus stock changes.

Apparent consumption, natural gas (international): The total of an individual nation's dry natural gas production plus imports less exports.

Apparent consumption, petroleum (international): Consumption that includes internal consumption, refinery fuel and loss, and bunkering. For countries in the Organization for Economic Cooperation and Development (OECD), apparent consumption is derived from refined product output plus refined product imports minus refined product exports plus refined product stock changes plus other oil consumption (such as direct use of crude oil). For countries outside the OECD, apparent consumption is either a reported figure or is derived from refined product output plus refined product imports minus refined product exports, with stock levels assumed to remain the same. Apparent consumption also includes, where available, liquefied petroleum gases sold directly from natural gas processing plants for fuel or chemical uses.

Appliance: A piece of equipment, commonly powered by electricity, used to perform a particular energy-driven function. Examples of common appliances are refrigerators, clothes washers and dishwashers, conventional ranges/ovens and microwave ovens, humidifiers and dehumidifiers, toasters, radios, and televisions.

Appliance efficiency index: A relative comparison of trends in new-model efficiencies for major appliances and energy-using equipment. The base year for relative comparisons was 1972 (1972=100). Efficiencies for each year were efficiencies of different model types that were weighted by their market shares.

Appliance efficiency standards: The National Appliance Energy Conservation Act of 1987 established minimum efficiency standards for major home appliances, including furnaces, central and room air conditioners, refrigerators, freezers, water heaters, dishwashers, and heat pumps. Most of the standards took effect in 1990. The standards for clothes washers, dishwashers, and ranges took effect in 1988, because they required only minor changes in product design, such as eliminating pilot lights and requiring cold water rinse options. The standards for central air conditioners and furnaces took effect in 1992, because it took longer to redesign these products. Appliance efficiency standards for refrigerators took effect in 1993.

Aromatics: Hydrocarbons characterized by unsaturated ring structures of carbon atoms. Commercial petroleum aromatics are benzene, toluene, and xylene (BTX).

Asbestos: A group of naturally occurring minerals that separate into long, thin fibers. Asbestos was used for many years to insulate and fireproof buildings. In the 1989 CBECS, information on asbestos in buildings was collected (Section R of the Buildings Questionnaire) for the U.S. Environmental Protection Agency (EPA) Asbestos treatment methods include removal, encapsulation or sealing, and enclosure behind a permanent barrier.

Ash: Impurities consisting of silica, iron, alumina, and other noncombustible matter that are contained in coal. Ash increases the weight of coal, adds to the cost of handling, and can affect its burning characteristics Ash content is measured as a percent by weight of coal on an "as received" or a "dry" (moisture-free, usually part of a laboratory analysis) basis.

Asphalt: A dark brown-to-black cement-like material obtained by petroleum processing and containing bitumens as the predominant component; used primarily for road construction. It includes crude asphalt as well as the following finished products: cements, fluxes, the asphalt content of emulsions (exclusive of water), and petroleum distillates blended with asphalt to make cutback asphalts.

Assembly identifier: A unique string of alphanumeric characters that identifies an assembly, bundle, or canister for a specific reactor in which it has been irradiated.

Assembly type: Each assembly is characterized by a fabricator, rod-array size, and model type. An eight-digit assembly type code is assigned to each assembly type based on certain distinguishing characteristics, such as the number of rods per assembly, fuel rod diameter, cladding type, materials used in fabrication, and other design features.

Assessment work: The annual or biennial work performed on a mining claim (or claims), after claim location and before patent, to benefit or develop the claim and to protect it from relocation by third parties.

Assistance for heating in winter: Assistance from the Low-Income Home Energy Assistance Program (LIHEAP). The purpose of LIHEAP is to assist eligible households to meet the costs of home energy, *i.e.*, a source of heating or cooling residential buildings.

Assistance for weatherization of residence: The household received services free, or at a reduced cost, from the Federal, State, or local Government. Any of the following services could have been received:

1. Insulation in the attic, outside wall, or basement/crawlspace below the floor of the house
2. Insulation around the hot water heater
3. Repair of broken windows or doors to keep out the cold or hot weather
4. Weather stripping or caulking around any windows or doors to the outside

5. Storm doors or windows added
6. Repair of broken furnace
7. Furnace tune up and/or modifications
8. Other home energy-saving devices.

Associated-dissolved natural gas: Natural gas that occurs in crude oil reservoirs either as free gas (associated) or as gas in solution with crude oil (dissolved gas).

Atmospheric crude oil distillation: The refining process of separating crude oil components at atmospheric pressure by heating to temperatures of about 600 degrees to 750 degrees Fahrenheit (depending on the nature of the crude oil and desired products) and subsequent condensing of the fractions by cooling.

Auger mine: A surface mine in which the coal bed is removed by means of a large diameter drill. Usually operated only when the overburden becomes too thick for economical strip mining.

Authorized cash distribution to municipality: The authorized cash distributions to the municipality from the earned surplus of the utility department.

Automatic set-back or clock thermostat: A thermostat that can be set to turn the heating/cooling system off and on at certain predetermined times.

Automobile and truck classifications: Vehicle classifications for automobiles and light duty trucks were obtained from the EPA (Environmental Protection Agency) mileage guide book. Almost every year there are small changes in the classifications, therefore the categories will change accordingly. The EPA mileage guide can be found at any new car dealership.

Available but not needed capability: Net capability of main generating units that are operable but not considered necessary to carry load and cannot be connected to load within 30 minutes.

Average daily production: The ratio of the total production at a mining operation to the total number of production days worked at the operation.

Average delivered price: The weighted average of all contract price commitments and market price settlements in a delivery year.

Average household energy expenditures: A ratio estimate defined as the total household energy expenditures divided by the total number of households.

Average mine price: The ratio of the total value of the coal produced at the mine to the total production tonnage.

Average production per miner per day: The product of the average production per miner per hour at a mining operation and the average length of a production shift at the operation.

Average production per miner per hour: The ratio of the total production at a mining operation to the total direct labor hours worked at the operation.

Average revenue per kilo watt hour: The average revenue per kilo watt hour of electricity sold by sector (residential, commercial, industrial, or other) and geographic area (State, Census division, and national) is calculated by dividing the total monthly revenue by the corresponding total monthly sales for each sector and geographic area.

Average stream flow: The rate, usually expressed in cubic feet per second, at which water passes a given point in a stream over a set period of time.

Average vehicle fuel consumption: A ratio estimate defined as total gallons of fuel consumed by all vehicles divided by: (1) the total number of vehicles (for average fuel consumption per vehicle) or (2) the total number of households (for average fuel consumption per household).

Average vehicle miles traveled: A ratio estimate defined as total miles traveled by all vehicles, divided by: (1) the total number of vehicles (for average miles traveled per vehicle) or (2) the total number of households (for average miles traveled per household).

Average water conditions: The amount and distribution of precipitation within a drainage basin and the run off conditions present as determined by reviewing the area water supply records over a long period of time.

Aviation gasoline (finished): A complex mixture of relatively volatile hydrocarbons with or without small quantities of additives, blended to form a fuel suitable for use in aviation reciprocating engines.

Aviation gasoline blending components: Naphthas that will be used for blending or compounding into finished aviation gasoline (*e.g.*, straight run gasoline, alkylate, reformate, benzene, toluene, and xylene). Excludes oxygenates (alcohols, ethers), butane, and pentanes plus. Oxygenates are reported as other hydrocarbons, hydrogen, and oxygenates.

Backup fuel: In a central heat pump system, the fuel used in the furnace that takes over the space heating when the outdoor temperature drops below that which is feasible to operate a heat pump.

Backup power: Electric energy supplied by a utility to replace power and energy lost during an unscheduled equipment outage.

Balancing item: Represents differences between the sum of the components of natural gas supply and the sum of the components of natural gas disposition. These differences may be due to quantities lost or to the effects of data reporting problems. Reporting problems include differences due to the net result of conversions of flow data metered at varying temperature and pressure bases and converted to a standard temperature and pressure base; the effect of variations in company accounting and billing practices; differences between billing cycle and calendar period time frames; and imbalances resulting from the merger of data reporting systems that vary in scope, format, definitions, and type of respondents.

Barrel: A unit of volume equal to 42 U.S. gallons.

Barrels per Calendar day: The amount of input that a distillation facility can process under usual operating conditions. The amount is expressed in terms of capacity during a 24-hour period and reduces the maximum processing capability of all units at the facility under continuous operation to account for the following limitations that may delay, interrupt, or slow down production.

1. the capability of downstream processing units to absorb the output of crude oil processing facilities of a given refinery. No reduction is necessary for intermediate streams that are distributed to other than downstream facilities as part of a refinery's normal operation;
2. the types and grades of inputs to be processed;
3. the types and grades of products expected to be manufactured;
4. the environmental constraints associated with refinery operations;
5. the reduction of capacity for scheduled downtime due to such conditions as routine inspection, maintenance, repairs, and turnaround; and

6. the reduction of capacity for unscheduled downtime due to such conditions as mechanical problems, repairs, and slowdowns.

Barrels per Stream day: The maximum number of barrels of input that a distillation facility can process within a 24-hour period when running at full capacity under optimal crude and product slate conditions with no allowance for downtime.

Base (cushion) gas: The volume of gas needed as a permanent inventory to maintain adequate reservoir pressures and deliverability rates throughout the withdrawal season. All native gas is included in the base gas volume.

Base bill: A charge calculated by taking the rate from the appropriate electric rate schedule and applying it to the level of consumption.

Base load: The minimum amount of electric power delivered or required over a given period of time at a steady rate.

Base load capacity: The generating equipment normally operated to serve loads on an around-the-clock basis.

Base load plant: A plant, usually housing high-efficiency steam-electric units, which is normally operated to take all or part of the minimum load of a system, and which consequently produces electricity at an essentially constant rate and runs continuously. These units are operated to maximize system mechanical and thermal efficiency and minimize system operating costs.

Base period: The period of time for which data used as the base of an index number, or other ratio, have been collected. This period is frequently one of a year but it may be as short as one day or as long as the average of a group of years. The length of the base period is governed by the nature of the material under review, the purpose for which the index number (or ratio) is being compiled, and the desire to use a period as free as possible from abnormal influences in order to avoid bias.

Base rate: A fixed kilo watt hour charge for electricity consumed that is independent of other charges and/or adjustments.

Baseboard heater: As a type of heating equipment, a system in which either electric resistance coils or finned tubes carrying steam or hot water are mounted behind shallow panels along baseboards. Baseboards rely on passive convection to distribute heated air in the space. Electric baseboards are an example of an "Individual Space Heater."

bbl: The abbreviation for barrel(s).

bbl/d: The abbreviation for barrel(s) per day.

bbl/sd: The abbreviation for barrel(s) per stream day

bcf: The abbreviation for billion cubic feet.

Benzene (C_6H_6): An aromatic hydrocarbon present in small proportion in some crude oils and made commercially from petroleum by the catalytic reforming of naphthenes in petroleum naphtha. Also made from coal in the manufacture of coke. Used as a solvent in the manufacture of detergents, synthetic fibers, petrochemicals, and as a component of high-octane gasoline.

Bi-fuel vehicle: A motor vehicle that operates on two different fuels, but not on a mixture of the fuels. Each fuel is stored in a separate tank.

Bilateral agreement: A written statement signed by two parties that specifies the terms for exchanging energy.

Bilateral electricity contract: A direct contract between an electric power producer and either a user or broker outside of a centralized power pool or power exchange.

Billing period: The time between meter readings. It does not refer to the time when the bill was sent or when the payment was to have been received. In some cases, the billing period is the same as the billing cycle that corresponds closely (within several days) to meter-reading dates. For fuel oil and LPG, the billing period is the number of days between fuel deliveries.

Biodiesel: Any liquid biofuel suitable as a diesel fuel substitute or diesel fuel additive or extender. Biodiesel fuels are typically made from oils such as soybeans, rapeseed, or sunflowers, or from animal tallow. Biodiesel can also be made from hydrocarbons derived from agricultural products such as rice hulls.

Biofuels: Liquid fuels and blending components produced from biomass (plant) feedstocks, used primarily for transportation.

Biomass: Organic non-fossil material of biological origin constituting a renewable energy source.

Biomass gas: A medium Btu gas containing methane and carbon dioxide, resulting from the action of microorganisms on organic materials such as a landfill.

Bitumen: A naturally occurring viscous mixture, mainly of hydrocarbons heavier than pentane, that may contain sulphur compounds and that, in its natural occurring viscous state, is not recoverable at a commercial rate through a well.

Bituminous coal: A dense coal, usually black, sometimes dark brown, often with well-defined bands of bright and dull material, used primarily as fuel in steam-electric power generation, with substantial quantities also used for heat and power applications in manufacturing and to make coke. Bituminous coal is the most abundant coal in active U.S. mining regions. Its moisture content usually is less than 20 percent. The heat content of bituminous coal ranges from 21 to 30 million Btu per ton on a moist, mineral-matter-free basis. The heat content of bituminous coal consumed in the United States averages 24 million Btu per ton, on the as-received basis (*i.e.*, containing both inherent moisture and mineral matter).

Black lung benefits: In the content of the coal operation statement of income, this term refers to all payments, including taxes, made by the company attributable to Black Lung.

Blast furnace: A furnace in which solid fuel (coke) is burned with an air blast to smelt ore.

Blast-furnace gas: The waste combustible gas generated in a blast furnace when iron ore is being reduced with coke to metallic iron. It is commonly used as a fuel within steel works.

Blending plant: A facility that has no refining capability but is either capable of producing finished motor gasoline through mechanical blending or blends oxygenates with motor gasoline.

Block-rate structure: An electric rate schedule with a provision for charging a different unit cost for various increasing blocks of demand for energy. A reduced rate may be charged on succeeding blocks.

BOE: The abbreviation for barrels of oil equivalent (used internationally).

Boiler: A device for generating steam for power, processing, or heating purposes; or hot water for heating purposes or hot water supply. Heat from an external combustion source is transmitted to a fluid contained within the tubes found in the boiler shell. This fluid is delivered to an end-use at a desired pressure, temperature, and quality.

Boiler fuel: An energy source to produce heat that is transferred to the boiler vessel in order to generate steam or hot water. Fossil fuel is the primary energy source used to produce heat for boilers.

Boiling-water reactor (BWR): A light-water reactor in which water, used as both coolant and moderator, is allowed to boil in the core. The resulting steam can be used directly to drive a turbine.

Bonded petroleum imports: Petroleum imported and entered into Customs bonded storage. These imports are not included in the import statistics until they are: (1) withdrawn from storage free of duty for use as fuel for vessels and aircraft engaged in international trade; or (2) withdrawn from storage with duty paid for domestic use.

Bone coal: Coal with a high ash content; it is dull in appearance, hard, and compact.

Book value: The portion of the carrying value (other than the portion associated with tangible assets) prorated in each accounting period, for financial reporting purposes, to the extracted portion of an economic interest in a wasting natural resource.

Booked costs: Costs allocated or assigned to inter-departmental or intracompany transactions, such as on-system or synthetic natural gas (SNG) production and company-owned gas used in gas operations and recorded in company books or records for accounting and / or regulatory purposes.

Borderline customer: A customer located in the service area of one utility, but supplied by a neighboring utility through an arrangement between the utilities.

Bottled gas, LPG, or propane: Any fuel gas supplied to a building in liquid form, such as liquefied petroleum gas, propane, or butane. It is usually delivered by tank truck and stored near the building in a tank or cylinder until used.

Bottom ash: Residue mainly from the coal burning process that falls to the bottom of the boiler for removal and disposal.

Bottom-hole contribution: A payment (either in cash or in acreage) that is required by agreement when a test well is drilled to a specified depth regardless of the outcome of the well and that is made in exchange for well and evaluation data.

bp: The abbreviation for boiling point.

Branded product: A refined petroleum product sold by a refiner with the understanding that the purchaser has the right to resell the product under a trademark, trade name, service mark, or other identifying symbol or names owned by such refiner.

Break-even cutoff grade: The lowest grade of material that can be mined and processed considering all applicable costs, without incurring a loss or gaining a profit.

Breccia: A coarse-grained clastic rock, composed of angular broken rock fragments held together by a mineral cement or in a fine-grained matrix.

Breeder reactor: A reactor that both produces and consumes fissionable fuel, especially one that creates more fuel than it consumes. The new fissionable material is created by a process known as breeding, in which neutrons from fission are captured in fertile materials.

Breeze: The fine screenings from crushed coke. Usually breeze will pass through a 1/2-inch or 3/4-inch screen opening. It is most often used as a fuel source in the process of agglomerating iron ore.

British thermal unit: The quantity of heat required to raise the temperature of 1 pound of liquid water by 1 degree Fahrenheit at the temperature at which water has its greatest density (approximately 39 degrees Fahrenheit).

Btu: The abbreviation for British thermal unit(s).

Btu conversion factors: Btu conversion factors for site energy are as follows:

Electricity	3,412 Btu/kilowatthour
Natural Gas	1,031 Btu/cubic foot
Fuel Oil No.1	135,000 Btu/gallon
Kerosene	135,000 Btu/gallon
Fuel Oil No.2	138,690 Btu/gallon
LPG (Propane)	91,330 Btu/gallon
Wood	20 million Btu/cord

Btu per cubic foot: The total heating value, expressed in Btu, produced by the combustion, at constant pressure, of the amount of the gas that would occupy a volume of 1 cubic foot at a temperature of 60 degrees F if saturated with water vapor and under a pressure equivalent to that of 30 inches of mercury at 32 degrees F and under standard gravitational force (980.665 cm. per sec. squared) with air of the same temperature and pressure as the gas, when the products of combustion are cooled to the initial temperature of gas and air when the water formed by combustion is condensed to the liquid state. (Sometimes called gross heating value or total heating value.)

BTX: The acronym for the commercial petroleum aromatics–benzene, toluene, and xylene.

Budget plan: An agreement between the household and the utility company or fuel supplier that allows the household to pay the same amount for fuel for each month for a number of months.

Building shell conservation feature: A building feature designed to reduce energy loss or gain through the shell or envelope of the building. Data collected on the following specific building shell energy conservation features: roof, ceiling, or wall insulation; storm windows or double- or triple-paned glass (multiple glazing); tinted or reflective glass or shading films; exterior or interior shadings or awnings; and weather stripping or caulking.

Building shell (envelope) DSM program: A DSM program that promotes reduction of energy consumption through improvements to the building envelope. Includes installations of insulation, weatherstripping, caulking, window film, and window replacement.

Built-in electric units: An individual-resistance electric-heating unit that is permanently installed in the floors, walls, ceilings, or baseboards and is part of the electrical installation of the building. Electric-heating devices that are plugged into an electric socket or outlet are not considered built in.

Bulk power transactions: The wholesale sale, purchase, and interchange of electricity among electric utilities. Bulk power transactions are used by electric utilities for many different aspects of electric utility operations, from maintaining load to reducing costs.

Bulk sales: Wholesale sales of gasoline in individual transactions which exceed the size of a truckload.

Bulk station: A facility used primarily for the storage and/or marketing of petroleum products, which has a total bulk storage capacity of less than 50,000 barrels and receives its petroleum products by tank car or truck.

Bulk terminal: A facility used primarily for the storage and/or marketing of petroleum products, which has a total bulk storage capacity of 50,000 barrels or more and/or receives petroleum products by tanker, barge, or pipeline.

Bundled utility service (electric): A means of operation whereby energy, transmission, and distribution services, as well as ancillary and retail services, are provided by one entity.

Bunker fuels: Fuel supplied to ships and aircraft, both domestic and foreign, consisting primarily of residual and distillate fuel oil for ships and kerosene-based jet fuel for aircraft. The term "international bunker fuels" is used to denote the consumption of fuel for international transport activities.

Burn days: The number of days the station could continue to operate by burning coal already on hand assuming no additional deliveries of coal and an average consumption rate.

Burnup: Amount of thermal energy generated per unit mass of fuel, expressed as Gigawatt-Days Thermal per Metric Ton of Initial Heavy Metal (GWDT/MTIHM), rounded to the nearest gigawatt day.

Bus: An electrical conductor that serves as a common connection for two or more electrical circuits.

Butane (C_4H_{10}): A normally gaseous straight-chain or branch-chain hydrocarbon extracted from hydrocarbon extracted from natural gas or refinery gas streams. It includes isobutane and normal butane and is designated in ASTM Specification D1835 and Gas Processors Association Specifications for commercial butane.

Butylene (C_4H_8): An olefinic hydrocarbon recovered from refinery processes.

Buy-back oil: Crude oil acquired from a host government whereby a portion of the government's ownership interest in the crude oil produced in that country may or should be purchased by the producing firm.

Bypassed footage: Bypassed footage is the footage in that section of hole that is abandoned as the result of remedial sidetrack drilling operations.

Byproduct: A secondary or additional product resulting from the feedstock use of energy or the processing of nonenergy materials. For example, the more common byproducts of coke ovens are coal gas, tar, and a mixture of benzene, toluene, and xylenes (BTX).

C_4H: A mixture of light hydrocarbons that have the general formula C_4H_n, where **n** is the number of hydrogen atoms per molecule. Examples include butane (C_4H_{10}) and butylene (C_4H_8).

Calcination: A process in which a material is heated to a high temperature without fusing, so that hydrates, carbonates, or other compounds are decomposed and the volatile material is expelled.

Calcium sulfate: A white crystalline salt, insoluble in water. Used in Keene's cement, in pigments, as a paper filler, and as a drying agent.

Calcium sulfite: A white powder, soluble in diluted sulfuric acid. Used in the sulfite process for the manufacture of wood pulp.

California Power Exchange: A State-chartered, non-profit corporation which provides day-ahead and hour-ahead markets for energy and ancillary services in accordance with the power exchange tariff. The power exchange is a scheduling coordinator and is independent of both the independent system operator and all other market participants.

Canadian deuterium uranium reactor (CANDU): Uses heavy water or deuterium oxide (D_2O), rather than light water (H_2O), as the coolant and moderator. Deuterium is an isotope of hydrogen that has a different neutron absorption spectrum from that of ordinary hydrogen. In a deuterium-moderated-reactor, fuel made from natural uranium (0.71 U-235) can sustain a chain reaction.

Cannel coal: A compact, tough variety of coal, originating from organic spore residues, that is noncaking, contains a high percentage of volatile matter, ignites easily, and burns with a luminous smoky flame.

Capable of being fueled: A vehicle is capable of being fueled by a particular fuel(s) if that vehicle has the engine components in place to make operation possible on the fuel(s). The vehicle does not necessarily have to run on the fuel(s) in order for that vehicle to be considered capable of being fueled by the fuel(s). For example, a vehicle that is equipped to operate on either gasoline or natural gas but normally operates on gasoline is considered to be capable of being fueled by gasoline and natural gas.

Capacity (purchased): The amount of energy and capacity available for purchase from outside the system.

Capacity charge: An element in a two-part pricing method used in capacity transactions (energy charge is the other element). The capacity charge, sometimes called *Demand Charge,* is assessed on the amount of capacity being purchased.

Capacity factor: The ratio of the electrical energy produced by a generating unit for the period of time considered to the electrical energy that could have been produced at continuous full power operation during the same period.

Capacity transaction: The acquisition of a specified quantity of generating capacity from another utility for a specified period of time. The utility selling the power is obligated to make available to the buyer a specified quantity of power.

Capacity utilization: Capacity utilization is computed by dividing production by productive capacity and multiplying by 100.

Capital cost: The cost of field development and plant construction and the equipment required for industry operations.

Capital stock: Property, plant and equipment used in the production, processing and distribution of energy resources.

Captive coal: Coal produced to satisfy the needs of the mine owner, or of a parent, subsidiary, or other affiliate of the mine owner (for example, steel companies and electricity generators), rather than for open market sale.

Captive refinery MTBE plants: MTBE (methyl tertiary butyl ether) production facilities primarily located within refineries. These integrated refinery units produce MTBE from Fluid Cat Cracker isobutylene with production dedicated to internal gasoline blending requirements.

Captive refinery oxygenate plants: Oxygenate production facilities located within or adjacent to a refinery complex.

Carbon black: An amorphous form of carbon, produced commercially by thermal or oxidative decomposition of hydrocarbons and used principally in rubber goods, pigments, and printer's ink.

Carbon budget: The balance of the exchanges (incomes and losses) of carbon between carbon sinks (*e.g.*, atmosphere and biosphere) in the carbon cycle.

Carbon cycle: All carbon sinks and exchanges of carbon from one sink to another by various chemical, physical, geological, and biological processes.

Carbon dioxide (CO_2): A colorless, odorless, non-poisonous gas that is a normal part of Earth's atmosphere. Carbon dioxide is a product of fossil-fuel combustion as well as other processes. It is considered a greenhouse gas as it traps heat (infrared energy) radiated by the Earth into the atmosphere and thereby contributes to the potential for global warming. The global warming potential (GWP) of other greenhouse gases is measured in relation to that of carbon dioxide, which by international scientific convention is assigned a value of one (1).

Carbon dioxide equivalent: The amount of carbon dioxide by weight emitted into the atmosphere that would produce the same estimated radiative forcing as a given weight of another radiatively active gas. Carbon dioxide equivalents are computed by multiplying the weight of the gas being measured (for example, methane) by its estimated global warming potential (which is 21 for methane). "Carbon equivalent units" are defined as carbon dioxide equivalents multiplied by the carbon content of carbon dioxide (*i.e.*, 12/44).

Carbon intensity: The amount of carbon by weight emitted per unit of energy consumed. A common measure of carbon intensity is weight of carbon per British thermal unit (Btu) of energy. When there is only one fossil fuel under consideration, the carbon intensity and the emissions coefficient are identical. When there are several fuels, carbon intensity is based on their combined emissions coefficients weighted by their energy consumption levels. **Carbon output rate:** The amount of carbon by weight per kilowatthour of electricity produced.

Carbon sequestration: The fixation of atmospheric carbon dioxide in a carbon sink through biological or physical processes.

Carbon sink: A reservoir that absorbs or takes up released carbon from another part of the carbon cycle. The four sinks, which are regions of the Earth within which carbon behaves in a systematic manner, are the atmosphere, terrestrial biosphere (usually including freshwater systems), oceans, and sediments (including fossil fuels).

Carburetor: A fuel delivery device for producing a proper mixture of gasoline vapor and air and for delivering it to the intake manifold of an internal combustion engine. Gasoline is gravity-fed from a reservoir bowl into a throttle bore, where it is allowed to evaporate into the stream of air being inducted by the engine.

Carrying costs: Costs incurred in order to retain exploration and property rights after acquisition but before production has occurred. Such costs include legal costs for title defense, ad valorem taxes on nonproducing mineral properties, shut-in royalties, and delay rentals.

Cash and carry: Kerosene, fuel oil, or bottled gas (tank or propane) purchased with cash, by check, or by credit card and taken home by the purchaser. The purchaser provides the container or pays extra for the container.

Casinghead gas (or oil well gas): Natural gas produced along with crude oil from oil wells. It contains either dissolved or associated gas or both.

Cast silicon: Crystalline silicon obtained by pouring pure molten silicon into a vertical mold and adjusting the temperature gradient along the mold volume during cooling to obtain slow, vertically advancing crystallization of the silicon. The polycrystalline ingot thus formed is composed of large, relatively parallel, interlocking crystals. The cast ingots are sawed into wafers for further fabrication into photovoltaic cells. Cast silicon wafers and ribbon silicon sheets fabricated into cells are usually referred to as polycrystalline photovoltaic cells.

Catalyst coke: In many catalytic operations (*e.g.*, catalytic cracking), carbon is deposited on the catalyst, thus deactivating the catalyst. The catalyst is reactivated by burning off the carbon, which is used as a fuel in the refining process. This carbon or coke is not recoverable in a concentrated form.

Catalytic converter: A device containing a catalyst for converting automobile exhaust into mostly harmless products.

Catalytic cracking: The refining process of breaking down the larger, heavier, and more complex hydrocarbon molecules into simpler and lighter molecules. Catalytic cracking is accomplished by the use of a catalytic agent and is an effective process for increasing the yield of gasoline from crude oil. Catalytic cracking processes fresh feeds and recycled feeds.

Catalytic hydrocracking: A refining process that uses hydrogen and catalysts with relatively low temperatures and high pressures for converting middle boiling or residual material to high octane gasoline, reformer charge stock, jet fuel, and /or high grade fuel oil. The process uses one or more catalysts, depending on product output, and can handle high sulfur feedstocks without prior desulfurization.

Catalytic hydrotreating: A refining process for treating petroleum fractions from atmospheric or vacuum distillation units (*e.g.*, naphthas, middle distillates, reformer feeds, residual fuel oil, and heavy gas oil) and other petroleum (*e.g.*, cat cracked naphtha, coker naphtha, gas oil, etc.) in the presence of catalysts and substantial quantities of hydrogen. Hydrotreating includes desulfurization, removal of substances (*e.g.*, nitrogen compounds) that deactivate catalysts, conversion of olefins to paraffins to reduce gum formation in gasoline, and other processes to upgrade the quality of the fractions.

Catalytic reforming: A refining process using controlled heat and pressure with catalysts to rearrange certain hydrocarbon molecules, thereby converting paraffinic and naphthenic type hydrocarbons (*e.g.*, low octane gasoline boiling range fractions) into petrochemical feedstocks and higher octane stocks suitable for blending into finished gasoline. Catalytic reforming is reported in two categories. They are:

1. Low Pressure. A processing unit operating at less than 225 pounds per square inch gauge (PSIG) measured at the outlet separator.
2. High pressure. A processing unit operating at either equal to or greater than 225 pounds per square inch gauge (PSIG) measured at the outlet separator.

Cells: Refers to the un-encapsulated semi-conductor components of the module that convert the solar energy to electricity.

Cells to OEM (non-PV): Cells shipped to non-photovoltaic original equipment manufacturers such as boat manufacturers, car manufacturers, etc.

Census division: Any of nine geographic areas of the United States as defined by the U.S. Department of Commerce, Bureau of the Census. The divisions, each consisting of several States, are defined as follows:

1. New England: Connecticut, Maine, Massachusetts, New Hampshire, Rhode Island, and Vermont;
2. Middle Atlantic: New Jersey, New York, and Pennsylvania;
3. East North Central: Illinois, Indiana, Michigan, Ohio, and Wisconsin;
4. West North Central: Iowa, Kansas, Minnesota, Missouri, Nebraska, North Dakota, and South Dakota;
5. South Atlantic: Delaware, District of Columbia, Florida, Georgia, Maryland, North Carolina, South Carolina, Virginia, and West Virginia;
6. East South Central: Alabama, Kentucky, Mississippi, and Tennessee;
7. West South Central: Arkansas, Louisiana, Oklahoma, and Texas;
8. Mountain: Arizona, Colorado, Idaho, Montana, Nevada, New Mexico, Utah, and Wyoming;
9. Pacific: Alaska, California, Hawaii, Oregon, and Washington.

In some cases, the Pacific division is subdivided into the Pacific Contiguous area (California, Oregon, and Washington) and the Pacific Noncontiguous area (Alaska and Hawaii).

Census Region: Any of four geographic areas of the United States as defined by the U.S. Department of Commerce, Bureau of the Census. The Regions, each consisting of various States selected according to population size and physical location, are defined as follows:

1. Northeast: Connecticut, Maine, Massachusetts, New Hampshire, New Jersey, New York, Pennsylvania, Rhode Island, and Vermont.
2. South: Alabama, Arkansas, Delaware, District of Columbia, Florida, Georgia, Kentucky, Louisiana, Maryland, Mississippi, North Carolina, Oklahoma, South Carolina, Tennessee, Texas, Virginia, and West Virginia.
3. Midwest: Illinois, Indiana, Iowa, Kansas, Michigan, Minnesota, Missouri, Nebraska, North Dakota, Ohio, South Dakota, and Wisconsin.
4. West: Alaska, Arizona, California, Colorado, Hawaii, Idaho, Montana, Nevada, New Mexico, Oregon, Utah, Washington, and Wyoming.

Census Region/division: An hierarchical organization of the **United States** according to geographic areas and sub-areas as follows:

Northeast Region

New England division

Middle Atlantic division

South Region

South Atlantic division

East South Central division

West South Central division

Midwest Region

East North Central division

West North Central division

West Region

Mountain division

Pacific division

Central chiller: Any centrally located air conditioning system that produces chilled water in order to cool air. The chilled water or cold air is then distributed throughout the building, using pipes or air ducts or both. These systems are also commonly known as "chillers," "centrifugal chillers," "reciprocating chillers," or "absorption chillers." Chillers are generally located in or just outside the building they serve. Buildings receiving district chilled water are served by chillers located at central physical plants.

Central cooling: Cooling of an entire building with a refrigeration unit to condition the air. Typically central chillers and ductwork are present in the centrally cooled building.

Central physical plant: A plant owned by, and on the grounds of, a multibuilding facility that provides district heating, district cooling, or electricity to other buildings on the same facility. To qualify as a central plant it must provide district heat, district chilled water, or electricity to at least one other building. The central physical plant may be by itself in a separate building or may be located in a building where other activities occur.

Central warm air furnace: A type of space heating equipment where a central combustor or resistance unit generally using gas, fuel oil, or electricity provides warm air through ducts leading to the various rooms. Heat pumps are not included in this category. A forced air furnace is one in which a fan is used to force the air through the ducts. In a gravity furnace, air is circulated by gravity, relying on the natural flow of warm air up and cold air down; the warm air rises through ducts and the cold air falls through ducts that return it to the furnace to be reheated and this completes the circulation cycle.

Centralized water heating system: Equipment, to heat and store water for other than space heating purposes, which provides hot water from a single location for distribution throughout a building. A residential type tank water heater is a good example of a centralized water heater.

Certificate: A type of permit for public convenience and necessity issued by a utility commission, which authorizes a utility or regulated company to engage in business, construct facilities, provide some services, or abandon service.

Certificate requirement: The maximum annual volume allowed for sales to resale or direct sale customers under certificate authorizations by the Federal Energy Regulatory Commission.

Cesspool: An underground reservoir for liquid waste, typically household sewage.

cfs: Cubic feet per second.

Chained dollars: A measure used to express real prices. Real prices are those that have been adjusted to remove the effect of changes in the purchasing power of the dollar; they usually reflect buying power relative to a reference year. Prior to 1996, real prices were expressed in constant dollars, a measure based on the weights of goods and services in a single year, usually a recent year. In 1996, the U.S. Department of Commerce introduced the chained-dollar measure. The new measure is based on the average weights of goods and services in successive pairs of years. It is "chained" because the second year in each pair, with its weights, becomes the first year of the next pair. The advantage of using the chained-dollar measure is that it is more closely related to any given period covered and is therefore subject to less distortion over time.

Characterization: Sampling, monitoring, and analysis activities to determine the extent and nature of contamination at a facility or site. Characterization provides the necessary technical information to develop, screen, analyze, and select appropriate cleanup techniques.

Charge capacity: The input (feed) capacity of the refinery processing facilities.

Chemical separation: A process for extracting uranium and plutonium from dissolved spent nuclear fuel and irradiated targets. The fission products that are left behind are high-level waste. Chemical separation is also known as reprocessing.

Chlorofluorocarbon (CFC): Any of various compounds consisting of carbon, hydrogen, chlorine, and flourine used as refrigerants. CFCs are now thought to be harmful to the earth's atmosphere.

Christmas tree: The valves and fittings installed at the top of a gas or oil well to control and direct the flow of well fluids.

CIF (cargo, insurance and freight): CIF refers to cargos for which the seller pays for the transportation and insurance up to the port of destination.

CIF (cost, insurance, freight): This term refers to a type of sale in which the buyer of the product agrees to pay a unit price that includes the f.o.b. value of the product at the point of origin plus all costs of insurance and transportation. This type of a transaction differs from a "delivered" purchase, in that the buyer accepts the quantity as determined at the loading port (as certified by the Bill of Lading and Quality Report) rather than pay based on the quantity and quality ascertained at the unloading port. It is similar to the terms of an f.o.b. sale, except that the seller, as a service for which he is compensated, arranges for transportation and insurance.

Circuit: A conductor or a system of conductors through which electric current flows.

Circuit-mile: The total length in miles of separate circuits regardless of the number of conductors used per circuit.

Citygate: A point or measuring station at which a distributing gas utility receives gas from a natural gas pipeline company or transmission system.

Class rate schedule: An electric rate schedule applicable to one or more specified classes of service, groups of businesses, or customer uses.

Classes of service: Customers grouped by similar characteristics in order to be identified for the purpose of setting a common rate for electric service. Usually classified into groups identified as residential, commercial, industrial, and other.

Clean Development Mechanism (CDM): A Kyoto Protocol program that enables industrialized countries to finance emissions-avoiding projects in developing countries and receive credit for reductions achieved against their own emissions limitation targets.

Climate change: A term used to refer to all forms of climatic inconsistency, but especially to significant change from one prevailing climatic condition to another. In some cases, "climate change" has been used synonymously with the term "global warming"; scientists, however, tend to use the term in a wider sense inclusive of natural changes in climate, including climatic cooling.

Clinker: Powdered cement, produced by heating a properly proportioned mixture of finely ground raw materials (calcium carbonate, silica, alumina, and iron oxide) in a kiln to a temperature of about 2,700 degrees Fahrenheit.

Cloud condensation nuclei: Aerosol particles that provide a platform for the condensation of water vapor, resulting in clouds with higher droplet concentrations and increased albedo.

CO control period ("seasons"): The portion of the year in which a CO nonattainment area is prone to high ambient levels of carbon monoxide. This portion of the year is to be specified by the Environmental Protection Agency but is to be not less than 4 months in length.

Coal: A readily combustible black or brownish-black rock whose composition, including inherent moisture, consists of more than 50 percent by weight and more than 70 percent by volume of carbonaceous material. It is formed from plant remains that have been compacted, hardened, chemically altered, and metamorphosed by heat and pressure over geologic time.

Coal analysis: Determines the composition and properties of coal so it can be ranked and used most effectively.

1. *Proximate analysis* determines, on an as-received basis, the moisture content, volatile matter (gases released when coal is heated), fixed carbon (solid fuel left after the volatile matter is driven off), and ash (impurities consisting of silica, iron, alumina, and other incombustible matter). The moisture content affects the ease with which coal can be handled and burned. The amount of volatile matter and fixed carbon provides guidelines for determining the intensity of the heat produced. Ash increases the weight of coal, adds to the cost of handling, and can cause problems such as clinkering and slagging in boilers and furnaces.
2. *Ultimate analysis* determines the amount of carbon, hydrogen, oxygen, nitrogen, and sulfur. Heating value is determined in terms of Btu, both on an as received basis (including moisture) and on a dry basis.
3. *Agglomerating* refers to coal that softens when heated and forms a hard gray coke; this coal is called caking coal. Not all caking coals are cooking coals. The agglomerating value is used to differentiate between coal ranks and also is a guide to determine how a particular coal reacts in a furnace.
4. *Agglutinating* refers to the binding qualities of a coal. The agglutinating value is an indication of how well a coke made from a particular coal will perform in a blast furnace. It is also called a caking index.
5. *Other tests* include the determination of the ash softening temperature, the ash fusion temperature (the temperature at which the ash forms clinkers or slag), the free swelling index (a guide to a coal's coking characteristics), the Gray King test (which determines the suitability of coal for making coke), and the Hardgrove grindability index (a measure of the ease with which coal can be pulverized). In a petrographic analysis, thin sections of coal or highly polished blocks of coal are studied with a microscope to determine the physical composition, both for scientific purposes and for estimating the rank and coking potential.

Coal bed: A bed or stratum of coal. Also called a coal seam.

Coal bed degasification: This refers to the removal of methane or coal bed gas from a coal mine before or during mining.

Coal bed methane: Methane is generated during coal formation and is contained in the coal microstructure. Typical recovery entails pumping water out of the coal to allow the gas to escape. Methane is the principal component of natural gas. Coal bed methane can be added to natural gas pipelines without any special treatment.

Coal briquets: Anthracite, bituminous, and lignite briquets comprise the secondary solid fuels manufactured from coal by a process in which the coal is partly dried, warmed to expel excess moisture, and then compressed into briquets, usually without the use of a binding substance. In the reduction of briquets to coal equivalent, different conversion factors are applied according to their origin from hard coal, peat, brown coal, or lignite.

Coal carbonized: The amount of coal decomposed into solid coke and gaseous products by heating in a coke oven in a limited air supply or in the absence of air.

Coal chemicals: Coal chemicals are obtained from the gases and vapor recovered from the manufacturing of coke. Generally, crude tar, ammonia, crude light oil, and gas are the basic products recovered. They are refined or processed to yield a variety of chemical materials.

Coal consumption: The quantity of coal burned for the generation of electric power (in short tons), including fuel used for maintenance of standby service.

Coal delivered: Coal which has been delivered from the coal supplier to any site belonging to the electric power company.

Coal exports: Amount of India coal shipped to foreign destinations, as reported in the Indian Department of Commerce, Bureau of Census.

Coal face: This is the exposed area from which coal is extracted.

Coal financial reporting regions: A geographic classification of areas with coal resources which is used for financial reporting of coal statistics.

1. Eastern Region. Consists of the Appalachian Coal Basin. The following comprise the Eastern Region: Alabama, eastern Kentucky, Georgia, Maryland, Mississippi, Ohio, Pennsylvania, Virginia, Tennessee, North Carolina, and West Virginia.
2. Midwest Region. Consists of the Illinois and Michigan Coal Basins. The following comprise the Midwest Region: Illinois, Indiana, Michigan, and western Kentucky.
3. Western Region. Consists of the Northern Rocky, Southern Rocky, West Coast Coal Basins and Western Interior. The following comprise the Western Region: Alaska, Arizona, Arkansas, California, Colorado, Idaho, Iowa, Kansas, Louisiana, Missouri, Montana, New Mexico, North Dakota, Oklahoma, Oregon, Texas, South Dakota, Utah, Washington, and Wyoming.

Coal fines: Coal with a maximum particle size usually less than one-sixteenth inch and rarely above one- eighth inch.

Coal gas: Substitute natural gas produced synthetically by the chemical reduction of coal at a coal gasification facility.

Coal gasification: The process of converting coal into gas. The basic process involves crushing coal to a powder, which is then heated in the presence of steam and oxygen to produce a gas. The gas is then refined to reduce sulfur and other impurities. The gas can be used as a fuel or processed further and concentrated into chemical or liquid fuel.

Coal grade: This classification refers to coal quality and use.

1. **Briquettes** are made from compressed coal dust, with or without a binding agent such as asphalt.
2. **Cleaned coal or prepared coal** has been processed to reduce the amount of impurities present and improve the burning characteristics.

3. **Compliance coal** is a coal, or a blend of coal, that meets sulfur dioxide emission standards for air quality without the need for flue-gas desulfurization.
4. **Culm and silt** are waste materials from preparation plants. In the anthracite region, culm consists of coarse rock fragments containing as much as 30 percent small-sized coal. Silt is a mixture of very fine coal particles (approximately 40 percent) and rock dust that has settled out from waste water from the plants. The terms culm and silt are sometimes used interchangeably and are sometimes called refuse. Culm and silt have a heat value ranging from 8 to 17 million Btu per ton.
5. **Low-sulfur coal** generally contains 1 percent or less sulfur by weight. For air quality standards, "low sulfur coal" contains 0.6 pounds or less sulfur per million Btu, which is equivalent to 1.2 pounds of sulfur dioxide per million Btu.
6. **Metallurgical coal (or coking coal)** meets the requirements for making coke. It must have a low ash and sulfur content and form a coke that is capable of supporting the charge of iron ore and limestone in a blast furnace. A blend of two or more bituminous coals is usually required to make coke.
7. **Pulverized coal** is a coal that has been crushed to a fine dust in a grinding mill. It is blown into the combustion zone of a furnace and burns very rapidly and efficiently.
8. **Slack coal** usually refers to bituminous coal one-half inch or smaller in size.
9. **Steam coal** refers to coal used in boilers to generate steam to produce electricity or for other purposes.
10. **Stoker coal** refers to coal that has been crushed to specific sizes (but not powdered) for burning on a grate in automatic firing equipment.

Coal imports: Amount of foreign coal shipped to the United States, as reported in the U.S. Department of Commerce, Bureau of the Census, "Monthly Report IM 145."

Coal liquefaction: A chemical process that converts coal into clean-burning liquid hydrocarbons, such as synthetic crude oil and methanol.

Coal mining productivity: Coal mining productivity is calculated by dividing total coal production by the total direct labor hours worked by all mine employees.

Coal preparation: The process of sizing and cleaning coal to meet market specifications by removing impurities such as rock, sulfur, etc. It may include crushing, screening, or mechanical cleaning.

Coal-producing regions:

1. Appalachian Region. Consists of Alabama, Georgia, Eastern Kentucky, Maryland, North Carolina, Ohio, Pennsylvania, Tennessee, Virginia, and West Virginia.
2. Interior Region (with Gulf Coast). Consists of Arkansas, Illinois, Indiana, Iowa, Kansas, Louisiana, Michigan, Mississippi, Missouri, Oklahoma, Texas, and Western Kentucky.
3. Western Region. Consists of Alaska, Arizona, Colorado, Montana, New Mexico, North Dakota, Utah, Washington, and Wyoming.

Coal producing districts: A classification of coal fields defined in the Bituminous Coal Act of 1937. The districts were originally established to aid in formulating minimum prices of bituminous and subbituminous coal and lignite. Because much statistical information was compiled in terms of these districts, their use for statistical purposes has continued since the abandonment of that legislation in 1943. District 24 was added for the anthracite-producing district in Pennslyvania.

Coal production: The sum of sales, mine consumption, issues to miners, and issues to coke, briquetting, and other ancillary plants at mines. Production data include quantities extracted from surface and underground mines, and normally exclude wastes removed at mines or associated reparation plants.

Coal rank: The classification of coals according to their degree of progressive alteration from lignite to anthracite. In the United States, the standard ranks of coal include lignite, subbituminous coal, bituminous coal, and anthracite and are based on fixed carbon, volatile matter, heating value, and agglomerating (or caking) properties.

Coal sampling: The collection and proper storage and handling of a relatively small quantity of coal for laboratory analysis. Sampling may be done for a wide range of purposes, such as: coal resource exploration and assessment, characterization of the reserves or production of a mine, to characterize the results of coal cleaning processes, to monitor coal shipments or receipts for adherence to coal quality contract specifications, or to subject a coal to specific combustion or reactivity tests related to the customer's intended use. During pre-development phases, such as exploration and resource assessment, sampling typically is from natural outcrops, test pits, old or existing mines in the region, drill cuttings, or drilled cores. Characterization of a mine's reserves or production may use sample collection in the mine, representative cuts from coal conveyors or from handling and loading equipment, or directly from stockpiles or shipments (coal rail cars or barges). Contract specifications rely on sampling from the production flow at the mining or coal handling facility or at the loadout, or from the incoming shipments at the receiver's facility. In all cases, the value of a sample taken depends on its being representative of the coal under consideration, which in turn requires that appropriate sampling procedures be carefully followed.

For coal resource and estimated reserve characterization, appropriate types of samples include:

1. **Face channel or channel sample:** a sample taken at the exposed coal in a mine by cutting away any loose or weathered coal then collecting on a clean surface a sample of the coal seam by chopping out a channel of uniform width and depth; a face channel or face sample is taken at or near the working face, the most freshly exposed coal where actual removal and loading of mined coal is taking place. Any partings greater than 3/8 inch and/or mineral concretions greater than 1/2 inch thick and 2 inches in maximum diameter are normally discarded from a channel sample so as better to represent coal that has been mined, crushed, and screened to remove at least gross non-coal materials.
2. **Column sample:** a channel or drill core sample taken to represent the entire geologic coalbed; it includes all partings and impurities that may exist in the coalbed.
3. **Bench sample:** a face or channel sample taken of just that contiguous portion of a coalbed that is considered practical to mine, also known as a "bench;" For example, bench samples may be taken of minable coal where impure coal that makes up part of the geologic coalbed is likely to be left in the mine, or where thick partings split the coal into two or more distinct minable seams, or where extremely thick coalbeds cannot be recovered by normal mining equipment, so that the coal is mined in multiple passes, or benches, usually defined along natural bedding planes.
4. **Composite sample:** a recombined coalbed sample produced by averaging together thickness-weighted coal analyses from partial samples of the coalbed, such as from one or more bench

samples, from one or more mine exposures or outcrops where the entire bed could not be accessed in one sample, or from multiple drill cores that were required to retrieve all local sections of a coal seam.

Coal stocks: Coal quantities that are held in storage for future use and disposition.

Coal sulfur: Coal sulfur occurs in three forms: organic, sulfate, and pyritic. Organic sulfur is an integral part of the coal matrix and cannot be removed by conventional physical separation. Sulfate sulfur is usually negligible. Pyritic sulfur occurs as the minerals pyrite and marcasite; larger sizes generally can be removed by cleaning the coal.

Coal type: The classification is based on physical characteristics or microscopic constituents. Examples of coal types are banded coal, bright coal, cannel coal, and splint coal. The term is also used to classify coal according to heat and sulfur content.

Coal zone: A series of laterally extensive and (or) lenticular coal beds and associated strata that arbitrarily can be viewed as a unit. Generally, the coal beds in a coal zone are assigned to the same geologic member or formation.

Code of Federal Regulations: A compilation of the general and permanent rules of the executive departments and agencies of the Federal Government as published in the Federal Register. The code is divided into 50 titles that represent broad areas subject to Federal regulation. Title 18 contains the FERC regulations.

Cofiring: The process of burning natural gas in conjunction with another fuel to reduce air pollutants.

Cogeneration: The production of electrical energy and another form of useful energy (such as heat or steam) through the sequential use of energy.

Cogeneration system: A system using a common energy source to produce both electricity and steam for other uses, resulting in increased fuel efficiency.

Cogenerator: A generating facility that produces electricity and another form of useful thermal energy (such as heat or steam), used for industrial, commercial, heating, or cooling purposes. To receive status as a qualifying facility (QF) under the Public Utility Regulatory Policies Act (PURPA), the facility must produce electric energy and "another form of useful thermal energy through the sequential use of energy" and meet certain ownership, operating, and efficiency criteria established by the Federal Energy Regulatory Commission (FERC).

Coincidental demand: The sum of two or more demands that occur in the same time interval.

Coincidental peak load: The sum of two or more peak loads that occur in the same time interval.

Coke (coal): A solid carbonaceous residue derived from low-ash, low-sulfur bituminous coal from which the volatile constituents are driven off by baking in an oven at temperatures as high as 2,000 degrees Fahrenheit so that the fixed carbon and residual ash are fused together. Coke is used as a fuel and as a reducing agent in smelting iron ore in a blast furnace. Coke from coal is grey, hard, and porous and has a heating value of 24.8 million Btu per ton.

Coke (petroleum): A residue high in carbon content and low in hydrogen that is the final product of thermal decomposition in the condensation process in cracking. This product is reported as marketable coke or catalyst coke. The conversion is 5 barrels (of 42 U.S. gallons each) per short ton. Coke from petroleum has a heating value of 6.024 million Btu per barrel.

Coke breeze: The term refers to the fine sizes of coke, usually less than one-half inch, that are recovered from coke plants. It is commonly used for sintering iron ore.

Coke button: A button-shaped piece of coke resulting from standard laboratory tests that indicates the coking or free-swelling characteristics of a coal; expressed in numbers and compared with a standard.

Coke oven gas: The mixture of permanent gases produced by the carbonization of coal in a coke oven at temperatures in excess of 1,000 degrees Celsius.

Coke plants: Plants where coal is carbonized for the manufacture of coke in slot or beehive ovens.

Coking: Thermal refining processes used to produce fuel gas, gasoline blendstocks, distillates, and petroleum coke from the heavier products of atomspheric and vacuum distillation. Includes:

1. **Delayed Coking.** A process by which heavier crude oil fractions can be thermally decomposed under conditions of elevated temperatures and pressure to produce a mixture of lighter oils and petroleum coke. The light oils can be processed further in other refinery units to meet product specifications. The coke can be used either as a fuel or in other applications such as the manufacturing of steel oraluminum.
2. **Flexicoking.** A thermal cracking process which converts heavy hydrocarbons such as crude oil, tar sands bitumen, and distillation residues into light hydrocarbons. Feedstocks can be any pumpable hydrocarbons including those containing high concentrations of sulfur and metals.
3. **Fluid Coking.** A thermal cracking process utilizing the fluidized-solids technique to remove carbon (coke) for continuous conversion of heavy, low-grade oils into lighter products.

Coking coal: Bituminous coal suitable for making coke.

Cold-deck imputation: A statistical procedure that replaces a missing value of an item with a constant value from an external source such as a value from a previous survey.

Combined cycle: An electric generating technology in which electricity is produced from otherwise lost waste heat exiting from one or more gas (combustion) turbines. The exiting heat is routed to a conventional boiler or to a heat recovery steam generator for utilization by a steam turbine in the production of electricity. This process increases the efficiency of the electric generating unit.

Combined cycle unit: An electric generating unit that consists of one or more combustion turbines and one or more boilers with a portion of the required energy input to the boiler(s) provided by the exhaust gas of the combustion turbine(s).

Combined heat and power (CHP) plant: A plant designed to produce both heat and electricity from a single heat source. CHP better describes the facilities because some of the plants included do not produce heat and power in a sequential fashion and, as a result, do not meet the legal definition of cogeneration specified in the Public Utility Regulatory Policies Act (PURPA).

Combined household energy expenditures: The total amount of funds spent for energy consumed in, or delivered to, a housing unit during a given period of time and for fuel used to operate the motor vehicles that are owned or used on a regular basis by the household. The total dollar amount for energy consumed in a housing unit includes state and local taxes but excludes merchandise repairs or special service charges. Electricity, and natural gas expenditures are for the amount of those energy sources consumed. Fuel oil, kerosene, and LPG expenditures are for the amount of fuel purchased, which may differ from the amount of fuel consumed. The total dollar amount of fuel spent for vehicles is the product of fuel consumption and price.

Combined hydroelectric plant: A hydroelectric plant that uses both pumped water and natural streamflow for the production of power.

Combined pumped-storage plant: A pumped-storage hydroelectric power plant that uses both pumped water and natural streamflow to produce electricity.

Combustion: Chemical oxidation accompanied by the generation of light and heat.

Combustion chamber: An enclosed vessel in which chemical oxidation of fuel occurs.

Commercial building: A building with more than 50 percent of its floor space used for commercial activities. Commercial buildings include, but are not limited to, stores, offices, schools, churches, gymnasiums, libraries, museums, hospitals, clinics, warehouses, and jails. Government buildings are included except for buildings on military bases or reservations.

Commercial facility: An economic unit that is owned or operated by one person or organization and that occupies two or more commercial buildings at a single location. A university and a large hospital complex are examples of a commercial multi-building facility.

Commercial operation (nuclear): The phase of reactor operation that begins when power ascension ends and the operating utility formally declares the nuclear power plant to be available for the regular production of electricity. This declaration is usually related to the satisfactory completion of qualification tests on critical components of the unit.

Commercial sector: An energy-consuming sector that consists of service-providing facilities and equipment of: businesses; Federal, State, and local governments; and other private and public organizations, such as religious, social, or fraternal groups. The commercial sector includes institutional living quarters. It also includes sewage treatment facilities. Common uses of energy associated with this sector include space heating, water heating, air conditioning, lighting, refrigeration, cooking, and running a wide variety of other equipment.

Commingling: The mixing of one utility's generated supply of electric energy with another utility's generated supply within a transmission system.

Commissioned agent: An agent who wholesales or retails a refined petroleum product under a commission arrangement. The agent does not take title to the product or establish the selling price, but receives a percentage of fixed fee for serving as an agent.

Common equity (book value): The retained earnings and common stock earnings plus the balances in common equity reserves and all other common stock accounts. This also includes the capital surplus, the paid-in surplus, the premium on common stocks, except those balances specifically related to preferred or preference stocks, less any common stocks held in the treasury.

Compact fluorescent bulbs: These are also known as "screw-in fluorescent replacements for incandescent" or "screw-ins." Compact fluorescent bulbs combine the efficiency of fluorescent lighting with the convenience of a standard incandescent bulb. There are many styles of compact fluorescent, including exit light fixtures and floodlights (lamps containing reflectors). Many screw into a standard light socket, and most produce a similar color of light as a standard incandescent bulb. Compact fluorescent bulbs come with ballasts that are electronic (lightweight, instant, no-flicker starting, and 10 to 15% more efficient) or magnetic (much heavier and slower starting). Other types of compact fluorescent bulbs include adaptive circulation and PL and SL lamps and ballasts. Compact fluorescent bulbs are designed for residential uses; they are also used in table lamps, wall sconces, and hall and ceiling fixtures of hotels, motels, hospitals, and other types of commercial buildings with residential-type applications.

Company automotive (retail) outlet: Any retail outlet selling motor fuel under the brand name of a company reporting in theFinancial Reporting System.

Company-lessee automotive outlet: One of three types of company automotive (retail) outlets. This type of outlet is operated by an independent marketer who leases the station and land and has use of tanks, pumps, signs, etc. A lessee dealer typically has a supply agreement with a refiner or a distributor and purchases products at dealer tank wagon prices. The term includes outlets operated by commissioned agents and is limited to those dealers who are supplied directly by a refiner or any affiliate or subsidiary company of a refiner.

Company-open automotive outlet: One of three types of company automotive (retail) outlets. This type of outlet is operated by an independent marketer who owns or leases (from a third party that is not a refiner) the station or land of a retail outlet and has use of tanks, pumps, signs, etc. An open dealer typically has a supply agreement with a refiner or a distributor and purchases products based on either rack or dealer tank wagon prices.

Company-operated automotive outlet: One of three types of company automotive (retail) outlets. This type of outlet is operated by salaried or commissioned personnel paid by the reporting company.

Company-operated retail outlet: Any retail outlet (*i.e.*, service station) which sells motor vehicle fuels and is under the direct control of a **firm** that sets the retail product price and directly collects all or part of the retail margin. The category includes retail outlets operated by (1) salaried employees of the firm and/or its subsidiaries and affiliates, (2) licensed or commissioned agents, and/or (3) personnel services contracted by the firm.

Competitive transition charge: A non-bypassable charge levied on each customer of the distribution utility, including those who are served under contracts with nonutility suppliers, for recovery of the utility's **stranded costs** that develop because of competition.

Completion (oil/gas production): The term refers to the installation of permanent equipment for the production of oil or gas. If a well is equipped to produce only oil or gas from one zone or reservoir, the definition of a "well" (classified as an oil well or gas well) and the definition of a "completion" are identical. However, if a well is equipped to produce oil and/or gas separately from more than one reservoir, a "well" is not synonymous with a "completion."

Completion date (oil/gas production): The date on which the installation of permanent equipment has been completed as reported to the appropriate regulatory agency. The date of completion of a dry hole is the date of abandonment as reported to the appropriate agency. The date of completion of a service well is the date on which the well is equipped to perform the service for which it was intended.

Compliance coal: A coal or a blend of coals that meets sulfur dioxide emission standards for air quality without the need for flue gas desulfurization.

Compressed natural gas (CNG): Natural gas which is comprised primarily of methane, compressed to a pressure at or above 2,400 pounds per square inch and stored in special high-pressure containers. It is used as a fuel for natural gas powered vehicles.

Compressor station: Any combination of facilities that supply the energy to move gas in transmission or distribution lines or into storage by increasing the pressure.

Concentrating solar power or solar thermal power system: A solar energy conversion system characterized by the optical concentration of solar rays through an arrangement of mirrors to generate a high temperature working fluid.

Concentrator: A reflective or refractive device that focuses incident insolation onto an area smaller than the reflective or refractive surface, resulting in increased insolation at the point of focus.

Concession: The operating right to explore for and develop petroleum fields in consideration for a share of production in kind (equity oil).

Concessionary purchases: The quantity of crude oil exported during a reporting period, which was acquired from the producing government under terms that arise from the firm's participation in a concession. It includes preferential crude where the reporting firm's access to such crude is derived from a former concessionary relationship.

Condensate (lease condensate): A natural gas liquid recovered from associated and nonassociated gas wells from lease separators or field facilities, reported in barrels of 42 U.S. gallons at atmospheric pressure and 60 degrees Fahrenheit.

Condenser cooling water: A source of water external to a boiler's feed system is passed through the steam leaving the turbine in order to cool and condense the steam. This reduces the steam's exit pressure and recaptures its heat, which is then used to preheat fluid entering the boiler, thereby increasing the plant's thermodynamic efficiency.

Conditional energy intensity: Total consumption of a particular energy source(s) or fuel(s) divided by the total floorspace of buildings that use the energy source(s) or fuel(s); *i.e.*, the ratio of consumption to energy source-specific floorspace.

Conditionally effective rates: An electric rate schedule that has been put into effect by the FERC subject to refund pending final disposition or refiling.

Conductor: Metal wires, cables, and bus-bar used for carrying electric current. Conductors may be solid or stranded, that is, built up by a assembly of smaller solid conductors.

Conference of the Parties (COP): The collection of nations that have ratified the Framework Convention on Climate Change (FCCC). The primary role of the COP is to keep implementation of the FCCC under review and make the decisions necessary for its effective implementation.

Congestion: A condition that occurs when insufficient transfer capacity is available to implement all of the preferred schedules for electricity transmission simultaneously.

Connected load: The sum of the continuous ratings or the capacities for a system, part of a system, or a customer's electric power consuming apparatus.

Connection: The physical connection (*e.g.*, transmission lines, transformers, switch gear, etc.) between two electric systems permitting the transfer of electric energy in one or both directions.

CO nonattainment area: Areas with carbon monoxide design values of 9.5 parts per million or more, generally based on data for 1988 and 1989.

Conservation and other DSM: This Demand-Side Management category represents the amount of consumer load reduction at the time of system peak due to utility programs that reduce consumer load during many hours of the year. Examples include utility rebate and shared savings activities for the installation of energy efficient appliances, lighting and electrical machinery, and weatherization materials. In addition, this category includes all other Demand-Side Management activities, such as thermal storage, time-of-use rates, fuel substitution, measurement and evaluation, and any other utility-administered Demand-Side Management activity designed to reduce demand and/or electricity use.

Conservation feature: A feature in the building designed to reduce the usage of energy.

Conservation program: A program in which a utility company furnishes home weatherization services free or at reduced cost or provides free or low cost devices for saving energy, such as energy efficient light bulbs, flow restrictors, weather stripping, and water heater insulation.

Consolidated Metropolitan Statistical Area (CMSA): An area that meets the requirements of a metropolitan statistical area, has a population of one million or more, and consists of two or more component parts that are recognized as **primary metropolitan statistical areas.**

Construction: An energy-consuming subsector of the **industrial sector** that consists of all facilities and equipment used to perform land preparation and construct, renovate, alter, install, maintain, or repair major infrastructure or individual systems therein. Infrastructure includes buildings; industrial plants; and other major structures, such as tanks, towers, monuments, roadways, tunnels, bridges, dams, pipelines, and transmission lines.

Construction costs (of the electric power industry): All direct and indirect costs incurred in acquiring and constructing electric utility plant and equipment and proportionate shares of common utility plants. Included are the cost of land and improvements, nuclear fuel and spare parts, allowance for funds used during construction, and general overheads capitalized, less the cost of acquiring plant and equipment previously operated in utility service.

Construction expenditures (of the electric power industry): The gross expenditures for construction costs (including the cost of replacing worn out plants), and electric construction costs, and land held for future use.

Construction pipeline (of a nuclear reactor): The various stages involved in the acquisition of a nuclear reactor by a utility. The events that define these stages are the ordering of a reactor, the licensing process, and the physical construction of the nuclear generating unit. A reactor is said to be "in the pipeline" when the reactor is ordered and "out of the pipeline" when it completes low power testing and begins operation toward full power.

Construction work in progress (CWIP): The balance shown on a utility's balance sheet for construction work not yet completed but in process. This balance line item may or may not be included in the rate base.

Constructive surplus or deficit: The amounts representing the exchange of services, supplies, etc., between the utility department and the municipality and its other departments without charge or at a reduced charge. Charges to this account include utility and other services, supplies, etc., furnished by the utility department to the municipality or its other departments without charge, or the amount of the reduction, if furnished at a reduced charge. Credits to the account consist of services, supplies, office space, etc., furnished by the municipality to the utility department without charge on the amount of the reduction, if furnished at a reduced charge.

Consumer (energy): Any individually metered dwelling, building, establishment, or location using natural gas, synthetic natural gas, and/or mixtures of natural and supplemental gas for feedstock or as fuel for any purpose other than in oil or gas lease operations; natural gas treating or processing plants; or pipeline, distribution, or storage compressors.

Consumer charge: An amount charged periodically to a consumer for such utility costs as billing and meter reading, without regard to demand or energy consumption.

Consumer Price Index (CPI): These prices are collected in 85 urban areas selected to represent all urban consumers about 80 percent of the total U.S. population. The service stations are selected initially and on a replacement basis, in such a way that they represent the purchasing habits of the CPI population. Service stations in the current sample include those providing all types of service (*i.e.*, full, mini, and self service).

Consumption per square foot: The aggregate ratio of total consumption for a particular set of buildings to the total floorspace of those buildings.

Continuous delivery energy sources: Those energy sources provided continuously to a building.

Continuous mining: A form of room pillar mining in which a continuous mining machine extracts and removes coal from the working face in one operation; no blasting is required.

Contract price: The delivery price determined when a contract is signed. It can be a fixed price or a base price escalated according to a given formula.

Contract receipts: Purchases based on a negotiated agreement that generally covers a period of 1 or more years.

Contracted gas: Any gas for which Interstate Pipeline has a contract to purchase from any domestic or foreign source that cannot be identified to a specific field or group. This includes tailgate plant purchases, single meter point purchases, pipeline purchases, natural gas imports, SNG purchases, and LNG purchases.

Contribution to net income: The FRS (Financial Reporting System survey) segment equivalent to net income. However, some consolidated items of revenue and expense are not allocated to the segments, and therefore they are not equivalent in a strict sense. The largest item not allocated to the segments is interest expense since this is regarded as a corporate level item for FRS purposes.

Control: Including the terms "controlling," "controlled by," and "under common control with," means the possession, direct or indirect, of the power to direct or cause the direction of the management and policies of a person, whether through the ownership of voting shares, by contract, or otherwise.

Control total: The number of elements in the population or a subset of the population. The sample weights for the observed elements in a survey are adjusted so that they add up to the control total. The value of a control total is obtained from an outside source. The control totals are given by the number of households in one of the 12 cells by categorizing households by the four Census regions and by three categories of metropolitan status (Metropolitan Statistical Area central city, Metropolitan Statistical Area outside central city, and non Metropolitan Statistical Area). The control totals are obtained from the Current Population Survey.

Conventional gasoline: Finished motor gasoline not included in the oxygenated or reformulated gasoline categories.

Conventional hydroelectric plant: A plant in which all of the power is produced from natural streamflow as regulated by available storage.

Conventional mill (uranium): A facility engineered and built principally for processing of uraniferous ore materials mined from the earth and the recovery, by chemical treatment in the mill's circuits, of uranium and/or other valued coproduct components from the processed one.

Conventional mining: The oldest form of room pillar mining, which consists of a series of operations that involve cutting the coal bed, so it breaks easily when blasted with explosives or high pressure air, and then loading the broken coal.

Conventional thermal electricity generation: Electricity generated by an electric power plant using coal, petroleum, or gas as its source of energy.

Conventionally fueled vehicle: A vehicle that runs on petroleum-based fuels such as motor gasoline or diesel fuel.

Conversion company: An organization that performs vehicle conversions on a commercial basis.

Conversion factor: A number that translates units of one measurement system into corresponding values of another measurement system.

Converted (alternative-fuel) vehicle: A vehicle originally designed to operate on gasoline/diesel that was modified or altered to run on an alternative fuel after its initial delivery to an end-user.

Cooling: Conditioning of room air for human comfort by a refrigeration unit (such as an air conditioner or heat pump) or by circulating chilled water through a central cooling or district cooling system. Use of fans or blowers by themselves, without chilled air or water, is not included in this definition of cooling.

Cooling degree-days: A measure of how warm a location is over a period of time relative to a base temperature, most commonly specified as 65 degrees Fahrenheit. The measure is computed for each day by subtracting the base temperature (65 degrees) from the average of the day's high and low temperatures, with negative values set equal to zero. Each day's cooling degree-days are summed to create a cooling degree-day measure for a specified reference period. Cooling degree-days are used in energy analysis as an indicator of air conditioning energy requirements or use.

Cooling pond: A natural or man made body of water that is used for dissipating waste heat from power plants.

Cooling system: An equipment system that provides water to the condensers and includes water intakes and outlets; cooling towers; and ponds, pumps, and pipes.

Cooperative electric utility: An electric utility legally established to be owned by and operated for the benefit of those using its service. The utility company will generate, transmit, and/or distribute supplies of electric energy to a specified area not being serviced by another utility. Such ventures are generally exempt from Federal income tax laws.

Coordination service: The sale, exchange, or transmission of electricity between two or more electric utilities that typically have sufficient generation and transmission capacity to supply their load requirements under normal conditions.

Coordination service pricing: The typical price components of a bulk power coordination sale are an energy charge, a capacity, or reservation charge, and an adder. The price for a particular sale may embody some or all of these components. The energy charge is made on a per-kilowatt basis and is intended to recover the seller's system incremental variable costs of making a sale. Because the nonfuel expenses are usually hard to quantify, and small relative to fuel expense, energy charges quoted are usually based on fuel cost. A capacity charge is set at a certain level per kilowatt and is normally paid whether or not energy is taken by the buyer. An adder is added to that energy charge to recover the hard quantify nonfuel variable costs. There are three types of adders: percentage, fixed, and split savings. A percentage adder increases the energy charge by a certain percentage. A fixed adder adds a fixed amount per kilowatthour to the energy charge. Split savings adders are used only in economy energy transactions. They split production costs savings between the seller and the buyer by adding one half of the savings to the energy cost.

Cord of wood: A cord of wood measures 4 feet by 4 feet by 8 feet, or 128 cubic feet.

Correlation (statistical term): In its most general sense, correlation denotes the interdependence between quanitative or qualitative data. It would include the association of dichotomized attributes and the contingency of multiple classified attributes. The concept is quite general and may be extended to more than two variates. The word is most frequently used in a somewhat narrower sense to denote the relationship between measurable variates or ranks.

Cost, insurance, freight (CIF): A type of sale in which the buyer of the product agrees to pay a unit price that includes the f.o.b. value of the product at the point of origin plus all costs of insurance and transportation. This type of transaction differs from a "delivered" purchase in that the buyer accepts the quantity as determined at the loading port (as certified by the Bill of Loading and Quality Report) rather than pay on the basis of the quantity and quality ascertained at the unloading port. It is similar to the terms of an f.o.b. sale except that the seller, as a service for which he is compensated, arranges for transportation and insurance.

Cost model for undiscovered resources: A computerized algorithm that uses the uranium endowment estimated for a given geological area and selected industry economic indexes to develop random variables that describe the undiscovered resources ultimately expected to be discovered in that area at chosen forward cost categories.

Cost of capital: The rate of return a utility must offer to obtain additional funds. The cost of capital varies with the leverage ratio, the effective income tax rate, conditions in the bond and stock markets, growth rate of the utility, its dividend strategy, stability of net income, the amount of new capital required, and other factors dealing with business and financial risks. It is a composite of the cost for debt interest, preferred stock dividends, and common stockholders' earnings that provide the facilities used in supplying utility service.

Cost of debt: The interest rate paid on new increments of debt capital multiplied by 1 minus the tax rate.

Cost of preferred stock: The preferred stock dividends divided by the net price of the preferred stock.

Cost of retained earnings: The residual of an entity's earnings over expenditures, including taxes and dividends, that are reinvested in its business. The cost of these funds is always lower than the cost of new equity capital, due to taxes and transactions costs. Therefore, the cost of retained earnings is the yield that retained earnings accrue upon reinvestment.

Cost of service: A rate making concept used for the design and development of rate schedules to ensure that the filed rate schedules recover only the cost of providing the electric service at issue. This concept attempts to correlate the utility's costs and revenue with the service provided to each of the various customer classes.

Cost-of-service regulation: A traditional electric utility regulation under which a utility is allowed to set rates based on the cost of providing service to customers and the right to earn a limited profit.

Costs (imports of natural gas): All expenses incurred by an importer up to the U.S. point of delivery for the reported quantity {of natural gas} imported.

Criteria pollutant: A pollutant determined to be hazardous to human health and regulated under EPA's National Ambient Air Quality Standards. The 1970 amendments to the Clean Air Act require EPA to describe the health and welfare impacts of a pollutant as the "criteria" for inclusion in the regulatory regime.

Crop residue: Organic residue remaining after the harvesting and processing of a crop.

Crude oil: A mixture of hydrocarbons that exists in liquid phase in natural underground reservoirs and remains liquid at atmospheric pressure after passing through surface separating facilities. Depending upon the characteristics of the crude stream, it may also include:

1. Small amounts of hydrocarbons that exist in gaseous phase in natural underground reservoirs but are liquid at atmospheric pressure after being recovered from oil well (casinghead) gas in lease separators and are subsequently commingled with the crude stream without being separately measured. Lease condensate recovered as a liquid from natural gas wells in lease or field separation facilities and later mixed into the crude stream is also included;
2. Small amounts of nonhydrocarbons produced with the oil, such as sulfur and various metals;
3. Drip gases, and liquid hydrocarbons produced from tar sands, gilsonite, and oil shale.

Liquids produced at natural gas processing plants are excluded. Crude oil is refined to produce a wide array of petroleum products, including heating oils; gasoline, diesel and jet fuels; lubricants; asphalt; ethane, propane, and butane; and many other products used for their energy or chemical content.

Crude oil acquisitions (unfinished oil acquisitions): The volume of crude oil either

1. acquired by the respondent for processing for his own account in accordance with accounting procedures generally accepted and consistently and historically applied by the refiner concerned, or
2. in the case of a processing agreement, delivered to another refinery for processing for the respondent's own account.

Crude oil that has not been added by a refiner to inventory and that is thereafter sold or otherwise disposed of without processing for the account of that refiner shall be deducted from its crude oil purchases at the time when the related cost is deducted from refinery inventory in accordance with accounting procedures generally applied by the refiner concerned.

Crude oil processed by the respondent for the account of another is not a crude oil acquisition.

Crude oil f.o.b. price: The crude oil price actually charged at the oil producing country's port of loading. Includes deductions for any rebates and discounts or additions of premiums, where applicable. It is the actual price paid with no adjustment for credit terms.

Crude oil input: The total crude oil put into processing units at refineries.

Crude oil landed cost: The price of crude oil at the port of discharge, including charges associated with purchasing, transporting, and insuring a cargo from the purchase point to the port of discharge. The cost does not include charges incurred at the discharge port (*e.g.*, import tariffs or fees, wharfage charges, and demurrage).

Crude oil less lease condensate: A mixture of hydrocarbons that exists in liquid phase in natural underground reservoirs and remains liquid at atmospheric pressure after passing through surface separating facilities. Such hydrocarbons as lease condensate and natural gasoline recovered as liquids from natural gas wells in lease or field separation facilities and later mixed into the crude stream are excluded. Depending upon the characteristics of the crude stream, crude oil may also include:

1. Small amounts of hydrocarbons that exist in gaseous phase in natural underground reservoirs but are liquid at atmospheric pressure after being recovered from oil well (casinghead) gas in lease separators and are subsequently commingled with the crude stream without being separately measured;
2. Small amounts of nonhydrocarbons produced with the oil, such as sulfur and various metals.

Crude oil losses: Represents the volume of crude oil reported by petroleum refineries as being lost in their operations. These losses are due to spills, contamination, fires, etc., as opposed to refining processing losses.

Crude oil production: The volume of crude oil produced from oil reservoirs during given periods of time. The amount of such production for a given period is measured as volumes delivered from lease storage tanks (*i.e.*, the point of custody transfer) to pipelines, trucks, or other media for transport to refineries or terminals with adjustments for (1) net differences between opening and closing lease inventories, and (2) basic sediment and water (BS&W).

Crude oil qualities: Refers to two properties of crude oil, the sulfur content, and API gravity, which affect processing complexity and product characteristics.

Crude oil refinery input: The total crude oil put into processing units at refineries.

Crude oil stocks: Stocks of crude oil and lease condensate held at refineries, in pipelines, at pipeline terminals, and on leases.

Crude oil used directly: Crude oil consumed as fuel by crude oil pipelines and on crude oil leases.

Crude oil, refinery receipts: Receipts of domestic and foreign crude oil at a refinery. Includes all crude oil in transit except crude oil in transit by pipeline. Foreign crude oil is reported as a receipt only after entry through customs. Crude oil of foreign origin held in bonded storage is excluded.

Crystalline fully refined wax: A light colored paraffin wax having the following characteristics: viscosity at 210 degrees Fahrenheit (D88)-59.9 SUS (10.18 centistokes) maximum; oil content (D721)-0.5 percent maximum; other +20 color, Saybolt minimum.

Crystalline other wax: A paraffin wax having the following characteristics: viscosity at 210 deg. F(D88)-59.9 SUS (10.18 centistokes) maximum; oil content (D721)-0.51 percent minimum to 15 percent maximum.

Cubic foot (cf), natural gas: The amount of natural gas contained at standard temperature and pressure (60 degrees Fahrenheit and 14.73 pounds standard per square inch) in a cube whose edges are one foot long.

Cull wood: Wood logs, chips, or wood products that are burned.

Culm: Waste from Pennsylvania anthracite preparation plants, consisting of coarse rock fragments containing as much as 30 percent small sized coal; sometimes defined as including very fine coal particles called silt. Its heat value ranges from 8 to 17 million Btu per short ton.

Cultivar: A horticulturally or agriculturally derived variety of a plant.

Cumulative depletion: The sum in tons of coal extracted and lost in mining as of a stated date for a specified area or a specified coal bed.

Current (electric): A flow of electrons in an electrical conductor. The strength or rate of movement of the electricity is measured in amperes.

Current assets: Cash and other assets that are expected to be turned into cash, sold, or exchanged within the normal operating cycle of the utility, usually one year. Current assets include cash, marketable securities, receivables, inventory, and current prepayments.

Current liabilities: A debt or other obligation that must be discharged within one year or the normal operating cycle of the utility by expending a current asset or the incurrence of another short-term obligation. Current liabilities include accounts payable, short-term notes payable, and accrued expenses payable such as taxes payable and salaries payable.

Current ratio: The ratio of current assets divided by current liabilities that shows the ability of a utility to pay its current obligations from its current assets. A measure of liquidity, the higher the ratio, the more assurance that current liabilities can be paid.

Customer choice: The right of customers to purchase energy from a supplier other than their traditional supplier or from more than one seller in the retail market.

Customs district (coal): Customs districts, as defined by the Bureau of the Census, U.S. Department of Commerce, " Monthly Report EM 545," are as follows:

1. Eastern: Bridgeport, CT, Washington, DC, Boston, MA, Baltimore, MD, Portland, ME, Buffalo, NY, New York City, NY, Ogdensburg, NY, Philadelphia, PA, Providence, RI, Norfolk, VA, St. Albans, VT.
2. Southern: Mobile, AL, Savannah, GA, Miami, FL, Tampa, FL, New Orleans, LA, Wilmington, NC, San Juan, PR, Charleston, SC, Dallas-Fort Worth, TX, El Paso, TX, Houston-Galveston, TX, Laredo, TX, Virgin Islands.
3. Western: Anchorage, AK, Nogales, AZ, Los Angeles, CA, San Diego, CA, San Francisco, CA, Honolulu, HI, Great Falls, MT, Portland, OR, Seattle, WA.
4. Northern: Chicago, IL, Detroit, MI, Duluth, MN, Minneapolis, MN, St. Louis, MO, Pembina, ND, Cleveland, OH, Milwaukee, WI.

Cut-off grade (uranium): The lowest grade, in percent U_3O_8, of uranium ore at a minimum specified thickness that can be mined at a specified cost.

Cycle: The time period running from the startup of one reactor cycle to the startup of the following cycle.

Cycle/reactor history: A group of assemblies that have been irradiated in the same cycles in an individual reactor and are said to have the same cycle/reactor history.

Cycling (natural gas): The practice of producing natural gas for the extraction of natural gas liquids, returning the dry residue to the producing reservoir to maintain reservoir pressure and increase the ultimate recovery of natural gas liquids. The reinjected gas is produced for disposition after cycling operations are completed.

Dam: A physical barrier constructed across a river or waterway to control the flow of or raise the level of water. The purpose of construction may be for flood control, irrigation needs, hydroelectric power production, and/or recreation usage.

Day-ahead and hour-ahead markets: Forward markets where electricity quantities and market clearing prices are calculated individually for each hour of the day on the basis of participant bids for energy sales and purchases.

Day-ahead schedule: A schedule prepared by a scheduling coordinator or the independent system operator before the beginning of a trading day. This schedule indicates the levels of generation and demand scheduled for each settlement period that trading day.

Daylighting controls: A system of sensors that assesses the amount of daylight and controls lighting or shading devices to maintain a specified lighting level. The sensors are sometimes referred to as "photocells."

Deadweight tons: The lifting capacity of a ship expressed in long tons (2,240 lbs.), including cargo, commodities, and crew.

Dealer tank wagon (DTW) sales: Wholesale sales of gasoline priced on a delivered basis to a retail outlet.

Decatherm: Ten therms or 1,000,000 Btu.

Decommissioning: Retirement of a nuclear facility, including decontamination and/or dismantlement.

Decontamination: Removal of unwanted radioactive or hazardous contamination by a chemical or mechanical process.

Dedicated reserves: The volume of recoverable, salable gas reserves committed to, controlled by, or possessed by the reporting pipeline company and used for acts and services for which both the seller and the company have received certificate authorization from the Federal Energy Regulatory Commission (FERC). Reserves include both company-owned reserves (including owned gas in underground storage), reserves under contract from independent producers, and short-term and emergency supplies from the intrastate market. Gas volumes under contract from other interstate pipelines are not included as reserves, but may constitute part or all of a company's gas supply.

Dedicated vehicle: A vehicle that operates only on an alternative fuel, as when a vehicle is configured to operate on compressed natural gas.

Deepest total depth: The deepest total depth of a given well is the distance from a surface reference point (usually the Kelly bushing) to the point of deepest penetration measured along the well bore. If a well is drilled from a platform or barge over water, the depth of the water is included in the total length of the well bore.

Deferred cost: An expenditure not recognized as a cost of operation of the period in which incurred, but carried forward to be written off in future periods.

Deferred fuel costs: An expenditure for fuel that is not recognized for bookkeeping practices as a cost in the operating period incurred, but carried forward to be written off in future periods.

Deferred income tax (liability): A liability in the balance sheet representing the additional Federal income taxes that would have been due if a utility had not been allowed to compute tax expenses differently for income tax reporting purposes than for rate making purposes.

eforestation: The net removal of trees from forested land.

Degasification system: The methods employed for removing methane from a coal seam that could not otherwise be removed by standard ventilation fans and thus would pose a substantial hazard to coal miners. These systems may be used prior to mining or during mining activities.

Degradable organic carbon: The portion of organic carbon present in such solid waste as paper, food waste, and yard waste that is susceptible to biochemical decomposition.

Delayed coking: A process by which heavier crude oil fractions can be thermally decomposed under conditions of elevated temperatures and pressure to produce a mixture of lighter oils and petroleum coke. The light oils can be processed further in other refinery units to meet product specifications. The coke can be used either as a fuel or in other applications such as the manufacturing of steel or aluminum.

Deliverability: Represents the number of future years during which a pipeline company can meet its annual requirements for its presently certificated delivery capacity from presently committed sources of supply. The availability of gas from these sources of supply shall be governed by the physical capabilities of these sources to deliver gas by the terms of existing gas-purchase contracts, and by limitations imposed by State or Federal regulatory agencies.

Delivered cost: The cost of fuel, including the invoice price of fuel, transportation charges, taxes, commissions, insurance, and expenses associated with leased or owned equipment used to transport the fuel.

Delivered energy: The amount of energy delivered to the site (building); no adjustment is made for the fuels consumed to produce electricity or district sources. This is also referred to as net energy.

Delivered (gas): The physical transfer of natural, synthetic, and/or supplemental gas from facilities operated by the responding company to facilities operated by others or to consumers.

Deliveries (electric): Energy generated by one system and delivered to another system through one or more transmission lines.

Demand bid: A bid into the power exchange indicating a quantity of energy or an ancillary service that an eligible customer is willing to purchase and, if relevant, the maximum price that the customer is willing to pay.

Demand charge: That portion of the consumer's bill for electric service based on the consumer's maximum electric capacity usage and calculated based on the billing demand charges under the applicable rate schedule.

Demand charge credit: Compensation received by the buyer when the delivery terms of the contract cannot be met by the seller.

Demand indicator: A measure of the number of energy-consuming units, or the amount of service or output, for which energy inputs are required.

Demand interval: The time period during which flow of electricity is measured (usually in 15-, 30-, or 60-minute increments.)

Demand-metered: Having a meter to measure peak demand (in addition to total consumption) during a billing period. Demand is not usually metered for other energy sources.

Demand-side management (DSM): The planning, implementation, and monitoring of utility activities designed to encourage consumers to modify patterns of electricity usage, including the timing and level of electricity demand. It refers to only energy and load-shape modifying activities that are undertaken in response to utility-administered programs. It does not refer to energy and load-shaped changes arising from the normal operation of the marketplace or from government-mandated energy-efficiency standards. Demand-Side Management covers the complete range of load-shape objectives, including strategic conservation and load management, as well as strategic load growth.

Demonstrated reserve base (coal): A collective term for the sum of coal in both measured and indicated resource categories of reliability, representing 100 percent of the in-place coal in those categories as of a certain date. Includes beds of bituminuous coal and anthracite 28 or more inches thick and beds of subbituminuous coal 60 or more inches thick that can occur at depths of up to 1,000 feet. Includes beds of lignite 60 or more inches thick that can be surface mined. Includes also thinner and/or deeper beds that presently are being mined or for which there is evidence that they could be mined commercially at a given time. Represents that portion of the identified coal resource from which reserves are calculated.

Demonstrated resources: Same qualifications as identified resources, but include measured and indicated degrees of geologic assurance and excludes the inferred.

Demonstration and test vehicles: Vehicles operated by a motor vehicle dealer solely for the purpose of promoting motor vehicle sales or permitting potential purchasers to drive the vehicle for pre-purchase or pre-lease evaluation; or a vehicle that is owned and operated by a motor vehicle manufacturer or motor vehicle component manufacturer, or owned or held by a university research department, independent testing laboratory, or other such evaluation facility, solely for the purpose of evaluating the performance of such vehicles for engineering, research and development, or quality control reasons.

Demurrage: The charge paid to the vessel owner or operator for detention of a vessel at the port(s) beyond the time allowed, usually 72 hours, for loading and unloading.

Dependable capacity: The load-carrying ability of a station or system under adverse conditions for a specified period of time.

Depletion (coal): The subtraction of both tonnage produced and the tonnage lost to mining from identified resources to determine the remaining tonnage as of a certain time.

Depletion allowance: A term for either (1) a periodic assignment to expense of recorded amounts or (2) an allowable income tax deduction that is related to the exhaustion of mineral reserves. Depletion is included as one of the elements of amortization. When used in that manner, depletion refers only to book depletion. Book. The portion of the carrying value (other than the portion associated with tangible assets) prorated in each accounting period, for financial reporting purposes, to the extracted portion of an economic interest in wasting natural resource. Tax-cost. A deduction (allowance) under U.S. Federal income taxation normally calculated under a formula whereby the adjusted basis of the mineral property is multiplied by a fraction, the numerator of which is the number of units of minerals sold during the tax year and the denominator of which is the estimated number of units of unextracted minerals remaining at the end of the tax year plus the number of units of minerals sold during the tax year. Tax-percentage (for Statutory). A deduction (allowance) allowed to certain mineral producers under U.S. Federal income taxation calculated on the basis of a specified percentage of gross revenue from the sale of minerals from each mineral property not to exceed the lesser of 50 percent of the taxable income from the property computed without allowance for depletion. (There are also other limits on percentage depletion on oil and gas production.) The taxpayer is entitled to a deduction representing the amount of tax-cost depletion or percentage (statutory) depletion, whichever is higher. Excess statutory depletion. The excess of estimated statutory depletion allowable as an income tax deduction over the amount of cost depletion otherwise allowable as a tax deduction, determined on a total enterprise basis.

Depleted resources: Resources that have been mined; include coal recovered, coal lost in mining, and coal reclassified as subeconomic because of mining.

Depleted storage field: A sub-surface natural geological reservoir, usually a depleted gas or oil field, used for storing natural gas.

Depletion factor: The multiplier applied to the tonnage produced to compute depletion. This multiplier takes into account both the tonnage recovered and the tonnage lost due to mining. The depletion factor is the reciprocal of the recovery factor in relation to a given quantity of production.

Depreciation and amortization of property, plant, and equipment: The monthly provision for depreciation and amortization (applicable to utility property other than electric plant, electric plant in service, and equipment).

Depth of deepest production: The depth of the deepest production is the length of the well bore measured (in feet) from the surface reference point to the bottom of the open hole or the deepest perforation in the casing of a producing well.

Deregulation: The elimination of some or all regulations from a previously regulated industry or sector of an industry.

Design electrical rating (capacity) net: The nominal net electrical output of a nuclear unit, as specified by the utility for the purpose of plant design.

Design head: The achieved river, pondage, or reservoir surface height (forebay elevation) that provides the water level to produce the full flow at the gate of the turbine in order to attain the manufacturer's installed nameplate rating for generation capacity.

Desulfurization: The removal of sulfur, as from molten metals, petroleum oil, or flue gases.

Development: The preparation of a specific mineral deposit for commercial production; this preparation includes construction of access to the deposit and of facilities to extract the minerals. The development process is sometimes further distinguished between a preproduction stage and a current stage, with the distinction being made on the basis of whether the development work is performed before or after production from the mineral deposit has commenced on a commercial scale.

Development costs: Costs incurred to obtain access to proved reserves and to provide facilities for extracting, treating, gathering, and storing the oil and gas. More specifically, development costs, depreciation and applicable operating costs of support equipment and facilities, and other costs of development activities, are costs incurred to: Gain access to and prepare well locations for drilling, including surveying well locations for the purpose of determing specific development drilling sites; clearing ground; draining; road building; and relocating public roads, gas lines, and power lines to the extent necessary in developing the proved reserves. Drill and equip development wells, development-type stratigraphic test wells, and service wells, including the costs of platforms and of well equipment such as casing, tubing, pumping equipment, and the wellhead assembly. Acquire, construct, and install production facilities such as lease flow lines, separators, treaters, heaters, manifolds, measuring devices, production storage tanks, natural gas cycling and processing plants, and utility waste disposal systems. Provide improved recovery systems.

Development drilling: Drilling done to determine more precisely the size, grade, and configuration of an ore deposit subsequent to when the determination is made that the deposit can be commercially developed. Not included are: (1) secondary development drilling, (2) solution-mining drilling for production, or (3) production-related underground and openpit drilling done for control of mining operations.

Development well: A well drilled within the proved area of an oil or gas reservoir to the depth of a stratigraphic horizon known to be productive.

Diesel-electric plant: A generating station that uses diesel engines to drive its electric generators.

Diesel fuel: A fuel composed of distillates obtained in petroleum refining operation or blends of such distillates with residual oil used in motor vehicles. The boiling point and specific gravity are higher for diesel fuels than for gasoline.

Diesel fuel system: Diesel engines are internal combustion engines that burn diesel oil rather than gasoline. Injectors are used to spray droplets of diesel oil into the combustion chambers, at or near the top of the compression stroke. Ignition follows due to the very high temperature of the compressed intake air, or to the use of "glow plugs," which retain heat from previous ignitions (spark plugs are not used). Diesel engines are generally more fuel-efficient than gasoline engines but must be stronger and heavier because of high compression ratios.

Diffusive transport: The process by which particles of liquids or gases move from an area of higher concentration to an area of lower concentration.

Direct access: The ability of a retail customer to purchase electricity or other energy sources directly from a supplier other than their traditional supplier.

Direct electricity load control: The utility installs a radio-controlled device on the HVAC equipment. During periods of particulary heavy use of electricity, the utility will send a radio signal to the building in its service territory with this device and turn off the HVAC for a certain period.

Direct labor hours: Direct labor hours worked by all mining employees at a mining operation during the year. Includes hours worked by those employees engaged in production, preparation, development, maintenance, repair, shop or yard work management, and technical or engineering work. Excludes office workers. Excludes vacation and leave hours.

Direct load control: This Demand-Side Management category represents the consumer load that can be interrupted at the time of annual peak load by direct control of the utility system operator. Direct Load Control does not include Interruptible Load. This type of control usually involves residential consumers.

Direct milling cost: Operating costs directly attributable to the processing of ores or other feed materials, including labor, supervision, engineering, power, fuel, supplies, reagents, and maintenance.

Direct mining cost: Operating cost directly attributable to the mining of ore, including costs for labor, supervision, engineering, power, fuel, supplies, equipment replacement, maintenance, and taxes on production.

Direct nonprocess end use: Those end uses that may be found on commercial, residential, or other sites, as well as at manufacturing establishments. They include heating, ventilation, and air conditioning (HVAC), facility lighting, facility support, onsite transportation, conventional electricity generation, and other nonprocess uses. "Direct" denotes that only the quantities of electricity or fossil fuel used in their original state (*i.e.*, not transformed) are included in the estimates.

Direct process end use: Those end uses that are specific to the carrying out of manufacturing. They include process heating, process cooling and refrigeration, machine drive, electrochemical processes, and other process uses. "Direct" denotes that only the quantities of electricity or fossil fuel used in their original state (*i.e.*, not transformed) are included in the estimates.

Directional (deviated) well: A well purposely deviated from the vertical, using controlled angles to reach an objective location other than directly below the surface location. A directional well may be the original hole or a directional "sidetrack" hole that deviates from the original bore at some point below the surface. The new footage associated with directional "sidetrack" holes should not be confused with footage resulting from remedial sidetrack drilling. If there is a common bore from which two or more wells are drilled, the first complete bore from the surface to the original objective is classified and reported as a well drilled. Each of the deviations from the common bore is reported as a separate well.

Discharged fuel: Irradiated fuel removed from a nuclear reactor during refueling.

Discrete-delivery energy sources: Energy sources that must be delivered to a site.

Dispatching: The operating control of an integrated electric system involving operations such as (1) the assignment of load to specific generating stations and other sources of supply to effect the most economical supply as the total or the significant area loads rise or fall (2) the control of operations and maintenance of high-voltage lines, substations, and equipment; (3) the operation of principal tie lines and switching; (4) the scheduling of energy transactions with connecting electric utilities.

Disposition, natural gas: The removal of natural, synthetic, and/or supplemental gas, or any components or gaseous mixtures contained therein, from the responding company's facilities within the report State by any means or for any purpose, including the transportation of such gas out of the report State.

Disposition, petroleum: A set of categories used to account for how crude oil and petroleum products are transferred, distributed, or removed from the supply stream. The categories include stock change, crude oil losses, refinery inputs, exports, and products supplied for domestic consumption.

Distillate fuel oil: A general classification for one of the petroleum fractions produced in conventional distillation operations. It includes diesel fuels and fuel oils. Products known as No. 1, No. 2, and No. 4 diesel fuel are used in on-highway diesel engines, such as those in trucks and automobiles, as well as off-highway engines, such as those in railroad locomotives and agricultural machinery. Products known as No. 1, No. 2, and No. 4 fuel oils are used primarily for space heating and electric power generation.

No. 1 Distillate: A light petroleum distillate that can be used as either a diesel fuel (see **No. 1 Diesel Fuel**) or a fuel oil. See **No. 1 Fuel Oil**.

No. 1 Diesel Fuel: A light distillate fuel oil that has distillation temperatures of 550 degrees Fahrenheit at the 90-percent point and meets the specifications defined in ASTM Specification D 975. It is used in high-speed diesel engines, such as those in city buses and similar vehicles.

No. 1 Fuel Oil: A light distillate fuel oil that has distillation temperatures of 400 degrees Fahrenheit at the 10-percent recovery point and 550 degrees Fahrenheit at the 90-percent point and meets the specifications defined in ASTM Specification D 396. It is used primarily as fuel for portable outdoor stoves and portable outdoor heaters. See **No. 1 Distillate** above.

No. 2 Distillate: A petroleum distillate that can be used as either a diesel fuel (see **No. 2 Diesel Fuel** definition below) or a fuel oil.

No. 2 Diesel Fuel: A fuel that has distillation temperatures of 500 degrees Fahrenheit at the 10-percent recovery point and 640 degrees Fahrenheit at the 90-percent recovery point and meets the specifications defined in ASTM Specification D 975. It is used in high-speed diesel engines, such as those in railroad locomotives, trucks, and automobiles.

Low Sulfur No. 2 Diesel Fuel: No. 2 diesel fuel that has a sulfur level no higher than 0.05 percent by weight. It is used primarily in motor vehicle diesel engines for on-highway use.

High Sulfur No. 2 Diesel Fuel: No. 2 diesel fuel that has a sulfur level above 0.05 percent by weight.

No. 2 Fuel oil (Heating Oil): A distillate fuel oil that has distillation temperatures of 400 degrees Fahrenheit at the 10-percent recovery point and 640 degrees Fahrenheit at the 90-percent recovery point and meets the specifications defined in ASTM Specification D 396. It is used in atomizing type burners for domestic heating or for moderate capacity commercial/industrial burner units.

No. 4 Fuel: A distillate fuel oil made by blending distillate fuel oil and residual fuel oil stocks. It conforms with ASTM Specification D 396 or Federal Specification VV-F-815C and is used extensively in industrial plants and in commercial burner installations that are not equipped with preheating facilities. It also includes No. 4 diesel fuel used for low- and medium-speed diesel engines and conforms to ASTM Specification D 975.

No. 4 Diesel Fuel and No. 4 Fuel Oil: See **No. 4 Fuel** above.

Distillation unit (atmospheric): The primary distillation unit that processes crude oil (including mixtures of other hydrocarbons) at approximately atmospheric conditions. It includes a pipe still for vaporizing the crude oil and a fractionation tower for separating the vaporized hydrocarbon components in the crude oil into fractions with different boiling ranges. This is done by continuously vaporizing and condensing the components to separate higher oiling point material. The selected boiling ranges are set by the processing scheme, the properties of the crude oil, and the product specifications.

Distributed/point-of-use water-heating system: A system for heating hot water, for other than space heating purposes, which is located at more than one space within a building. A point-of-use water heater is located at the faucet and heats water only as required for immediate use. Because water is not heated until it is required, this equipment is more energy-efficient.

Distribution: The delivery of energy to retail customers.

Distribution system: The portion of the transmission and facilities of an electric system that is dedicated to delivering electric energy to an end-user.

Distribution use: Natural gas used as fuel in the respondent's operations.

Distributor: A company primarily engaged in the sale and delivery of natural and/or supplemental gas directly to consumers through a system of mains.

District chilled water: Chilled water from an outside source used as an energy source for cooling in a building. The water is chilled in a central plant and piped into the building. Chilled water may be purchased from a utility or provided by a central physical plant in a separate building that is part of the same multibuilding facility (for example, a hospital complex or university).

District heat: Steam or hot water from an outside source used as an energy source in a building The steam or hot water is produced in a central plant and piped into the building. The district heat may be purchased from a utility or provided by a physical plant in a separate building that is part of the same facility (for example, a hospital complex or university).

Diversity: The electric utility system's load is made up of many individual loads that make demands upon the system usually at different times of the day. The individual loads within the customer classes follow similar usage patterns, but these classes of service place different demands upon the facilities and the system grid. The service requirements of one electrical system can differ from another by time-of-day usage, facility usage, and/or demands placed upon the system grid.

Diversity exchange: An exchange of capacity or energy, or both, between systems whose peak loads occur at different times.

Divestiture: The stripping off of one utility function from the others by selling (spinning-off) or in some other way changing the ownership of the assets related to that function. Stripping off is most commonly associated with spinning-off generation assets so they are no longer owned by the shareholders that own the transmission and distribution assets.

Docket: A formal record of a Federal Energy Regulatory Commission proceeding. These records are available for inspection and copying by the public. Each individual case proceeding is identified by an assigned number.

DOE: Department of Energy.

Domestic inland consumption: Domestic inland consumption is the sum of all refined petroleum products supplied for domestic use (excludes international marine bunkers). Consumption is calculated by product by adding production, imports, crude oil burned directly, and refinery fuel and losses, and then subtracting exports and charges in primary stocks (net withdrawals is a plus quantity and net additions is a minus quantity).

Domestic uranium industry: Collectively, those businesses (whether U.S. or foreign-based) that operate under the laws and regulations pertaining to the conduct of commerce within the United States and its territories and possessions and that engage in activities within the United States, its territories, and possessions specifically directed toward uranium exploration, development, mining, and milling; marketing of uranium materials; enrichment; fabrication; or acquisition and management of uranium materials for use in commercial nuclear power plants.

Domestic vehicle producer: An Original Vehicle Manufacturer that assembles vehicles in the United States for domestic use. The term "domestic" pertains to the fifty states, the District of Columbia, commonwealths, territories, and possessions of the United States.

Double circuit line: A transmission line having two separate circuits. Because each carries three-phase power, at least six conductors, three per circuit, are required.

Drainage basin: The land drained by a river system.

Drawdown: The lowering of the water level of a reservoir as a result of withdrawing water.

Drawdown (maximum): The distance that the water surface of the reservoir is lowered from the normal full elevation to the lowest allowable elevation as the result of the withdrawal of water for the purposes of generating electricity.

Dredge mining: A method of recovering coal from rivers or streams.

Drift mine: A mine that opens horizontally into the coal bed or coal outcrop.

Drilling: The act of boring a hole (1) to determine whether minerals are present in commercially recoverable quantities and (2) to accomplish production of the minerals (including drilling to inject fluids).

1. Exploratory. Drilling to locate probable mineral deposits or to establish the nature of geological structures; such wells may not be capable of production if minerals are discovered.
2. Developmental. Drilling to delineate the boundaries of a known mineral deposit to enhance the productive capacity of the producing mineral property.
3. Directional. Drilling that is deliberately made to depart significantly from the vertical.

Drilling and equipping of wells: The drilling and equipping of wells through completion of the "christmas tree."

Drilling arrangement: A contractual agreement under which a working interest owner (assignor) assigns a part of a working interest in a property to another party (the assignee) in exchange for which the assignee agrees to develope the property. The term may also be applied to an agreement under which an operator assigns fractional shares in production from a property to participants for cash considerations as a means of acquiring cash for developing the property. Under a "disproportionate cost" drilling arrangement, the participants normally pay a greater total share of costs than the total value of the fractional shares of the property received in the arrangement.

Dry bottom boiler: No slag tanks at furnace throat area. The throat area is clear. Bottom ash drops through the throat to the bottom ash water hoppers. This design is used where the ash melting temperature is greater than the temperature on the furnace wall, allowing for relatively dry furnace wall conditions.

Dry (coal) basis: Coal quality data calculated to a theoretical basis in which no moisture is associated with the sample. This basis is determined by measuring the weight loss of a sample when its inherent moisture is driven off under controlled conditions of low temperature air-drying followed by heating to just above the boiling point of water (104 to 110 degrees Centigrade).

Dry hole: An exploratory or development well found to be incapable of producing either oil or gas in sufficient quantities to justify completion as an oil or gas well.

Dry hole charge: The charge-off to expense of a previously capitalized cost upon the conclusion of an unsuccessful drilling effort.

Dry hole contribution: A payment (either in cash or acreage) that is required by agreement only if a test well is unsuccessful and that is made in exchange for well test and evaluation data.

Dry natural gas: Natural gas which remains after: 1) the liquefiable hydrocarbon portion has been removed from the gas stream (*i.e.*, gas after lease, field, and/or plant separation); and 2) any volumes of nonhydrocarbon gases have been removed where they occur in sufficient quantity to render the gas unmarketable. Note: Dry natural gas is also known as consumer-grade natural gas. **Dry natural gas production:** The process of producing consumer-grade natural gas. Natural gas withdrawn from reservoirs is reduced by volumes used at the production (lease) site and by processing losses. Volumes used at the production site include (1) the volume returned to reservoirs in cycling, repressuring of oil reservoirs, and conservation operations; and (2) gas vented and flared. Processing losses include (1) nonhydrocarbon gases (*e.g.*, water vapor, carbon dioxide, helium, hydrogen sulfide, and nitrogen) removed from the gas stream; and (2) gas converted to liquid form, such as lease condensate and plant liquids. Volumes of dry gas withdrawn from gas storage reservoirs are not considered part of production. Dy natural gas production equals marketed production less extraction loss.

Dual fuel vehicle (1): A motor vehicle that is capable of operating on an alternative fuel and on gasoline or diesel fuel. These vehicles have at least two separate fuel systems which inject each fuel simultaneously into the engine combustion chamber.

Dual fuel vehicle (2): A motor vehicle that is capable of operating on an alternative fuel and on gasoline or diesel fuel. This term is meant to represent all such vehicles whether they operate on the alternative fuel and gasoline/diesel simultaneously (*e.g.*, flexible-fuel vehicles) or can be switched to operate on gasoline/diesel or an alternative fuel (*e.g.*, bi-fuel vehicles).

Dual-fired unit: A generating unit that can produce electricity using two or more input fuels. In some of these units, only the primary fuel can be used continuously; the alternate fuel(s) can be used only as a start-up fuel or in emergencies.

Dump energy: Energy generated in a hydroelectric plant by water that cannot be stored or conserved and which energy is in excess of the needs of the system producing the energy.

E85: A fuel containing a mixture of 85 percent ethanol and 15 percent gasoline.

E95: A fuel containing a mixture of 95 percent ethanol and 5 percent gasoline

Economy of scale: The principle that larger production facilities have lower unit costs than smaller facilities.

Effective full-power days: The number of effective full-power days produced by a unit is a measure of the unit's energy generation. It is determined using the following ratio: Heat generation (planned or actual) in megawatt days thermal (MWdt)(divided by) Licensed thermal power in megawatts thermal (MWt).

Electric baseboard: An individual space heater with electric resistance coils mounted behind shallow panels along baseboards. Electric baseboards rely on passive convection to distribute heated air to the space.

Electric current: The flow of electric charge. The preferred unit of measure is the ampere.

Electric energy: The ability of an electric current to produce work, heat, light, or other forms of energy. It is measured in kilowatthours.

Electric expenses: The cost of labor, material, and expenses incurred in operating a facility's prime movers, generators, auxiliary apparatus, switching gear, and other electric equipment for each of the points where electricity enters the transmission or distribution grid.

Electric generation industry: Stationary and mobile generating units that are connected to the electric power grid and can generate electricity. The electric generation industry includes the "electric power sector" (utility generators and independent power producers) and industrial and commercial power generators, including combined-heat-and-power producers, but excludes units at single-family dwellings.

Electric generator: A facility that produces only electricity, commonly expressed in kilowatthours (kWh) or megawatthours (MWh). Electric generators include electric utilities and independent power producers.

Electric hybrid vehicle: An electric vehicle that either (1) operates solely on electricity, but contains an internal combustion motor that generates additional electricity (series hybrid); or (2) contains an electric system and an internal combustion system and is capable of operating on either system (parallel hybrid).

Electric industry reregulation: The design and implementation of regulatory practices to be applied to the remaining traditional utilities after the electric power industry has been restructured. Reregulation applies to those entities that continue to exhibit characteristics of a natural monopoly. Reregulation could employ the same or different regulatory practices as those used before restructuring.

Electric industry restructuring: The process of replacing a monopolistic system of electric utility suppliers with competing sellers, allowing individual retail customers to choose their supplier but still receive delivery over the power lines of the local utility. It includes the reconfiguration of vertically-integrated electric utilities.

Electric motor vehicle: A motor vehicle powered by an electric motor that draws current from rechargeable storage batteries, fuel cells, photovoltaic arrays, or other sources of electric current.

Electric operating expenses: Summation of electric operation-related expenses, such as operation expenses, maintenance expenses, depreciation expenses, amortization, taxes other than income taxes, Federal income taxes, other income taxes, provision for deferred income taxes, provision for deferred income-credit, and investment tax credit adjustment.

Electric plant (physical): A facility containing prime movers, electric generators, and auxiliary equipment for converting mechanical, chemical, and/or fission energy into electric energy.

Electric plant acquisition adjustment: The difference between (a) the cost to the respondent utility of an electric plant acquired as an operating unit or system by purchase and (b) the depreciated original cost, estimated if not known, of such property.

Electric power: The rate at which electric energy is transferred. Electric power is measured by capacity and is commonly expressed in megawatts (MW).

Electric power grid: A system of synchronized power providers and consumers connected by transmission and distribution lines and operated by one or more control centers. In the continental United States, the electric power grid consists of three systems: the Eastern Interconnect, the Western Interconnect, and the Texas Interconnect. In Alaska and Hawaii, several systems encompass areas smaller than the State (*e.g.*, the interconnect serving Anchorage, Fairbanks, and the Kenai Peninsula; individual islands).

Electric power plant: A station containing prime movers, electric generators, and auxiliary equipment for converting mechanical, chemical, and/or fission energy into electric energy.

Electric power sector: An energy-consuming sector that consists of electricity only and combined heat and power(CHP) plants whose primary business is to sell electricity, or electricity and heat, to the public–*i.e.*, North American Industry Classification System 22 plants.

Electric power system: An individual electric power entity–a company; an electric cooperative; a public electric supply corporation as the Tennesse Valley Authority; a similar Federal department or agency such as the Bonneville Power Administration; the Bureau of Reclamation or the Corps of Engineers; a municipally owned electric department offering service to the public; or an electric public utility district (a "PUD"); also a jointly owned electric supply project such as the Keystone.

Electric pump for well water: This pump forces the water from a well below ground level up into the water pipes that circulate through the house. When this pump is not working, there is a limited supply of running water in the house.

Electric rate: The price set for a specified amount and type of electricity by class of service in an electric rate schedule or sales contract.

Electric rate schedule: A statement of the electric rate and the terms and conditions governing its application, including attendant contract terms and conditions that have been accepted by a regulatory body with appropriate oversight authority.

Electric system loss: Total energy loss from all causes for an electric utility.

Electric system reliability: The degree to which the performance of the elements of the electrical system results in power being delivered to consumers within accepted standards and in the amount desired. Reliability encompasses two concepts, adequacy and security. Adequacy implies that there are sufficient generation and transmission resources installed and available to meet projected electrical demand plus reserves for contingencies. Security implies that the system will remain intact operationally (*i.e.*, will have sufficient available operating capacity) even after outages or other equipment failure. The degree of reliability may be measured by the frequency, duration, and magnitude of adverse effects on consumer service.

Electric utility: A corporation, person, agency, authority, or other legal entity or instrumentality aligned with distribution facilities for delivery of electric energy for use primarily by the public. Included are investor-owned electric utilities, municipal and State utilities, Federal electric utilities, and rural electric cooperatives. A few entities that are tariff based and corporately aligned with companies that own distribution facilities are also included.

Electric utility divestiture: The separation of one electric utility function from others through the selling of the management and ownership of the assets related to that function. It is most commonly associated with selling generation assets so they are no longer owned or controlled by the shareholders that own the company's transmission and distribution assets.

Electric utility restructuring: The introduction of competition into at least the generation phase of electricity production, with a corresponding decrease in regulatory control.

Electric utility sector: The electric utility sector consists of privately and publicly owned establishments that generate, transmit, distribute, or sell electricity primarily for use by the public and that meet the defintion of an electric utility. Nonutility power producers are not included in the electric sector.

Electric zone: A portion of the grid controlled by the independent system operator.

Electrical system energy losses: The amount of energy lost during generation, transmission, and distribution of electricity, including plant and unaccounted for use.

Electricity: A form of energy characterized by the presence and motion of elementary charged particles generated by friction, induction, or chemical change.

Electricity broker: An entity that arranges the sale and purchase of electric energy, the transmission of electricity, and/or other related services between buyers and sellers but does not take title to any of the power sold.

Electricity congestion: A condition that occurs when insufficient transmission capacity is available to implement all of the desired transactions simultaneously.

Electricity demand: The rate at which energy is delivered to loads and scheduling points by generation, transmission, and distribution facilities.

Electricity demand bid: A bid into the power exchange indicating a quantity of energy or an ancillary service that an eligible customer is willing to purchase and, if relevant, the maximum price that the customer is willing to pay.

Electricity generation: The process of producing electric energy or the amount of electric energy produced by transforming other forms of energy, commonly expressed in kilowatthours (kWh) or megawatthours (MWh).

Electricity only plant: A plant designed to produce electricity only.

Electricity paid by household: The household paid the electric utility company directly for all household uses of electricity (such as water heating, space heating, air-conditioning, cooking, lighting, and operating appliances.) Bills paid by a third party are not counted as paid by the household.

Electricity sales: The amount of kilowatthours sold in a given period of time; usually grouped by classes of service, such as residential, commercial, industrial, and other. "Other" sales include sales for public street and highway lighting and other sales to public authorities, sales to railroads and railways, and interdepartmental sales.

Electrochemical process: The direct process end use in which electricity is used to cause a chemical ransformation. Major uses of electrochemical process occur in the aluminum industry in which alumina is reduced to molten aluminum metal and oxygen, and in the alkalies and chlorine industry, in which brine is separated into caustic soda, chlorine, and hydrogen.

Elution: Activities of removing "elutes" a material (uranium) adsorbed on ion exchange resin from the "eluant" solution.

Emergency: The failure of an electric power system to generate or deliver electric power as normally intended, resulting in the cutoff or curtailment of service.

Emergency backup generation: The use of electric generators only during interruptions of normal power supply.

Emergency energy: Electric energy provided for a limited duration, intended only for use during emergency conditions.

Emissions: Anthropogenic releases of gases to the atmosphere. In the context of global climate change, they consist of radiatively important greenhouse gases (*e.g.*, the release of carbon dioxide during fuel combustion).

Emissions coefficient: A unique value for scaling emissions to activity data in terms of a standard rate of emissions per unit of activity (*e.g.*, pounds of carbon dioxide emitted per Btu of fossil fuel consumed).

End user: A firm or individual that purchases products for its own consumption and not for resale (*i.e.*, an ultimate consumer).

Ending stocks: Primary stocks of crude oil and petroleum products held in storage as of 12 midnight on the last day of the month. Primary stocks include crude oil or petroleum products held in storage at (or in) leases, refineries, natural gas processing plants, pipelines, tank farms, and bulk terminals that can store at least 50,000 barrels of petroleum products or that can receive petroleum products by tanker, barge, or pipeline. Crude oil that is in-transit by water from Alaska or that is stored on Federal leases or in the Strategic Petroleum Reserve is included. Primary Stocks exclude stocks of foreign origin that are held in bonded warehouse storage.

Energy: The capacity for doing work as measured by the capability of doing work (potential energy) or the conversion of this capability to motion (kinetic energy). Energy has several forms, some of which are easily convertible and can be changed to another form useful for work. Most of the world's convertible energy comes from fossil fuels that are burned to produce heat that is then used as a transfer medium to mechanical or other means in order to accomplish tasks. Electrical energy is usually measured in kilowatthours, while heat energy is usually measured in British thermal units (Btu).

Energy audit: A program carried out by a utility company in which an auditor inspects a home and suggests ways energy can be saved.

Energy broker system: Introduced into Florida by the Public Service Commission, the energy broker system is a system for exchanging information that allows utilities to efficiently exchange hourly quotations of prices at which each is willing to buy and sell electric energy. For the broker system to operate, utility systems must have in place bilateral agreements between all potential parties, must have transmission arrangements between all potential parties, and must have transmission arrangements that allow the exchanges to take place.

Energy charge: That portion of the charge for electric service based upon the electric energy (kWh) consumed or billed.

Energy conservation features: This includes building shell conservation features, HVAC conservation features, lighting conservation features, any conservation features, and other conservation features incorporated by the building. However, this category does not include any demand-side management (DSM) program participation by the building. Any DSM program participation is included in the DSM Programs.

Energy consumption: The use of energy as a source of heat or power or as a raw material input to a manufacturing process.

Energy demand: The requirement for energy as an input to provide products and/or services.

Energy deliveries: Energy generated by one electric utility system and delivered to another system through one or more transmission lines.

Energy effects: The changes in aggregate electricity use (measured in megawatthours) for consumers that participate in a utility DSM (demand-side management) program. Energy effects represent changes at the consumer's meter (*i.e.*, exclude ransmission and distribution effects) and reflect only activities that are undertaken specifically in response to utility-administered programs, including those activities implemented by third parties under contract to the utility. To the extent possible, Energy effects should exclude non-program related effects such as changes in energy usage attributable to nonparticipants, government-mandated energy-efficiency standards that legislate improvements in building and appliance energy usage, changes in consumer behavior that result in greater energy use after initiation in a DSM program, the natural operations of the marketplace, and weather and business-cycle adjustments.

Energy efficiency: Refers to programs that are aimed at reducing the energy used by specific end-use devices and systems, typically without affecting the services provided. These programs reduce overall electricity consumption (reported in megawatthours), often without explicit consideration for the timing of program-induced savings. Such savings are generally achieved by substituting technically more advanced equipment to produce the same level of end-use services (*e.g.* lighting, heating, motor drive) with less electricity. Examples include high-efficiency appliances, efficient lighting programs, high-efficiency heating, ventilating and air conditioning (HVAC) systems or control modifications, efficient building design, advanced electric motor drives, and heat recovery systems.

Energy efficient motors: Are also known as "high-efficiency motors" and "premium motors." They are virtually interchangeable with standard motors, but differences in construction make them more energy efficient.

Energy exchange: Any transaction in which quantities of energy are received or given up in return for similar energy products.

Energy expenditures: The money directly spent by consumers to purchase energy. Expenditures equal the amount of energy used by the consumer multiplied by the price per unit paid by the consumer.

Energy information: Includes (A) all information in whatever form on fuel reserves, extraction, and energy resources (including petrochemical feedstocks) wherever located; production, distribution, and consumption of energy and fuels wherever carried on; and (B) matters relating to energy and fuels, such as corporate structure and proprietary relationships, costs, prices, capital investment, and assets, and other matters directly related thereto, wherever they exist.

Energy intensity (Commercial Buildings Energy Consumption Survey): The ratio of consumption to floor space.

Energy loss: Deleted because there is no need for a general term to encompass all forms of energy loss. Terms referring to losses specific to particular energy sources are defined separately.

Energy management and control system (EMCS): An energy conservation feature that uses mini/microcomputers, instrumentation, control equipment, and software to manage a building's use of energy for heating, ventilation, air conditioning, lighting, and/or business-related processes. These systems can also manage fire control, safety, and security. Not included as EMCS are time-clock thermostats.

Energy management practices: Involvement, as a part of the building's normal operations, in energy efficiency programs that are designed to reduce the energy used by specific end-use systems. This includes the following: EMCS, DSM Program Participation, Energy Audit, and a Building Energy Manager.

Energy Policy Act of 1992 (EPACT): This legislation creates a new class of power generators, exempt wholesale generators, that are exempt from the provisions of the Public Holding Company Act of 1935 and grants the authority to the Federal Energy Regulatory Commission to order and condition access by eligible parties to the interconnected transmission grid.

Energy receipts: Energy brought into a site from another location.

Energy reserves: Estimated quantities of energy sources that are demonstrated to exist with reasonable certainty on the basis of geologic and engineering data (proved reserves) or that can reasonably be expected to exist on the basis of geologic evidence that supports projections from proved reserves (probable/indicated reserves). Knowledge of the location, quantity, and grade of probable/indicated reserves is generally incomplete or much less certain than it is for proved energy reserves.

Energy sale(s): The transfer of title to an energy commodity from a seller to a buyer for a price or the quantity transferred during a specified period.

Energy savings: A reduction in the amount of electricity used by end users as a result of participation in energy efficiency programs and load management programs.

Energy service provider: An energy entity that provides service to a retail or end-use customer.

Energy source: Any substance or natural phenomenon that can be consumed or transformed to supply heat or power. Examples include petroleum, coal, natural gas, nuclear, biomass, electricity, wind, sunlight, geothermal, water movement, and hydrogen in fuel cells.

Energy supply: Energy made available for future disposition. Supply can be considered and measured from the point of view of the energy provider or the receiver.

Energy supplier: Fuel companies supplying electricity, natural gas, fuel oil, kerosene, or LPG (liquefied petroleum gas) to the household.

Energy used in the home: For electricity or natural gas, the quantity is the amount used by the household during the 365- or 366-day period. For fuel oil, kerosene, and liquefied petroleum gas (LPG), the quantity consists of fuel purchased, not fuel consumed. If the level of fuel in the storage tank was the same at the beginning and end of the annual period, then the quantity consumed would be the same as the quantity purchased.

Energy-use sectors: A group of major energy-consuming components of U.S. society developed to measure and analyze energy use. The sectors most commonly referred are: residential, commercial, industrial, transportation, and electric power.

Energy-weighted industrial output: The weighted sum of real output for all two-digit Standard Industrial Classification (SIC) manufacturing industries plus agriculture, construction, and mining. The weight for each industry is the ratio between the quantity of end-use energy consumption to the value of real output.

Engine size: The total volume within all cylinders of an engine when pistons are at their lowest positions. The engine is usually measured in "liters" or "cubic inches of displacement (CID)." Generally, larger engines result in greater engine power, but less fuel efficiency. There are 61.024 cubic inches in a liter.

Enriched uranium: Uranium in which the U-235 isotope concentration has been increased to greater than the 0.711 percent U-235 (by weight) present in natural uranium.

Enrichment feed deliveries: Uranium that is shipped under contract to a supplier of enrichment services for use in preparing enriched uranium product to a specified U-235 concentration and that ultimately will be used as fuel in a nuclear reactor.

Enrichment tails assay: A measure of the amount of fissile uranium (U-235) remaining in the waste stream from the uranium enrichment process. The natural uranium "feed" that enters the enrichment process generally contains 0.711 percent (by weight) U-235. The "product stream" contains enriched uranium (more than 0.711 percent U-235) and the "waste" or "tails" stream contains depleted uranium (less than 0.711 percent U-235). At the historical enrichment tails assay of 0.2 percent, the waste stream would contain 0.2 percent U-235. A higher enrichment tails assay requires more uranium feed (thus permitting natural uranium stockpiles to be decreased), while increasing the output of enriched material for the same energy expenditure.

Environmental impact statement: A report that documents the information required to evaluate the environmental impact of a project. It informs decision makers and the public of the reasonable alternatives that would avoid or minimize adverse impacts or enhance the quality of the environment.

Environmental Protection Agency (EPA) certification files: Computer files produced by EPA for analysis purposes. For each vehicle make, model and year, the files contain the EPA test MPGs (city, highway, and 55/45 composite). These MPG's are associated with various combinations of engine and drive-train technologies (*e.g.*, number of cylinders, engine size, gasoline or diesel fuel, and automatic or manual transmission). These files also contain information similar to that in the DOE/EPA Gas Mileage Guide, although the MPGs in that publication are adjusted for shortfall.

Environmental restoration: Although usually described as "cleanup," this function encompasses a wide range of activities, such as stabilizing contaminated soil; treating ground water; decommissioning process buildings, nuclear reactors, chemical separations plants, and many other facilities; and exhuming sludge and buried drums of waste.

Environmental restrictions: In reference to coal accessibility, land-use restrictions that constrain, postpone, or prohibit mining in order to protect environmental resources of an area; for example, surface- or groundwater quality, air quality affected by mining, or plants or animals or their habitats.

EPA certification: A permanent label on fireplace inserts and freestanding wood stoves manufactured after July 1, 1988, indicating that the equipment meets EPA standards for clean burning.

EPA composite MPG: The harmonic mean of the EPA city and highway MPG (miles per gallon), weighted under the assumption of 55 percent city driving and 45 percent highway driving.

Equilibrium cycle: An analytical term that refers to fuel cycles that occur after the initial one or two cycles of a reactor's operation. For a given type of reactor, equilibrium cycles have similar fuel characteristics.

Equity (financial): Ownership of shareholders in a corporation represented by stock.

Equity capital: The sum of capital from retained earnings and the issuance of stock.

Equity crude oil: The proportion of production that a concession owner has the legal and contractual right to retain.

Equity in earnings of unconsolidated affiliates: A company's proportional share (based on ownership) of the net earnings or losses of an unconsolidated affiliate.

Establishment: An economic unit, generally, at a single physical location where business is conducted or where services or industrial operations are performed. However, "establishment" is not synonymous with "building."

Estimated additional resources (EAR): The uranium in addition to reasonable assured resources (RAR) that is expected to occur, mostly on the basis of direct geological evidence, in extensions of well-explored deposits, little-explored deposits, and undiscovered deposits believed to exist along a well-defined geologic trend with known deposits, such that the uranium can subsequently be recovered within the given cost ranges. Estimates of tonnage and grade are based on available sampling data and on knowledge of the deposit characteristics as determined in the best known parts of the deposit or in similar deposits. EAR correspond to DOE's Probable Potential Resource Category.

ETBE (ethyl tertiary butyl ether): $(CH_3)_3COC2H$: An oxygenate blend stock formed by the catalytic etherification of isobutylene with ethanol.

Ethane (C_2H_6): A normally gaseous straight-chain hydrocarbon. It is a colorless paraffinic gas that boils at a temperature of -127.48 degrees Fahrenheit. It is extracted from natural gas and refinery gas streams.

Ethanol (CH_3-CH_2OH): A clear, colorless, flammable oxygenated hydrocarbon. Ethanol is typically produced chemically from ethylene, or biologically from fermentation of various sugars from carbohydrates found in agricultural crops and cellulosic residues from crops or wood. It is used in the United States as a gasoline octane enhancer and oxygenate (blended up to 10 percent concentration). Ethanol can also be used in high concentrations (E85) in vehicles designed for its use.

Ether: A generic term applied to a group of organic chemical compounds composed of carbon, hydrogen, and oxygen, characterized by an oxygen atom attached to two carbon atoms (*e.g.*, methyl tertiary butyl ether).

Ethylene: An olefinic hydrocarbon recovered from refinery processes or petrochemical processes. Ethylene is used as a petrochemical feedstock for numerous chemical applications and the production of consumer goods.

Ethylene dichloride: A colorless, oily liquid used as a solvent and fumigant for organic synthesis, and for ore flotation.

Evacuated-tube collector: A collector in which solar thermal heat is captured by use of a collector fluid that flows through an absorber tube contained inside an evacuated glass tube.

Evaporation pond: A containment pond (that preferably has an impermeable lining of clay or synthetic material such as hypalon) to hold liquid wastes and to concentrate the waste through evaporation.

Evaporative cooler (swamp cooler): An air-cooling unit that turns air into moist, cool air by saturating the air with water vapor. It does not cool air by use of a refrigeration unit.

Excess statutory depletion: The excess of estimated statutory depletion allowable as an income tax deduction over the amount of cost depletion otherwise allowable as a tax deduction, determined on a total enterprise basis.

Exchange agreement: A contractual agreement in which quantities of crude oil, petroleum products, natural gas, or electricity are delivered, either directly or through intermediaries, from one company to another company, in exchange for the delivery by the second company to the first company of an equivalent volume or heat content. The exchange may take place at the same time and location or at different times and/or locations. Such agreements may also involve the payment of cash.

Exchange, electricity: A type of energy exchange in which one electric utility agrees to supply electricity to another. Electricity received is returned in kind at a later time or is accumulated as an energy balance until the end of a specified period, after which settlement may be made by monetary payment.

Exchange, natural gas: A type of energy exchange in which one company agrees to deliver gas, either directly or through intermediaries, to another company at one location or in one time period in exchange for the delivery by the second company to the first company of an equivalent volume or heat content at a different location or time period. **Exchange, petroleum:** A type of energy exchange in which quantities of crude oil or any petroleum product(s) are received or given up in return for other crude oil or petroleum products. It includes reciprocal sales and purchases.

Exchange, power: Delete in favor of the already-defined term **exchange energy**, which should be renamed **exchange electricity** or **exchange, electricity**.

Exempt wholesale generator (EWG): Wholesale generators created under the 1992 Energy Policy Act that are exempt from certain financial and legal restrictions stipulated in the Public Utilities Holding Company Act of 1935.

Exhaust fan: Small fans located in the wall or ceiling that exhaust air, odors, and moisture from the bathroom, kitchen, or basement to the outside.

Expenditure: The incurrence of a liability to obtain an asset or service.

Expenditures per million Btu: The aggregate ratio of a group of buildings' total expenditures for a given fuel to the total consumption of that fuel.

Expenditures per square foot: The aggregate ratio of a group of buildings' total expenditures for a given fuel to the total floor space in those buildings.

Exploration drilling: Drilling done in search of new mineral deposits, on extensions of known ore deposits, or at the location of a discovery up to the time when the company decides that sufficient ore reserves are present to justify commercial exploration. Assessment drilling is reported as exploration drilling.

Exploratory well: A hole drilled: a) to find and produce oil or gas in an area previously considered unproductive area; b) to find a new reservoir in a known field, *i.e.*, one previously producing oil and gas from another reservoir, or c) to extend the limit of a known oil or gas reservoir.

Exports: Shipments of goods from within the 50 States and the District of Columbia to U.S. possessions and territories or to foreign countries.

Extensions: Any new reserves credited to a previously producing reservoir because of enlargement of its proved area. This enlargement in proved area is usually due to new well drilling outside of the previously known productive limits of the reservoir.

Extensions, discoveries, and other additions: Additions to an enterprise's proved reserves that result from (1) extension of the proved acreage of previously discovered (old) reserves through additional drilling in periods subsequent to discovery and (2) discovery of new fields with proved reserves or of new reservoirs of proved reserves in old fields.

Externalities: Benefits or costs, generated as a byproduct of an economic activity, that do not accrue to the parties involved in the activity. Environmental externalities are benefits or costs that manifest themselves through changes in the physical or biological environment.

Extraction loss: The reduction in volume of natural gas due to the removal of natural gas liquid constituents such as ethane, propane, and butane at natural gas processing plants.

Extractive industries: Industries involved in the activities of (1) prospecting and exploring for wasting (non-regenerative) natural resources, (2) acquiring them, (3) further exploring them, (4) developing them, and (5) producing (extracting) them from the earth. The term does not encompass the industries of forestry, fishing, agriculture, animal husbandry, or any others that might be involved with resources of a regenerative nature.

Extraordinary income deductions (electric utility): Those items related to transactions of a nonrecurring nature that are not typical or customary business activities of the utility and that would significantly distort the current year's net income if reported other than as extraordinary items.

Fabricated fuel: Fuel assemblies composed of an array of fuel rods loaded with pellets of enriched uranium dioxide.

Facilities charge: An amount to be paid by the customer in a lump sum, or periodically as reimbursement for facilities furnished. The charge may include operation and maintenance as well as fixed costs.

Facility: An existing or planned location or site at which prime movers, electric generators, and/or equipment for converting mechanical, chemical, and/or nuclear energy into electric energy are situated or will be situated. A facility may contain more than one generator of either the same or different prime mover type. For a cogenerator, the facility includes the industrial or commercial process.

Fahrenheit: A temperature scale on which the boiling point of water is at 212 degrees above zero on the scale and the freezing point is at 32 degrees above zero at standard atmospheric pressure.

Failure or hazard: Any electric power supply equipment or facility failure or other event that, in the judgement of the reporting entity, constitutes a hazard to maintaining the continuity of the bulk electric power supply system such that a load reduction action may become necessary and reportable outage may occur. Types of abnormal conditions that should be reported include the imposition of a special operating procedure, the extended purchase of emergency power, other bulk power system actions that may be caused by a natural disaster, a major equipment failure that would impact the bulk power supply, and an environmental and/or regulatory action requiring equipment outages.

Farm out (in) arrangement: An arrangement, used primarily in the oil and gas industry, in which the owner or lessee of mineral rights (the first party) assigns a working interest to an operator (the second party), the consideration for which is specified exploration and/or development activities. The first party retains an overriding royalty or other type of economic interest in the mineral production. The arrangement from the viewpoint of the second party is termed a "farm-in arrangement."

Farm use: Energy use at establishments where the primary activity is growing crops and/or raising animals. Energy use by all facilities and equipment at these establishments is included, whether or not it is directly associated with growing crops and/or raising animals. Common types of energy-using equipment include tractors, irrigation pumps, crop dryers, smudge pots, and milking machines. Facility energy use encompasses all structures at the establishment, including the farm house.

f.a.s. value: Free alongside ship value. The value of a commodity at the port of exportation, generally including the purchase price plus all charges incurred in placing the commodity alongside the carrier at the port of exportation in the country of exportation.

Fast breeder reactor (FBR): A reactor in which the fission chain reaction is sustained primarily by fast neutrons rather than by thermal or intermediate neutrons. Fast reactors require little or no moderator to slow down the neutrons from the speeds at which they are ejected from fissioning nuclei. This type of reactor produces more fissile material than it consumes.

Federal coal lease: A lease granted to a mining company to produce coal from land owned and administered by the Federal Government in exchange for royalties and other revenues.

Federal electric utility: A utility that is either owned or financed by the Federal Government.

Federal Energy Regulatory Commission (FERC): The Federal agency with jurisdiction over interstate electricity sales, wholesale electric rates, hydroelectric licensing, natural gas pricing, oil pipeline rates, and gas pipeline certification. FERC is an independent regulatory agency within the Department of Energy and is the successor to the Federal Power Commission.

Federal Power Act: Enacted in 1920, and amended in 1935, the Act consists of three parts. The first part incorporated the Federal Water Power Act administered by the former Federal Power Commission, whose activities were confined almost entirely to licensing non-Federal hydroelectric projects. Parts II and III were added with the passage of the Public Utility Act. These parts extended the Act's jurisdiction to include regulating the interstate transmission of electrical energy and rates for its sale as wholesale in interstate commerce. The Federal Energy Regulatory Commission is now charged with the administration of this law.

Federal Power Commission (FPC): The predecessor agency of the Federal Energy Regulatory Commission. The Federal Power Commission was created by an Act of Congress under the Federal Water Power Act on June 10, 1920. It was charged originally with regulating the electric power and natural gas industries. It was abolished on September 30, 1977, when the Department of Energy was created. Its functions were divided between the Department of Energy and the Federal Energy Regulatory Commission, an independent regulatory agency.

Federal region: In a Presidential directive issued in 1969, various Federal agencies (among them the currently designated Department of Health and Human Services, the Department of Labor, the Office of Economic Opportunity, and the Small Business Administration) were instructed to adopt a uniform field system of 10 geographic regions with common boundaries and headquarters cities. The action was taken to correct the evolution of fragmented Federal field organization structures that each agency or component created independently, usually with little reference to other agencies' arrangements. Most Federal domestic agencies or their components have completed realignments and relocations to conform to the Standard Federal Administration Regions (SFARs).

Fee interest: The absolute, legal possession and ownership of land, property, or rights, including mineral rights. A fee interest can be sold (in its entirety or in part) or passed on to heirs or successors.

Feeder line: An electrical line that extends radially from a distribution substation to supply electrical energy within an electric area or sub-area.

FERC: The Federal Energy Regulatory Commission.

FERC guidelines: A compilation of the Federal Energy Regulatory Commission's enabling statutes; procedural and program regulations; and orders, opinions, and decisions.

Fertile material: Material that is not itself fissionable by thermal neutrons but can be converted to fissile material by irradiation. The two principal fertile materials are uranium-238 and thorium-232.

Field: An area consisting of a single reservoir or multiple reservoirs all grouped on, or related to, the same individual geological structural feature and/or stratigraphic condition. There may be two or more reservoirs in a field that are separated vertically by intervening impervious strata or laterally by local geologic barriers, or by both.

Field area: A geographic area encompassing two or more pools that have a common gathering and metering system, the reserves of which are reported as a single unit. This concept applies primarily to the Appalachian region.

Field discovery year: The calendar year in which a field was first recognized as containing economically recoverable accumulations of oil and / or gas.

Field production: Represents crude oil production on leases, natural gas liquids production at natural gas processing plants, new supply of other hydrocarbons / oxygenates and motor gasoline blending components, and fuel ethanol blended into finished motor gasoline.

Field separation facility: A surface installation designed to recover lease condensate from a produced natural gas stream usually originating from more than one lease and managed by the operator of one or more of these leases.

File rate schedule: The rate for a particular electric service, including attendant contract terms and conditions, accepted for filing by a regulatory body with appropriate oversight authority.

Filing: Any written application, complaint, declaration, petition, protest, answer, motion, brief, exception, rate schedule, or other pleading, amendment to a pleading, document, or similar paper that is submitted to a utility commission.

Final order: A final ruling by FERC that terminates an action, decides some matter litigated by the petitioning parties, operates to some right, or completely disposes of the subject matter.

Financial Accounting Standards Board (FASB): An independent board responsible, since 1973, for establishing generally accepted accounting principles. Its official pronouncement are called "Statements of Financial Accounting Standards" and "Interpretations of Financial Accounting Standards."

Finished leaded gasoline: Contains more than 0.05 gram of lead per gallon or more than 0.005 gram of phosphorus per gallon. Premium and regular grades are included, depending on the octane rating. Includes leaded gasohol. Blendstock is excluded until blending has been completed. Alcohol that is to be used in the blending of gasohol is also excluded.

Finished motor gasoline: A complex mixture of relatively volatile hydrocarbons, with or without small quantities of additives, blended to form a fuel suitable for use in spark-ignition engines. Specification for motor gasoline, as given in ASTM Specification D439-88 or Federal Specification VV-G-1690B, include a boiling range of 122 degrees to 158 degrees Fahrenheit at the 10-percent point to 365 degrees to 374 degrees Fahrenheit at the 90-percent point and a Reid vapor pressure range from 9 to 15 psi. "Motor gasoline" includes finished leaded gasoline, finished unleaded gasoline, and gasohol. Blendstock is excluded until blending has been completed. (Alcohol that is to be used in the blending of gasohol is also excluded.)

Finished unleaded gasoline: Contains not more than 0.05 gram of lead per gallon and not more than 0.005 gram of phosphorus per gallon. Premium and regular grades are included, depending on the octane rating. Includes unleaded gasohol. Blendstock is excluded until blending has been completed. Alcohol that is to be used in the blending of gasohol is also excluded.

Firm: An association, company, corporation, estate, individual, joint venture, partnership, or sole proprietorship, or any other entity, however organized, including: (a) charitable or educational institutions; (b) the Federal Government, including corporations, departments, Federal agencies, and other instrumentalities; and State and local governments. A firm may consist of (1) a parent entity, including the consolidated and unconsolidated entities (if any) that it directly or indirectly controls; (2) a parent and its consolidated entities only; (3) an unconsolidated entity; or (4) any part or combination of the above.

Firm power: Power or power-producing capacity, intended to be available at all times during the period covered by a guaranteed commitment to deliver, even under adverse conditions.

First purchase (of crude oil): An equity (not custody) transaction commonly associated with a transfer of ownership of crude oil associated with the physical removal of the crude oil from a property for the first time (also referred to as a lease sale). A first purchase normally occurs at the time and place of ownership transfer where the crude oil volume sold is measured and recorded on a run ticket or other similar physical evidence of purchase. The volume purchased and the cost of such transaction shall not be measured farther from the wellhead than the point at which the value for landowner royalties is established, if there was a separate landowner.

First purchase price: The marketed first sales price of domestic crude oil, consistent with the removal price defined by the provisions of the Windfall Profits Tax on Domestic Crude Oil (Public Law 96-223, Sec. 4998 (c)).

Fiscal year: The U.S. Government's fiscal year runs from October 1 through September 30. The fiscal year is designated by the calendar year in which it ends; *e.g.*, fiscal year 2004 begins on October 1, 2003 and ends on September 30, 2004

Fissile material: Material that can be caused to undergo atomic fission when bombarded by neutrons. The most important fissionable materials are uranium-235, plutonium-239, and uranium-233.

Fission: The process whereby an atomic nucleus of appropriate type, after capturing a neutron, splits into (generally) two nuclei of lighter elements, with the release of substantial amounts of energy and two or more neutrons.

Fixed asset turnover: A ratio of revenue to fixed assets which is a measure of the productivity and efficiency of property, plant, and equipment in generating revenue. A high turnover reflects positively on the entity's ability to utilize properly its fixed assets in business operations.

Fixed assets: Tangible property used in the operations of an entity, but not expected to be consumed or converted into cash in the ordinary course of events. With a life in excess of one year, not intended for resale to customers, and subject to depreciation (with the exception of land), they are usually referred to as property, plant, and equipment.

Fixed carbon: The nonvolatile matter in coal minus the ash. Fixed carbon is the solid residue other than ash obtained by prescribed methods of destructive distillation of a coal. Fixed carbon is the part of the total carbon that remains when coal is heated in a closed vessel until all matter is driven off.

Fixed charge coverage: The ratio of earnings available to pay so-called fixed charges to such fixed charges. Fixed charges include interest on funded debt, including leases, plus the related amortizations of debt discount, premium, and expense. Earnings available for fixed charges may be computed before or after deducting income taxes. Occasionally credits for the "allowance for funds used during construction" are excluded from the earnings figures. The precise procedures followed in calculating fixed charge or interest coverages vary widely.

Fixed cost (expense): An expenditure or expense that does not vary with volume level of activity.

Fixed operating costs: Costs other than those associated with capital investment that do not vary with the operation, such as maintenance and payroll.

Flared: Gas disposed of by burning in flares usually at the production sites or at gas processing plants.

Flat and meter rate schedule: An electric rate schedule consisting of two components, the first of which is a service charge and the second a price for the energy consumed.

Flat demand rate schedule: An electric rate schedule based on billing demand that provides no charge for energy.

Flat plate pumped: A medium-temperature solar thermal collector that typically consists of a metal frame, glazing, absorbers (usually metal), and insulation and that uses a pumped liquid as the heat-transfer medium: predominant use is in water-heating applications.

Fleet vehicle: Any motor vehicle a company owns or leases that is in the normal operations of a company. Vehicles which are used in the normal operation of a company, but are owned by company employees are not fleet vehicles. If a company provides services in addition to providing natural gas, only those vehicles that are used by the natural gas provider portion of a company should be counted as fleet vehicles. Vehicles that are considered "off-road" (*e.g.*, farm or construction vehicles) or demonstration vehicles are not to be counted as fleet vehicles. Fleet vehicles include gasoline/diesel powered vehicles and alternative-fuel vehicles.

Flexible fuel vehicle: A vehicle that can operate on:

1. alternative fuels (such as M85 or E85)
2. 100-percent petroleum-based fuels
3. any mixture of an alternative fuel (or fuels) and a petroleum-based fuel.

Flexible fuel vehicles have a single fuel system to handle alternative and petroleum-based fuels. Flexible fuel vehicle and variable fuel vehicle are synonymous terms.

Flexicoking: A thermal cracking process which converts heavy hydrocarbons such as crude oil, tar sands bitumen, and distillation residues into light hydrocarbons. Feedstocks can be any pumpable hydrocarbons, including those containing high concentrations of sulfur and metals.

Floor (coal): The upper surface of the stratum underlying a coal seam. In coals that were formed in persistent swamp environments, the floor is typically a bed of clay, known as "underclay," representing the soil in which the trees or other coal-forming swamp vegetation was rooted.

Floor price: A price specified in a market-price contract as the lowest purchase price of the uranium, even if the market price falls below the specified price. The floor price may be related to the seller's production costs.

Floor space: The area enclosed by exterior walls of a building, including parking areas, basements, or other floors below ground level. It is measured in square feet.

Floor, wall, or pipeless furnace: Space-heating equipment consisting of a ductless combustor or resistance unit, having an enclosed chamber where fuel is burned or where electrical-resistance heat is generated to warm the rooms of a building. A floor furnace is located below the floor and delivers heated air to the room immediately above or (if under a partition) to the room on each side. A wall furnace is installed in a partition or in an outside wall and delivers heated air to the rooms on one or both sides of the wall. A pipeless furnace is installed in a basement and delivers heated air through a large register in the floor of the room or hallway immediately above.

Flow control: The laws, regulations, and economic incentives or disincentives used by waste managers to direct waste generated in a specific geographic area to a designated landfill, recycling, or waste-to-energy facility.

Flue: An enclosed passageway for directing products of combustion to the atmosphere.

Flue gas desulfurization: Equipment used to remove sulfur oxides from the combustion gases of a boiler plant before discharge to the atmosphere. Also referred to as scrubbers. Chemicals such as lime are used as scrubbing media.

Flue-gas desulfurization unit (scrubber): Equipment used to remove sulfur oxides from the combustion gases of a boiler plant before discharge to the atmosphere. Chemicals such as lime are used as the scrubbing media.

Flue-gas particulate collector: Equipment used to remove fly ash from the combustion gases of a boiler plant before discharge to the atmosphere. Particulate collectors include electrostatic precipitators, mechanical collectors (cyclones), fabric filters (baghouses), and wet scrubbers.

Fluid catalytic cracking: The refining process of breaking down the larger, heavier, and more complex hydrocarbon molecules into simpler and lighter molecules. Catalytic cracking is accomplished by the use of a catalytic agent and is an effective process for increasing the yield of gasoline from crude oil. Catalytic cracking processes fresh feeds and recycled feeds.

Fluid coking: A thermal cracking process utilizing the fluidized-solids technique to remove carbon (coke) for continuous conversion of heavy, low-grade oils into lighter products.

Fluidized-bed combustion: A method of burning particulate fuel, such as coal, in which the amount of air required for combustion far exceeds that found in conventional burners. The fuel particles are continually fed into a bed of mineral ash in the proportions of 1 part fuel to 200 parts ash, while a flow of air passes up through the bed, causing it to act like a turbulent fluid.

Fluorescent lamp: A glass enclosure in which light is produced when electricity is passed through mercury vapor inside the enclosure. The electricity creates a radiation discharge that strikes a coating on the inside surface of the enclosure, causing the coating to glow.

Fluorescent light bulbs: These are usually long, narrow, white tubes made of glass coated on the inside with fluorescent material, which is connected to a fixture at both ends of the light bulb; some are circular tubes. The light bulb produces light by passing electricity through mercury vapor, which causes the fluorescent coating to glow or fluoresce.

Fluorescent lighting other than compact fluorescent bulbs: In fluorescent lamps, energy is converted to light by using an electric charge to "excite" gaseous atoms within a fluorescent tube. Common types are "cool white," "warm white," etc. Special energy efficient fluorescent lights have been developed that produce the same amount of light while consuming less energy.

Flux material: A substance used to promote fusion, *e.g.*, of metals or minerals.

Fly ash: Particulate matter mainly from coal ash in which the particle diameter is less than 1×10^4 meter. This ash is removed from the flue gas using flue gas particulate collectors such as fabric filters and electrostatic precipitators.

FME: Free Market Economies. Countries that are members of the Council for Mutual Economic Assistance (CMEA) are not included.

Free on board (f.o.b.): A sales transaction in which the seller makes the product available for pick up at a specified port or terminal at a specified price and the buyer pays for the subsequent transportation and insurance.

f.o.b. price: The price actually charged at the producing country's port of loading. The reported price should be after deducting any rebates and discounts or adding premiums where applicable and should be the actual price paid with no adjustment for credit terms.

f.o.b. value (coal): Free-on-board value. This is the value of coal at the coal mine or of coke and breeze at the coke plant without any insurance or freight transportation charges added.

Foreign-controlled firms (coal): Foreign-controlled firms are U.S. coal producers with more than 50 percent of their stock or assets owned by a foreign firm.

Forward costs (uranium): The operating and capital costs that will be incurred in any future production of uranium from in-place reserves. Included are costs for labor, materials, power and fuel, royalties, payroll taxes, insurance, and general and administrative costs that are dependent upon the quantity of production and, thus, applicable as variable costs of production. Excluded from forward costs are prior expenditures, if any, incurred for property acquisition, exploration, mine development, and mill construction, as well as income taxes, profit, and the cost of money.

Footage drilled: Total footage for wells in various categories, as reported for any specified period, includes (1) the deepest total depth (length of well bores) of all wells drilled from the surface, (2) the total of all bypassed footage drilled in connection with reported wells, and (3) all new footage drilled for directional sidetrack wells. Footage reported for directional sidetrack wells does not include footage in the common bore that is reported as footage for the original well. In the case of old wells drilled deeper, the reported footage is that which was drilled below the total depth of the old well.

Forced outage: The shutdown of a generating unit, transmission line, or other facility for emergency reasons or a condition in which the generating equipment is unavailable for load due to unanticipated breakdown.

Foreign access: Refers to proved reserves of crude, condensate, and natural gas liquids applicable to long-term supply agreements with foreign governments or authorities in which the company or one of its affiliates acts as producer.

Foreign currency transaction gains and losses: Gains or losses resulting from the effect of exchange rate changes on transactions denominated in currencies other than the functional currency (for example, a U.S. enterprise may borrow Swiss francs or a French subsidiary may have a receivable denominated in kroner from a Danish customer). Gains and losses on those foreign currency transactions are generally included in determining net income for the period in which exchange rates change unless the transaction hedges a foreign currency commitment or a net investment in a foreign entity. Intercompany transactions of a long-term investment nature are considered part of a parent's net investment and hence do not give rise to gains or losses.

Foreign currency translation effects: Gains or losses resulting from the process of expressing amounts denominated or measured in one currency in terms of another currency by use of the exchange rate between the two currencies. This process is generally required to consolidate the financial statements of foreign affiliates into the total company financial statements and to recognize the conversion of foreign currency or the settlement of a receivable or payable denominated in foreign currency at a rate different from that at which the item is recorded. Translation adjustments are not included in determining net income, but are disclosed as separate components of consolidated equity.

Foreign operations: These are operations that are located outside the United States. Determination of whether an enterprise's mobile assets, such as offshore drilling rigs or ocean-going vessels, constitute foreign operations should depend on whether such assets are normally identified with operations located outside the United States.

Forward cost (1): Forward costs are those operating and capital costs yet to be incurred at the time an estimate of reserves is made. Profits and "sunk" costs, such as past expenditures for property acquisition, exploration, and mine development, are not included. Therefore, the various forward-cost categories are independent of the market price at which uranium produced from the reserves would be sold.

Forward cost (2): The operating and capital costs still to be incurred in the production of uranium from in-place reserves. By using forward costing, estimates for reserves for ore deposits in differing geological settings and status of development can be aggregated and reported for selected cost categories. Included are costs for labor, materials, power and fuel, royalties, payroll taxes, insurance, and applicable general and administrative costs. Excluded from forward cost estimates are prior expenditures, if any, incurred for property acquisition, exploration, mine development, and mill construction, as well as income taxes, profit, and the cost of money. Forward costs are neither the full costs of production nor the market price at which the uranium, when produced, might be sold.

Forward coverage: Amount of uranium required to assure uninterrupted operation of nuclear power plants.

Fossil fuel: An energy source formed in the earth crust from decayed organic material. The common fossil fuels are petroleum, coal, and natural gas.

Fossil-fuel electric generation: Electric generation in which the prime mover is a turbine rotated by high-pressure steam produced in a boiler by heat from burning fossil fuels.

Fossil fuel plant: A plant using coal, petroleum, or gas as its source of energy.

Fossil fuel steam-electric power plant: An electricity generation plant in which the prime mover is a turbine rotated by high-pressure steam produced in a boiler by heat from burning fossil fuels.

Foundry: An operation where metal castings are produced, using coke as a fuel.

Foundry coke: This is a special coke that is used in furnaces to produce cast and ductile iron products. It is a source of heat and also helps maintain the required carbon content of the metal product. Foundry coke production requires lower temperatures and longer times than blast furnace coke.

Fractionation: The process by which saturated hydrocarbons are removed from natural gas and separated into distinct products, or "fractions," such as propane, butane, and ethane.

Framework Convention on Climate Change (FCCC): An agreement opened for signature at the "Earth Summit" in Rio de Janeiro, Brazil, on June 4, 1992, which has the goal of stabilizing greenhouse gas concentrations in the atmosphere at a level that would prevent significant anthropogenically forced climate change.

Free alongside ship (f.a.s.): The value of a commodity at the port of exportation, generally including the purchase price plus all charges incurred in placing the commodity alongside the carrier at the port of exportation.

Free on board (f.o.b.): A sales transaction in which the seller makes the product available for pick up at a specified port or terminal at a specified price and the buyer pays for the subsequent transportation and insurance.

Free well: A well drilled and equipped by an assignee as consideration for the assignment of a fractional share of the working interest, commonly under a farm-out agreement.

Fresh feed input: Represents input of material (crude oil, unfinished oils, natural gas liquids, other hydrocarbons and oxygenates or finished products) to processing units at a refinery that is being processed (input) into a particular unit for the first time. Examples:

1. Unfinished oils coming out of a crude oil distillation unit that are put into a catalytic cracking unit are considered fresh feed to the catalytic cracking unit.
2. Unfinished oils coming out of a catalytic cracking unit being looped back into the same catalytic cracking unit to be reprocessed are not considered fresh feed.

Fresh feeds: Crude oil or petroleum distillates that are being fed to processing units for the first time.

FRS: Financial Reporting System Survey.

Fuel: Any material substance that can be consumed to supply heat or power. Included are petroleum, coal, and natural gas (the fossil fuels), and other consumable materials, such as uranium, biomass, and hydrogen.

Fuel cell: A device capable of generating an electrical current by converting the chemical energy of a fuel directly into electrical energy. Fuel cells differ from conventional electrical cells in that the active materials such as fuel and oxygen are not contained within the cell but are supplied from outside. It does not contain an intermediate heat cycle, as do most other electrical generation techniques.

Fuel cycle: The entire set of sequential processes or stages involved in the utilization of fuel, including extraction, transformation, transportation, and combustion. Emissions generally occur at each stage of the fuel cycle.

Fuel emergencies: An emergency that exists when supplies of fuels or hydroelectric storage for generation are at a level or estimated to be at a level that would threaten the reliability or adequacy of bulk electric power supply. The following factors should be taken into account to determine that a fuel emergency exists:

1. Fuel stock or hydroelectric project water storage levels are 50 percent or less of normal for that particular time of the year and a continued downward trend in fuel stock or hydroelectric project water storage level is estimated; or
2. Unscheduled dispatch or emergency generation is causing an abnormal use of a particular fuel type, such that the future supply of stocks of that fuel could reach a level that threatens the reliability or adequacy of bulk electric power supply.

Fuel ethanol (CH_3-CH_2OH): An anhydrous denatured aliphatic alcohol intended for gasoline blending as described in the definition of oxygenates. It is also used in high concentrations to produce E85.

Fuel expenses: These costs include the fuel used in the production of steam or driving another prime mover for the generation of electricity. Other associated expenses include unloading the shipped fuel and all handling of the fuel up to the point where it enters the first bunker, hopper, bucket, tank, or holder in the boiler-house structure.

Fuel/fabricator assembly identifier: Individual assembly identifier based on a numbering scheme developed by individual fuel fabricators. Most fuel fabricator assembly identifiers schemes closely match the scheme developed by the American National Standards Institute (ANSI) and are therefore unique.

Fuel injection: A fuel delivery system whereby gasoline is pumped to one or more fuel injectors under high pressure. The fuel injectors are valves that, at the appropriate times, open to allow fuel to be sprayed or atomized into a throttle bore or into the intake manifold ports. The fuel injectors are usually solenoid operated valves under the control of the vehicle's on-board computer (thus the term "electronic fuel injection"). The fuel efficiency of fuel injection systems is less temperature-dependent than carburetor systems. Diesel engines always use injectors.

Fuel oil: A liquid petroleum product less volatile than gasoline, used as an energy source. Fuel oil includes distillate fuel oil (No. 1, No. 2, and No. 4), and residual fuel oil (No. 5 and No. 6).

Fuel purchase agreement: An agreement between a company and a fuel provider which stipulates that the company agrees to purchase its fuel from the fuel provider. If the company has a credit card for use at a fuel provider's locations, but is not bound by an additional agreement to purchase fuel from that provider, the credit card agreement alone is not considered a fuel purchase agreement.

Fuel ratio: The ratio of fixed carbon to volatile matter in coal.

Fuel switching capability: The short-term capability of a manufacturing establishment to have used substitute energy sources in place of those actually consumed. Capability to use substitute energy sources means that the establishment's combustors (for example, boilers, furnaces, ovens, and blast furnaces) had the machinery or equipment either in place or available for installation so that substitutions could actually have been introduced within 30 days without extensive modifications. Fuel-switching capability does not depend on the relative prices of energy sources; it depends only on the characteristics of the equipment and certain legal constraints.

Fuel-switching DSM program assistance: DSM program assistance where the sponsor encourages consumers to change from one fuel to another for a particular end-use service. For example, utilities might encourage consumers to replace electric water heaters with gas units or encourage industrial consumers to use electric microwave heaters instead of natural gas-heaters.

Fuel wood: Wood and wood products, possibly including scrubs and branches, etc, bought or gathered, and used by direct combustion.

Fuels solvent deasphalting: A refining process for removing asphalt compounds from petroleum fractions, such as reduced crude oil. The recovered stream from this process is used to produce fuel products.

Fugitive emissions: Unintended leaks of gas from the processing, transmission, and/or transportation of fossil fuels.

Full forced outage: The net capability of main generating units that are unavailable for load for emergency reasons.

Full power day: The equivalent of 24 hours of full power operation by a reactor. The number of full power days in a specific cycle is the product of the reactor's capacity factor and the length of the cycle.

Full power operation: Operation of a unit at 100 percent of its design capacity. Full-power operation precedes commercial operation.

Full requirements consumer: A wholesale consumer without other generating resources whose electric energy seller is the sole source of long-term firm power for the consumer's service area. The terms and conditions of sale are equivalent to the seller's obligations to its own retail service, if any.

Fumarole: A vent from which gas or steam issue; a geyser or spring that emits gases.

Furnace: The part of a boiler or warm-air space-heating plant in which combustion takes place.

Furnace coke plant: A coke plant whose coke production is used primarily by the producing company.

Furnaces that heat directly, without using steam or hot water (similar to a residential furnace): Furnaces burn natural gas, fuel oil, propane/ butane (bottled gas), or electricity to warm the air. The warmed air is then distributed throughout the building through ducts. Many people use the words "boilers" and "furnaces" interchangeably. They are not the same. We mean that warm air is produced directly by burning some fuel.

Warm-air furnaces typically rely on air ducts to carry the warm air throughout the building. Warm-air furnaces are often built in combination with central air-conditioning systems, so that they can use the same air ducts for both heating or air-conditioning (depending on the season).

Other terms for describing this type of equipment include: "central system," "split system," and "forced air/forces air furnace."

Futures market: A trade center for quoting prices on contracts for the delivery of a specified quantity of a commodity at a specified time and place in the future.

Gallon: A volumetric measure equal to 4 quarts (231 cubic inches) used to measure fuel oil. One barrel equals 42 gallons.

Gas: A non-solid, non-liquid combustible energy source that includes natural gas, coke-oven gas, blast-furnace gas, and refinery gas.

Gas cooled fast breeder reactor (GCFB): A fast breeder reactor that is cooled by a gas (usually helium) under pressure.

Gas oil: European and Asian designation for No. 2 heating oil and No. 2 diesel fuel.

Gas plant operator: Any firm, including a gas plant owner, which operates a gas plant and keeps the gas plant records. A gas plant is a facility in which natural gas liquids are separated from natural gas or in which natural gas liquids are fractionated or otherwise separated into natural gas liquid products or both.

Gas processing unit: A facility designed to recover natural gas liquids from a stream of natural gas that may or may not have passed through lease separators and/or field separation facilties. Another function of natural gas processing plants is to control the quality of the processed natural gas stream. Cycling plants are considered natural gas processing plants.

Gas to liquids (GTL): A process that combines the carbon and hydrogen elements in natural gas molecules to make synthetic liquid petroleum products, such as diesel fuel.

Gas turbine plant: A plant in which the prime mover is a gas turbine. A gas turbine consists typically of an axial-flow air compressor and one or more combustion chambers where liquid or gaseous fuel is burned and the hot gases are passed to the turbine and where the hot gases expand drive the generator and are then used to run the compressor.

Gas well: A well completed for production of natural gas from one or more gas zones or reservoirs. Such wells contain no completions for the production of crude oil.

Gas well productivity: Derived annually by dividing gross natural gas withdrawals from gas wells by the number of producing gas wells on December 31 and then dividing the quotient by the number of days in the year.

Gasification: A method for converting coal, petroleum, biomass, wastes, or other carbon-containing materials into a gas that can be burned to generate power or processed into chemicals and fuels.

Gasohol: A blend of finished motor gasoline containing alcohol (generally ethanol but sometimes methanol) at a concentration between 5.7 percent and 10 percent by volume.

Gasoline blending components: Naphthas which will be used for blending or compounding into finished aviation or motor gasoline (*e.g.*, straight-run gasoline, alkylate, reformate, benzene, toluene, and xylene). Excludes oxygenates (alcohols, ethers), butane, and pentanes plus.

Gasoline grades: The classification of gasoline by octane ratings. Each type of gasoline (conventional, oxygenated, and reformulated) is classified by three grades–Regular, Midgrade, and Premium. In general, automotive octane requirements are lower at high altitudes. Therefore, in some areas of the United States, such as the Rocky Mountain States, the octane ratings for the gasoline grades may be 2 or more octane points lower.

1. **Regular gasoline:** Gasoline having an antiknock index, *i.e.*, octane rating, greater than or equal to 85 and less than 88.
2. **Midgrade gasoline:** Gasoline having an antiknock index, *i.e.*, octane rating, greater than or equal to 88 and less than or equal to 90.
3. **Premium gasoline:** Gasoline having an antiknock index, *i.e.*, octane rating, greater than 90.

Gasoline motor, (leaded): Contains more than 0.05 grams of lead per gallon or more than 0.005 grams of phosphorus per gallon. The actual lead content of any given gallon may vary. Premium and regular grades are included, depending on the octane rating. Includes leaded gasohol. Blendstock is excluded until blending has been completed. Alcohol that is to be used in the blending of gasohol is also excluded.

Gasoline motor, (unleaded): Contains not more than 0.05 grams of lead per gallon and not more than 0.005 grams of phosphorus per gallon. Premium and regular grades are included, depending on the octane rating. Includes unleaded gasohol. Blendstock is excluded until blending has been completed. Alcohol that is to be used in the blending of gasohol is also excluded.

Gate station: Location where the pressure of natural gas being transferred from the transmission system to the distribution system is lowered for transport through small diameter, low pressure pipelines.

Gatherer: A company primarily engaged in the gathering of natural gas from well or field lines for delivery, for a fee, to a natural gas processing plant or central point. Gathering companies may also provide compression, dehydration, and/or treating services.

Generally accepted accounting principles (GAAP): Defined by the FASB as the conventions, rules, and procedures necessary to define accepted accounting practice at a particular time, includes both broad guidelines and relatively detailed practices and procedures.

Generating facility: An existing or planned location or site at which electricity is or will be produced.

Generating station: A station that consists of electric generators and auxiliary equipment for converting mechanical, chemical, or nuclear energy into electric energy.

Generating unit: Any combination of physically connected generators, reactors, boilers, combustion turbines, and other prime movers operated together to produce electric power.

Generation: The process of producing electric energy by transforming other forms of energy; also, the amount of electric energy produced, expressed in kilowatthours.

Generation company: An entity that owns or operates generating plants. The generation company may own the generation plants or interact with the short-term market on behalf of plant owners.

Generator capacity: The maximum output, commonly expressed in megawatts (MW), that generating equipment can supply to system load, adjusted for ambient conditions.

Generator nameplate capacity (installed): The maximum rated output of a generator, prime mover, or other electric power production equipment under specific conditions designated by the manufacturer. Installed generator nameplate capacity is commonly expressed in megawatts (MW) and is usually indicated on a nameplate physically attached to the generator.

Geologic assurance: State of sureness, confidence, or certainty of the existence of a quantity of resources based on the distance from points where coal is measured or sampled and on the abundance and quality of geologic data as related to thickness of overburden, rank, quality, thickness of coal, areal extent, geologic history, structure, and correlations of coal beds and enclosing rocks. The degree of assurance increases as the nearness to points of control, abundance, and quality of geologic data increases.

Geologic considerations: Conditions in the coal deposit or in the rocks in which it occurs that may complicate or preclude mining. Geologic considerations are evaluated in the context of the current state of technology and regulations, so the impact on mining may change with time.

Geological and geophysical (G&G) costs: Costs incurred in making geological and geophysical studies, including, but not limited to, costs incurred for salaries, equipment, obtaining rights of access, and supplies for scouts, geologists, and geophysical crews.

Geological repository: A mined facility for disposal of radioactive waste that uses waste packages and the natural geology as barriers to provide waste isolation.

Geopressured: A type of geothermal resource occurring in deep basins in which the fluid is under very high pressure.

Geyser: A special type of thermal spring that periodically ejects water with great force.

Geothermal energy: Hot water or steam extracted from geothermal reservoirs in the earth's crust. Water or steam extracted from geothermal reservoirs can be used for geothermal heat pumps, water heating, or electricity generation.

Geothermal plant: A plant in which the prime mover is a steam turbine. The turbine is driven either by steam produced from hot water or by natural steam that derives its energy from heat found in rocks or fluids at various depths beneath the surface of the earth. The energy is extracted by drilling and/or pumping.

Giga: One billion.

Gigawatt (GW): One billion watts or one thousand megawatts.

Gigawatt-electric (GWe): One billion watts of electric capacity.

Gigawatthour (GWh): One billion watthours.

Gilsonite: Trademark name for uintaite (or uintahite), a black, brilliantly lustrous natural variety of asphalt found in parts of Utah and western Colorado.

Global warming: An increase in the near surface temperature of the Earth. Global warming has occurred in the distant past as the result of natural influences, but the term is today most often used to refer to the warming some scientists predict will occur as a result of increased anthropogenic emissions of greenhouse gases.

Global warming potential (GWP): An index used to compare the relative radiative forcing of different gases without directly calculating the changes in atmospheric concentrations. GWPs are calculated as the ratio of the radiative forcing that would result from the emission of one kilogram of a greenhouse gas to that from the emission of one kilogram of carbon dioxide over a fixed period of time, such as 100 years.

Government-owned stocks: Oil stocks owned by the national government and held for national security. In the United States, these stocks are known as the Strategic Petroleum Reserve.

Greenhouse effect: The result of water vapor, carbon dioxide, and other atmospheric gases trapping radiant (infrared) energy, thereby keeping the earth's surface warmer than it would otherwise be. Greenhouse gases within the lower levels of the atmosphere trap this radiation, which would otherwise escape into space, and subsequent re-radiation of some of this energy back to the Earth maintains higher surface temperatures than would occur if the gases were absent.

Greenhouse gases: Those gases, such as water vapor, carbon dioxide, nitrous oxide, methane, hydrofluorocarbons (HFCs), perfluorocarbons (PFCs) and sulfur hexafluoride, that are transparent to solar (short-wave) radiation but opaque to long-wave (infrared) radiation, thus preventing long-wave radiant energy from leaving Earth's atmosphere. The net effect is a trapping of absorbed radiation and a tendency to warm the planet's surface.

Green pricing: In the case of renewable electricity, green pricing represents a market solution to the various problems associated with regulatory valuation of the nonmarket benefits of renewables. Green pricing programs allow electricity customers to express their willingness to pay for renewable energy development through direct payments on their monthly utility bills.

Grid: The layout of an electrical distribution system.

Gross additions to construction work in progress for the month: This amount should include the monthly gross additions for an electric plant in the process of construction.

Gross company-operated production: Total production from all company-operated properties, including all working and nonworking interests.

Gross domestic product (GDP): The total value of goods and services produced by labor and property located in the United States. As long as the labor and property are located in the United States, the supplier (that is, the workers and, for property, the owners) may be either U.S. residents or residents of foreign countries.

Gross domestic product (GDP) implicit price deflator: The implicit price deflator, published by the U.S. Department of Commerce, Bureau of Economic Analysis, is used to convert nominal figures to real figures.

Gross energy intensity: Total consumption of a particular energy source(s) or fuel(s) by a group of buildings, divided by the total floor space of those buildings, including buildings and floor space where the energy source or fuel is not used, *i.e.*, the ratio of consumption to gross floor space.

Gross gas withdrawal: The full-volume of compounds extracted at the wellhead, including nonhydrocarbon gases and natural gas plant liquids.

Gross generation: The total amount of electric energy produced by generating units and measured at the generating terminal in kilowatthours (kWh) or megawatthours (MWh).

Gross head: A dam's maximum allowed vertical distance between the upstream's surface water (headwater) forebay elevation and the downstream's surface water (tailwater) elevation at the tail-race for reaction wheel dams or the elevation of the jet at impulse wheel dams during specified operation and water conditions.

Gross inputs: The crude oil, unfinished oils, and natural gas plant liquids put into atmospheric crude oil distillation units.

Gross input to atmospheric crude oil distillation units: Total input to atmospheric crude oil distillation units. Includes all crude oil, lease condensate, natural gas plant liquids, unfinished oils, liquefied refinery gases, slop oils, and other liquid hydrocarbons produced from tar sands, gilsonite, and oil shale.

Gross national product (GNP): The total value of goods and services produced by the nation's economy before deduction of depreciation charges and other allowances for capital consumption. It includes the total purchases of goods and services by private consumers and government, gross private domestic capital investment, and net foreign trade.

Gross vehicle weight rating (GVWR): Vehicle weight plus carrying capacity.

Gross withdrawals: Full well stream volume, including all natural gas plant liquid and nonhydrocarbon gases, but excluding lease condensate. Also includes amounts delivered as royalty payments or consumed in field operations.

Gross working interest ownership basis: Gross working interest ownership is the respondent's working interest in a given property plus the proportionate share of any royalty interest, including overriding royalty interest, associated with the working interest.

Group: A group is a logical grouping of assemblies with similar characteristics. All assemblies in a group have the same initial average enrichment, the same cycle/reactor history, the same current location, the same burnup, the same owner, and the same assembly type.

Group name: The DOE/EIA-assigned name identifying a composite supply source (*i.e.*, commonly metered gas streams from more than one field), which is often the case in contract areas, field areas, and plants. A group name can also be a pipeline purchase (*i.e.*, FERC Gas Tariff, Canadian Gas, Mexican Gas, and Algerian LNG). Emergency purchases and short term purchases are also group names. Group Code–The DOE/EIA-assigned code identifying a composite supply source.

Group quarters: Living arrangement for institutional groups containing ten or more unrelated persons. Group quarters are typically found in hospitals, nursing or rest homes, military barracks, ships, halfway houses, college dormitories, fraternity and sorority houses, convents, monasteries, shelters, jails, and correctional institutions. Group quarters may also be found in houses or apartments shared by ten or more unrelated persons. Group quarters are often equipped with a dining area for residents.

Gypsum: Calcium sulfate dihydrate (C_aSO_4 $2H_2O$) a sludge constituent from the conventional lime scrubber process, obtained as a byproduct of the dewatering operation and soid for commercial use.

Half-life: The time it takes for an isotope to lose half of its radioactivity.

Halogen lamp: A type of incandescent lamp that lasts much longer and is more efficient than the common incandescent lamp. The lamp uses a halogen gas, usually iodine or bromine, that causes the evaporating tungsten to be redeposited on the filament, thus prolonging its life.

Halogenated substances: A volatile compound containing halogens, such as chlorine, fluorine or bromine.

Hand loading: An underground loading method by which coal is removed from the working face by manual labor through the use of a shovel for conveyance to the surface. Though rapidly disappearing, it is still used in small-tonnage mines.

Haulage cost: Cost of loading ore at a mine site and transporting it to a processing plant.

Head: The product of the water's weight and a usable difference in elevation gives a measurement of the potential energy possessed by water.

Heap leach solutions: The separation, or dissolving-out from mined rock of the soluble uranium constituents by the natural action of percolating a prepared chemical solution through mounded (heaped) rock material. The mounded material usually contains low grade mineralized material and/or waste rock produced from openpit or underground mines. The solutions are collected after percolation is completed and processed to recover the valued components.

Heat content: Measurement: The gross heat content (or heating value), is the number of British thermal units (Btu) produced by the combustion, at constant pressure, of the amount of the gas that would occupy a volume of one cubic foot at a temperature of 60 degrees Fahrenheit, if saturated with water vapor and under a pressure equivalent to 30 inches of mercury at 32 degrees Fahrenheit and under standard gravitational force (980.665 cm per sec.^{2}), with air of the same temperature and pressure as the gas, when the products of combustion are cooled to the initial temperature of gas and air and when the water formed by combustion is condensed to the liquid state.

Heat pump: Heating and/or cooling equipment that, during the heating season, draws heat into a building from outside and, during the cooling season, ejects heat from the building to the outside. Heat pumps are vapor-compression refrigeration systems whose indoor/outdoor coils are used reversibly as condensers or evaporators, depending on the need for heating or cooling.

Heat pump (air source): An air-source heat pump is the most common type of heat pump. The heat pump absorbs heat from the outside air and transfers the heat to the space to be heated in the heating mode. In the cooling mode the heat pump absorbs heat from the space to be cooled and rejects the heat to the outside air. In the heating mode when the outside air approaches 32o F or less, air-source heat pumps loose efficiency and generally require a back-up (resistance) heating system.

Heat pump (geothermal): A heat pump in which the refrigerant exchanges heat (in a heat exchanger) with a fluid circulating through an earth connection medium (ground or ground water). The fluid is contained in a variety of loop (pipe) configurations depending on the temperature of the ground and the ground area available. Loops may be installed horizontally or vertically in the ground or submersed in a body of water.

Heat pump efficiency: The efficiency of a heat pump, that is, the electrical energy to operate it, is directly related to temperatures between which it operates. Geothermal heat pumps are more efficient than conventional heat pumps or air conditioners that use the outdoor air since the ground or ground water a few feet below the earth's surface remains relatively constant throughout the year. It is more efficient in the winter to draw heat from the relatively warm ground than from the atmosphere where the air temperature is much colder, and in summer transfer waste heat to the relatively cool ground than to hotter air. Geothermal heat pumps are generally more expensive ($2,000-$5,000) to install than outside air heat pumps. However, depending on the location geothermal heat pumps can reduce energy consumption (operating cost) and correspondingly, emissions by more than 20 percent compared to high-efficiency outside air heat pumps. Geothermal heat pumps also use the waste heat from air-conditioning to provide free hot water heating in the summer.

Heat rate: A measure of generating station thermal efficiency, generally expressed in Btu per net kilowatthour. It is computed by dividing the total Btu content of fuel burned for electric generation by the resulting net kilowatthour generation.

Heated floorspace: The area within a buildingt hat is space heated.

Heating degree-days (HDD): A measure of how cold a location is over a period of time relative to a base temperature, most commonly specified as 65 degrees Fahrenheit. The measure is computed for each day by subtracting the average of the day's high and low temperatures from the base temperature (65 degrees), with negative values set equal to zero. Each day's heating degree-days are summed to create a heating degree-day measure for a specified reference period. Heating degree-days are used in energy analysis as an indicator of space heating energy requirements or use.

Heating equipment: Any equipment designed and/or specifically used for heating ambient air in an enclosed space. Common types of heating equipment include: central warm air furnace, heat pump, plug-in or built-in room heater, boiler for steam or hot water heating system, heating stove, and fireplace.

Heating intensity: The ratio of space-heating consumption or expenditures to square footage of heated floor space and heating degree-days (base 65 degrees Fahrenheit). This ratio provides a way of comparing different types of housing units and households by controlling for differences in housing unit size and weather conditions. The square footage of heated floor space is based on the measurements of the floor space that is heated. The ratio is calculated on a weighted, aggregate basis according to the following formula: Heating Intensity = Btu for Space Heating / (Heated Square Feet * Heating Degree-Days).

Heating stove burning wood, coal, or coke: Any free-standing box or controlled-draft stove; or a stove installed in a fireplace opening, using the chimney of the fireplace. Stoves are made of cast iron, sheet metal, or plate steel. Free-standing fireplaces that can be detached from their chimneys are considered heating stoves.

Heating value: The average number of British thermal units per cubic foot of natural gas as determined from tests of fuel samples.

Heavy gas oil: Petroleum distillates with an approximate boiling range from 651 degrees to 1000 degrees Fahrenheit.

Heavy metals: Metallic elements, including those required for plant and animal nutrition, in trace concentration but which become toxic at higher concentrations. Examples are mercury, chromium, cadmium, and lead.

Heavy oil: The fuel oils remaining after the lighter oils have been distilled off during the refining process. Except for start-up and flame stabilization, virtually all petroleum used in steam plants is heavy oil. Includes fuel oil numbers 4, 5, and 6; crude; and topped crude.

Heavy rail: An electric railway with the capacity for a "heavy volume" of traffic and characterized by exclusive rights-of-way, multi-car trains, high speed and rapid acceleration, sophisticated signaling, and high platform loading. Also known as "subway," elevated (railway), "metropolitan railway (metro)."

Heavy water: Water containing a significantly greater proportion of heavy hydrogen (deuterium) atoms to ordinary hydrogen atoms than is found in ordinary (light) water. Heavy water is used as a moderator in some reactors because it slows neutrons effectively and also has a low cross section for absorption of neutrons.

Heavy-water-moderated reactor: A reactor that uses heavy water as its moderator. Heavy water is an excellent moderator and thus permits the use of inexpensive natural (unenriched) uranium as fuel.

Hedging: The buying and selling of futures contracts so as to protect energy traders from unexpected or adverse price fluctuations.

Hedging contracts: Contracts which establish future prices and quantities of electricity independent of the short-term market. Derivatives may be used for this purpose.

Heliostat: A mirror that reflects solar rays onto a central receiver. A heliostat automatically adjusts its position to track daily or seasonal changes in the sun's position. The arrangement of heliostats around a central receiver is also called a solar collector field.

Henry Hub: A pipeline hub on the Louisiana Gulf coast. It is the delivery point for the natural gas futures contract on the New York Mercantile Exchange (NYMEX).

High efficiency ballast: A lighting conservation feature consisting of an energy-efficient version of a conventional electromagnetic ballast. The ballast is the transformer for fluorescent and high-intensity discharge (HID) lamps, which provides the necessary current, voltage, and wave-form conditions to operate the lamp. A high-efficiency ballast requires lower power input than a conventional ballast to operate HID and fluorescent lamps.

High efficiency lighting: Lighting provided by high-intensity discharge (HID) lamps and/or fluorescent lamps.

High-intensity discharge (HID) lamp: A lamp that produces light by passing electricity through gas, which causes the gas to glow. Examples of HID lamps are mercury vapor lamps, metal halide lamps, and high-pressure sodium lamps. HID lamps have extremely long life and emit far more lumens per fixture than do fluorescent lights.

High-mileage households: Households with estimated aggregate annual vehicle mileage that exceeds 12,500 miles.

High-temperature collector: A solar thermal collector designed to operate at a temperature of 180 degrees Fahrenheit or higher.

Highwall: The unexcavated face of exposed over-burden and coal in a surface mine.

Hinshaw pipeline: A pipeline or local distribution company that has received exemptions from regulations pursuant to the Natural Gas Act. These companies transport interstate natural gas not subject to regulations under NGA.

Holding company: A company that confines its activities to owning stock in and supervising management of other companies. The Securities and Exchange Commission, as administrator of the Public Utility Holding Company Act of 1935, defines a holding company as "a company which directly or indirectly owns, controls or holds 10 percent or more of the outstanding voting securities of a holding company" (15 USC 79b, par. a (7)).

Holding pond: A structure built to contain large volumes of liquid waste to ensure that it meets environmental requirements prior to release.

Horizontal axis wind turbine: The most common type of wind turbine where the axis of rotation is oriented horizontally.

Horsepower: A unit for measuring the rate of work (or power) equivalent to 33,000 foot-pounds per minute or 746 watts.

Host government: The government (including any government-controlled firm engaged in the production, refining, or marketing of crude oil or petroleum products) of the foreign country in which the crude oil is produced.

Hot dry rock: Heat energy residing in impermeable, crystalline rock. Hydraulic fracturing may be used to create permeability to enable circulation of water and removal of the heat.

Hot tub: Water-filled wood, plastic, or ceramic container in which up to 12 people can lounge. Normally equipped with a heater that heats the water from 80 degrees to 106 degrees Fahrenheit. It may also have jets to bubble the water. The water is not drained after each use. An average-size hot tub holds 200 to 400 gallons of water. All reported hot tubs are assumed to include an electric pump. These are also called spas or jacuzzis.

Hours under load: The hours the boiler is operating to drive the generator producing electricity.

Household: A family, an individual, or a group of up to nine unrelated persons occupying the same housing unit. "Occupy" means that the housing unit is the person's usual or permanent place of residence.

Household energy expenditures: The total amount of funds spent for energy consumed in, or delivered to, a housing unit during a given period of time.

Housing unit: A house, an apartment, a group of rooms, or a single room if it is either occupied or intended for occupancy as separate living quarters by a family, an individual, or a group of one to nine unrelated persons. Separate living quarters means the occupants (1) live and eat separately from other persons in the house or apartment and (2) have direct access from the outside of the buildings or through a common hall–that is, they can get to it without going through someone else's living quarters. Housing units do not include group quarters such as prisons or nursing homes where ten or more unrelated persons live. A common dining area used by residents is an indication of group quarters. Hotel and motel rooms are considered housing units if occupied as the usual or permanent place of residence.

Hub height: In a horizontal-axis wind turbine, the distance from the turbine platform to the rotor shaft.

Humidifier: A humidifier adds moisture to the air (often needed in winter when indoor air is very dry). It may be a portable unit or attached to the heating system.

Humidity: The moisture content of air. Relative humidity is the ratio of the amount of water vapor actually present in the air to the greatest amount possible at the same temperature.

HVAC: An abbreviation for the heating, ventilation, and air-conditioning system; the system or systems that condition air in a building.

HVAC conservation feature: A building feature designed to reduce the amount of energy consumed by the heating, cooling, and ventilating equipment.

HVAC DSM program: A DSM (demand-side management) program designed to promote the efficiency of the heating or cooling delivery system, including replacement. Includes ventilation (economizers; heat recovery from exhaust air), cooling (evaporative cooling, cool storage; heat recovery from chillers; high-efficiency air conditioning), heating, and automatic energy management systems.

Hybrid transmission line: A double-circuit line that has one alternating current and one direct circuit. The AC circuit usually serves local loads along the line.

Hydraulic fracturing: Fracturing of rock at depth with fluid pressure. Hydraulic fracturing at depth may be accomplished by pumping water into a well at very high pressures. Under natural conditions, vapor pressure may rise high enough to cause fracturing in a process known as hydrothermal brecciation.

Hydraulic head: The distance between the respective elevations of the upstream water surface (headwater) above and the downstream surface water (tailwater) below a hydroelectric power plant.

Hydrocarbon: An organic chemical compound of hydrogen and carbon in the gaseous, liquid, or solid phase. The molecular structure of hydrocarbon compounds varies from the simplest (methane, a constituent of natural gas) to the very heavy and very complex.

Hydrochlorofluorocarbons (HCFCs): Chemicals composed of one or more carbon atoms and varying numbers of hydrogen, chlorine, and fluorine atoms.

Hydroelectric power: The use of flowing water to produce electrical energy.

Hydrofluorocarbons (HFCs): A group of man-made chemicals composed of one or two carbon atoms and varying numbers of hydrogen and fluorine atoms. Most HFCs have 100-year Global Warming Potentials in the thousands.

Hydrogen: A colorless, odorless, highly flammable gaseous element. It is the lightest of all gases and the most abundant element in the universe, occurring chiefly in combination with oxygen in water and also in acids, bases, alcohols, petroleum, and other hydrocarbons.

Hydroxyl radical (OH): An important chemical scavenger of many trace gases in the atmosphere that are greenhouse gases. Atmospheric concentrations of OH affect the atmospheric lifetimes of greenhouse gases, their abundance, and, ultimately, the effect they have on climate.

Hypothetical resources (coal): Undiscovered coal resources in beds that may reasonably be expected to exist in known mining districts under known geologic conditions. In general, hypothetical resources are in broad areas of coalfields where points of observation are absent and evidence is from distant outcrops, drill holes, or wells. Exploration that confirms their existence and better defines their quantity and quality would permit their reclassification as identified resources. Quantitative estimates are based on a broad knowledge of the geologic character of coalbed or region. Measurements of coal thickness are more than 6 miles apart. The assumption of continuity of coalbed is supported only by geologic evidence.

Identified resources: Coal deposits whose location, rank, quality, and quantity are known from geologic evidence supported by engineering measurements. Included are beds of bituminous coal and anthracite (14 or more inches thick) and beds of subbituminous coal and lignite (30 or more inches thick) that occur at depths to 6,000 feet. The existence and quantity of these beds have been delineated within specified degrees of geologic assurance as measured, indicated, or inferred. Also included are thinner and / or deeper beds that presently are being mined or for which there is evidence that they could be mined commercially.

Idle capacity: The component of operable capacity that is not in operation and not under active repair, but capable of being placed in operation within 30 days; and capacity not in operation but under active repair that can be completed within 90 days.

Impedance: The opposition to power flow in an AC circuit. Also, any device that introduces such opposition in the form of resistance, reactance, or both. The impedance of a circuit or device is measured as the ratio of voltage to current, where a sinusoidal voltage and current of the same frequency are used for the measurement; it is measured in ohms.

Implicit price deflator: The implicit price deflator, published by the U.S. Department of Commerce, Bureau of Economic Analysis, is used to convert nominal figures to real figures.

Imported crude oil burned as fuel: The amount of foreign crude oil burned as a fuel oil, usually as residual fuel oil, without being processed as such. Imported crude oil burned as fuel includes lease condensate and liquid hydrocarbons produced from tar sands, gilsonite, and oil shale.

Imported Refiners' Acquisition Cost (IRAC): The average price for imported oil paid by U.S. refiners.

Imports: Receipts of goods into the 50 States and the District of Columbia from U.S. possessions and territories or from foreign countries.

Improved recovery: Extraction of crude oil or natural gas by any method other than those that rely primarily on natural reservoir pressure, gas lift, or a system of pumps.

Inadvertent power exchange: An unintended power exchange among utilities that is either not previously agreed upon or in an amount different from the amount agreed upon.

Incandescent lamp: A glass enclosure in which light is produced when a tungsten filament is electrically heated so that it glows. Much of the energy is converted into heat; therefore, this class of lamp is a relatively inefficient source of light. Included in this category are the familiar screw-in light bulbs, as well as somewhat more efficient lamps, such as tungsten halogen lamps, reflector or r-lamps, parabolic aluminized reflector (PAR) lamps, and ellipsoidal reflector (ER) lamps.

Incandescent light bulbs, including regular or energy-efficient light bulbs: An incandescent bulb is a type of electric light in which light is produced by a filament heated by electric current. The most common example is the type you find in most table and floor lamps. In commercial buildings, incandescent lights are used for display lights in retail stores, hotels and motels. This includes the very small, high-intensity track lights used to display merchandise or provide spot illumination in restaurants. Energy efficient light bulbs, known as "watt-savers," use less energy than a standard incandescent bulb. "Long-life" bulbs, bulbs that last longer than standard incandescent but produce considerably less light, are not considered energy-efficient bulbs. This category also includes halogen lamps. Halogen lamps are a special type of incandescent lamp containing halogen gas to produce a brighter, whiter light than standard incandescent. Halogen lamps come in three styles: bulbs, models with reflectors, and infrared models with reflectors. Halogen lamps are especially suited to recessed or "canned fixtures," track lights, and outdoor lights.

Incentives Demand-Side Management (DSM) program assistance: This DSM program assistance offers monetary or non-monetary awards to encourage consumers to buy energy-efficient equipment and to participate in programs designed to reduce energy usage. Examples of incentives are zero or low-interest loans, rebates, and direct installation of low cost measures, such as water heater wraps or duct work for distributing the cool air; the units condition air only in the room or areas where they are located.

Incremental effects: The annual changes in energy use (measured in megawatthours) and peak load (measured in kilowatts) caused by new participants in existing DSM (Demand-Side Management) programs and all participants in new DSM programs during a given year. Reported Incremental Effects are annualized to indicate the program effects that would have occurred had these participants been initiated into the program on January 1 of the given year. Incremental effects are not simply the Annual Effects of a given year minus the Annual Effects of the prior year, since these net effects would fail to account for program attrition, equipment degradation, building demolition, and participant dropouts. Please note that Incremental Effects are not a monthly disaggregate of the Annual Effects, but are the total year's effects of only the new participants and programs for that year.

Incremental energy costs: The additional cost of producing and/or transmitting electric energy above some previously determined base cost.

Independent power producer: A corporation, person, agency, authority, or other legal entity or instrumentality that owns or operates facilities for the generation of electricity for use primarily by the public, and that is not an **electric utility**.

Independent system operator (ISO): An independent, Federally regulated entity established to coordinate regional transmission in a non-discriminatory manner and ensure the safety and reliability of the electric system.

Indian coal lease: A lease granted to a mining company to produce coal from Indian lands in exchange for royalties and other revenues; obtained by direct negotiation with Indian tribal authorities, but subject to approval and administration by the U.S. Department of the Interior.

Indicated resources, coal: Coal for which estimates of the rank, quality, and quantity are based partly on sample analyses and measurements and partly on reasonable geologic projections. Indicated resources are computed partly from specified measurements and partly from projection of visible data for a reasonable distance on the basis of geologic evidence. The points of observation are 1/2 to 1-1/2 miles apart. Indicated coal is projected to extend as a 1/2-mile-wide belt that lies more than 1/4 mile from the outcrop, points of observation, or measurement.

Indirect cost: Costs not directly related to mining or milling operations, such as overhead, insurance, security, office expenses, property taxes, and similar administrative expenses.

Indirect uses (end-use category): The end-use category that handles boiler fuel. Fuel in boilers is transformed into another useful energy source, steam or hot water, which is in turn used in other end uses, such as process or space heating or electricity generation. Manufacturers find measuring quantities of steam as it passes through to various end uses especially difficult because variations in both temperature and pressure affect energy content. Thus, the MECS does not present end-use estimates of steam or hot water and shows only the amount of the fuel used in the boiler to produce those secondary energy sources.

Industrial production: The Federal Reserve Board calculates this index by compiling indices of physical output from a variety of agencies and trade groups, weighting each index by the Census' value added, and adding it to the cost of materials. When physical measures are not available, the Federal Reserve Board uses the number of production workers or amount of electricity consumed as the basis for the index. To convert industrial production into dollars, multiply by the "real value added" estimate used by the Federal Reserve Board.

Industrial restrictions (coal): Land-use restrictions that constrain, postpone, or prohibit mining in order to meet other industrial needs or goals; for example, resources not mined due to safety concerns or due to industrial or societal priorities, such as to preserve oil or gas wells that penetrate the coal reserves; to protect surface features such as pipelines, power lines, or company facilities; or to preserve public or private assets, such as highways, railroads, parks, or buildings.

Industrial sector: An energy-consuming sector that consists of all facilities and equipment used for producing, processing, or assembling goods The industrial sector encompasses the following types of activity: manufacturing (NAICS codes 31-33); agriculture, forestry, fishing and hunting (NAICS code 11); mining, including oil and gas extraction (NAICS code 21); natural gas distribution (NAICS code 2212); and construction (NAICS code 23). Overall energy use in this sector is largely for process heat and cooling and powering machinery, with lesser amounts used for facility heating, air conditioning, and lighting. Fossil fuels are also used as raw material inputs to manufactured products.

Inferred resources: Coal in unexplored extensions of demonstrated resources for which estimates of the quality and size are based on geologic evidence and projection. Quantitative estimates are based largely on broad knowledge of the geologic character of the bed or region and where few measurements of bed thickness are available. The estimates are based primarily on an assumed continuation from demonstrated coal for which there is geologic evidence. The points of observation are 1-1/2 to 6 miles apart. Inferred coal is projected to extend as a 2-1/4-mile wide belt that lies more than 3/4 mile from the outcrop, points of observation, or measurement.

In-house Demand-Side Management (DSM) program sponsor: The building's owner or management encourages consumers in the building to improve energy efficiency, reduce energy costs, change timing or energy usage, or promote the use of a different energy source by sponsoring its own DSM programs.

Initial enrichment: Average enrichment for a fresh fuel assembly as specified and ordered in fuel cycle planning. This average should include axial blankets and axially and radially zoned enrichments.

Initial operation: First availability of a newly constructed unit to provide power to the grid. For a nuclear unit, this time is when the Full Power Operating License for the unit is received.

Injections: Natural gas injected into storage reservoirs.

In situ leach mining (ISL): The recovery, by chemical leaching, of the valuable components of a mineral deposit without physical extraction of the mineralized rock from the ground. Also referred to as "solution mining."

Instantaneous peak demand: The maximum demand at the instant of greatest load.

Instantaneous water heater: Also called a "tankless" or "point-of-use" water heater. The water is heated at the point of use as it is needed.

Institutional living quarters: Space provided by a business or organization for long-term housing of individuals whose reason for shared residence is their association with the business or organization. Such quarters commonly have both individual and group living spaces, and the business or organization is responsible for some aspects of resident life beyond the simple provision of living quarters. Examples include prisons; nursing homes and other long-term medical care facilities; military barracks; college dormitories; and convents and monasteries.

Insulation: Any material or substance that provides a high resistance to the flow of heat from one surface to another. The different types include blanket or batt, foam, or loose fill, which are used to reduce heat transfer by conduction. Dead air space is an insulating medium in storm windows and storms as it reduces passage of heat through conduction and convection. Reflective materials are used to reduce heat transfer by radiation.

Insulation around heating and/or cooling ducts: Extra insulation around the heating and / or cooling ducts intended to reduce the loss of hot or cold air as it travels to different parts of the residence.

Insulation around hot-water pipes: Wrapping of insulating material around hot-water pipes to reduce the loss of heat through the pipes.

Insulation around water heater: Blanket insulation wrapped around the water heater to reduce loss of heat. To qualify under this definition, this wrapping must be in addition to any insulation provided by the manufacturer.

Insulator: A material that is a very poor conductor of electricity. The insulating material is usually a ceramic or fiberglass when used in the transmission line and is designed to support a conductor physically and to separate it electrically from other conductors and supporting material.

Intangible drilling and development costs (IDC): Costs incurred in preparing well locations, drilling and deepening wells, and preparing wells for initial production up through the point of installing control valves. None of these functions, because of their nature, have salvage value. Such costs would include labor, transportation, consumable supplies, drilling tool rentals, site clearance, and similar costs.

Integral collector storage (ICS): A solar thermal collector in which incident solar radiation is absorbed directly by the storage medium.

Integrated demand: The summation of the continuously varying instantaneous demand averaged over a specified interval of time. The information is usually determined by examining a demand meter.

Integrated gasification-combined cycle technology: Coal, water, and oxygen are fed to gasifier, which produces syngas. This medium-Btu gas is cleaned (particulates and sulfur compounds removed) and is fed to a gas turbine. The hot exhaust of the gas turbine and heat recovered from the gasification process are routed through a heat-recovery routed through a heat-recovery generator to produce steam, which drives a steam turbine to produce electricity.

Intensity: The amount of a quantity per unit floor space. This method adjusts either the amount of energy consumed or expenditures spent, for the effects of various building characteristics, such as size of the building, number of workers, or number of operating hours, to facilitate comparisons of energy across time, fuels, and buildings.

Intensity per hour: Total consumption of a particular fuel(s) divided by the total floor space of buildings that use the fuel(s) divided by total annual hours of operation.

Interchange energy: Kilowatthours delivered to or received by one electric utility or pooling system from another. Settlement may be payment, returned in kind at a later time, or accumulated as energy balances until the end of the stated period.

Intercity bus: A bus designed for high speed, long distance travel; equipped with front doors only, high backed seats, and usually restroom facilities.

Interconnected system: A system consisting of two or more individual power systems normally operating with connecting tie lines.

Interconnection: Two or more electric systems having a common transmission line that permits a flow of energy between them. The physical connection of the electric power transmission facilities allows for the sale or exchange of energy.

Interdepartmental sales: Includes amounts charged by the electric department at tariff or other specified rates for electricity supplied by it to other utility departments.

Interdepartmental service (electric): Interdepartmental service includes amounts charged by the electric department at tariff or other specified rates for electricity supplied by it to other utility departments.

Interest coverage ratio: The number of times that fixed interest charges were earned. It indicates the margin of safety of interest on fixed debt. The times-interest-earned ratio is calculated using net income before and after income taxes; and the credits of interest charged to construction being treated as other income. The interest charges include interest on long-term debt, interest on debt of associated companies, and other interest expenses.

Intergovernmental Panel on Climate Change (IPCC): A panel established jointly in 1988 by the World Meteorological Organization and the United Nations Environment Program to assess the scientific information relating to climate change and to formulate realistic response strategies.

Interlocking directorates: The holding of a significant position in management or a position on the corporate board of a utility while simultaneously holding a comparable position with another utility, or with a firm doing business with the utility.

Intermediate grade gasoline: A grade of unleaded gasoline with an octane rating intermediate between "regular" and "premium." Octane boosters are added to gasolines to control engine pre-ignition or "knocking" by slowing combustion rates.

Intermediate load (electric system): The range from base load to a point between base load and peak. This point may be the midpoint, a percent of the peak load, or the load over a specified time period.

Intermittent electric generator or intermittent resource: An electric generating plant with output controlled by the natural variability of the energy resource rather than dispatched based on system requirements. Intermittent output usually results from the direct, non-stored conversion of naturally occurring energy fluxes such as solar energy, wind energy, or the energy of free-flowing rivers (that is, run-of-river hydroelectricity).

Internal Collector Storage (ICS): A solar thermal collector in which incident solar radiation is absorbed by the storage medium.

Internal combustion plant: A plant in which the prime mover is an internal combustion engine. An internal combustion engine has one or more cylinders in which the process of combustion takes place, converting energy released from the rapid burning of a fuel-air mixture into mechanical energy. Diesel or gas-fired engines are the principal types used in electric plants. The plant is usually operated during periods of high demand for electricity.

Interruptible gas: Gas sold to customers with a provision that permits curtailment or cessation of service at the discretion of the distributing company under certain circumstances, as specified in the service contract.

Interruptible load: This Demand-Side Management category represents the consumer load that, in accordance with contractual arrangements, can be interrupted at the time of annual peak load by the action of the consumer at the direct request of the system operator. This type of control usually involves large-volume commercial and industrial consumers. Interruptible Load does not include Direct Load Control.

Interruptible or curtailable rate: A special electricity or natural gas arrangement under which, in return for lower rates, the customer must either reduce energy demand on short notice or allow the electric or natural gas utility to temporarily cut off the energy supply for the utility to maintain service for higher priority users. This interruption or reduction in demand typically occurs during periods of high demand for the energy (summer for electricity and winter for natural gas).

Interruptible power: Power and usually the associated energy made available by one utility to another. This transaction is subject to curtailment or cessation of delivery by the supplier in accordance with a prior agreement with the other party or under specified conditions.

Interstate companies: Natural gas pipeline companies subject to Federal Energy Regulatory Commission (FERC) jurisdiction.

Interstate pipeline: Any person engaged in natural gas transportation subject to the jurisdiction of Federal Energy Regulatory Commission (FERC) under the Natural Gas Act.

Interstate pipeline purchase: Any gas supply contracted from and volumes purchased from other interstate pipelines, overland natural gas import purchases, and LNG, SNG, or coal gas purchases from domestic or foreign sources. Purchases from intrastate pipelines to section 311 (b) of the NGPA of 1978 and from independent producers are not included with interstate pipelines purchase.

Intransit deliveries: Redeliveries to a foreign country of foreign gas received for transportation across U.S. territory, and deliveries of U.S. gas to a foreign country for transportation across its territory and redelivery to the United States.

Intransit receipts: Receipts of foreign gas for transportation across U.S. territory and redelivery to a foreign country, and redeliveries to the United States of U.S. gas transported across foreign territory.

Intrastate companies: Companies not subject to Federal Energy Regulatory Commission (FERC) jurisdiction.

Intrastate pipeline: Any person engaged in natural gas transportation (not including gathering) that is not subject to the jurisdiction of the Commission under the Natural Gas Act (other than any such pipeline that is not subject to the jurisdiction of the Commission soley by reason of Section 1(c) of the Natural Gas Act).

In-use (vehicles): Implies that a vehicle is:

1. Registered with the Government of one or more States, the District of Columbia, the Commonwealth of Puerto Rico, or the Virgin Islands; or

2. The vehicle is owned or operated by a Government or military organization within the United States that is not required to register vehicles with the Government agencies listed under 1 above. For example, civilian Federal vehicles are generally not required to register with the State Government in which they are assigned.

Investment of municipality: The investment of the municipality in its utility department, when such investment is not subject to cash settlement on demand or at a fixed future time. Include the cost of debt-free utility plant constructed or acquired by the municipality and made available for the use of the utility department, cash transferred to the utility department for working capital, and other expenditures of an investment nature.

Investments and advances to unconsolidated affiliates: The balance sheet account representing the cost of investments and advances to unconsolidated affiliates. Generally, affiliates that are less than 50-percent owned by a company may not be consolidated into the company's financial statements.

Investor-owned utility (IOU): A privately-owned electric utility whose stock is publicly traded. It is rate regulated and authorized to achieve an allowed rate of return.

Ion exchange: Reversible exchange of ions adsorbed on a mineral or synthetic polymer surface with ions in solution in contact with the surface. A chemical process used for recovery of uranium from solution by the interchange of ions between a solution and a solid, commonly a resin.

Iron and steel industry: Steel Works, Blast Furnaces (Including Coke Ovens), and Rolling Mills: Establishments primarily engaged in manufacturing hot metal, pig iron, and silvery pig iron from iron ore and iron and steel scrap; converting pig iron, scrap iron, and scrap steel into steel; and in hot-rolling iron and steel into basic shapes, such as plates, sheets, strips, rods, bars, and tubing.

Irradiated nuclear fuel: Nuclear fuel that has been exposed to radiation in the reactor core at any power level.

Isobutane (C_4H_{10}): A normally gaseous branch-chain hydrocarbon. It is a colorless paraffinic gas that boils at a temperature of 10.9 degrees Fahrenheit. It is extracted from natural gas or refinery gas streams.

Isobutylene (C_4H_8): An olefinic hydrocarbon recovered from refinery processes or petrochemical processes.

Isohexane (C_6H_{14}): A saturated branch-chain hydrocarbon. It is a colorless liquid that boils at a temperature of 156.2 degrees Fahrenheit.

Isomerization: A refining process that alters the fundamental arrangement of atoms in the molecule without adding or removing anything from the original material. Used to convert normal butane into isobutane (C_4), an alkylation process feedstock, and normal pentane and hexane into isopentane (C_5) and isohexane (C_6), high-octane gasoline components.

Isopach: A line on a map drawn through points of equal thickness of a designated unit (such as a coal bed).

Isopentane: A saturated branched-chain hydrocarbon (C_5H_{12}) obtained by fractionation of natural gasoline or isomerization of normal pentane.

Isotopes: Forms of the same chemical element that differ only by the number of neutrons in their nucleus. Most elements have more than one naturally occurring isotope. Many isotopes have been produced in reactors and scientific laboratories.

Jacket: The enclosure on a water heater, furnace, or boiler.

Jet fuel: A refined petroleum product used in jet aircraft engines. It includes kerosene-type jet fuel and naphtha-type jet fuel.

Joint Implementation (JI): Agreements made between two or more nations under the auspices of the Framework Convention on Climate Change (FCCC) whereby a developed country can receive "emissions reduction units" when it helps to finance projects that reduce net emissions in another developed country (including countries with economies in transition).

Joint-use facility: A multiple-purpose hydroelectric plant. An example is a dam that stores water for both flood control and power production.

Joule (J): The meter-kilogram-second unit of work or energy, equal to the work done by a force of one newton when its point of application moves through a distance of one meter in the direction of the force; equivalent to 107 ergs and one watt-second.

Joule's Law: The rate of heat production by a steady current in any part of an electrical circuit that is proportional to the resistance and to the square of the current, or, the internal energy of an ideal gas depends only on its temperature.

Junction: A region of transition between semiconductor layers, such as a p/n junction, which goes from a region that has a high concentration of acceptors (p-type) to one that has a high concentration of donors (n-type).

Kaplan turbine: A type of turbine that that has two blades whose pitch is adjustable. The turbine may have gates to control the angle of the fluid flow into the blades.

Kerosene: A light petroleum distillate that is used in space heaters, cook stoves, and water heaters and is suitable for use as a light source when burned in wick-fed lamps. Kerosene has a maximum distillation temperature of 400 degrees Fahrenheit at the 10-percent recovery point, a final boiling point of 572 degrees Fahrenheit, and a minimum flash point of 100 degrees Fahrenheit. Included are No. 1-K and No. 2-K, the two grades recognized by ASTM Specification D 3699 as well as all other grades of kerosene called range or stove oil, which have properties similar to those of No. 1 fuel oil.

Kerosene-type jet fuel: A kerosene-based product having a maximum distillation temperature of 400 degrees Fahrenheit at the 10-percent recovery point and a final maximum boiling point of 572 degrees Fahrenheit and meeting ASTM Specification D 1655 and Military Specifications MIL-T-5624P and MIL-T-83133D (Grades JP-5 and JP-8). It is used for commercial and military turbojet and turboprop aircraft engines.

Ketone-alcohol (cyclohexanol): An oily, colorless, hygroscopic liquid with a camphor-like odor. Used in soapmaking, dry cleaning, plasticizers, insecticides, and germicides.

Kilovolt-Ampere (kVa): A unit of apparent power, equal to 1,000 volt-amperes; the mathematical product of the volts and amperes in an electrical circuit.

Kilowatt (kW): One thousand watts.

Kilowatt-electric (kWe): One thousand watts of electric capacity.

Kilowatthour (kWh): A measure of electricity defined as a unit of work or energy, measured as 1 kilowatt (1,000 watts) of power expended for 1 hour. One kWh is equivalent to 3,412 Btu.

Kinetic energy: Energy available as a result of motion that varies directly in proportion to an object's mass and the square of its velocity.

Kyoto Protocol: The result of negotiations at the third Conference of the Parties (COP-3) in Kyoto, Japan, in December of 1997. The Kyoto Protocol sets binding greenhouse gas emissions targets for countries that sign and ratify the agreement. The gases covered under the Protocol include carbon dioxide, methane, nitrous oxide, hydrofluorocarbons (HFCs), perfluorocarbons (PFCs) and sulfur hexafluoride.

Lamp: A term generally used to describe artificial light. The term is often used when referring to a "bulb" or "tube."

Land use: The ultimate uses to be permitted for currently contaminated lands, waters, and structures at each Department of Energy installation. Land-use decisions will strongly influence the cost of environmental management.

Land-use restrictions: Constraints placed upon mining by societal policies to protect surface features or entities that could be affected by mining. Because laws and regulations may be modified or repealed, the restrictions, including industrial and environmental restrictions, are subject to change.

Langley: A unit or measure of solar radiation; 1 calorie per square centimeter or 3.69 Btu per square foot.

Large passenger car: A passenger car with more than 120 cubic feet of interior passenger and luggage volume.

Large pickup truck: A pickup truck weighing between 4,500-8,500 lbs gross vehicle weight (GVW).

Latitude and longitude: The distance on the earth's surface measured, respectively, north or south of the equator and east or west of the standard meridian, expressed in angular degrees, minutes, and seconds.

Leachate: The liquid that has percolated through the soil or other medium.

Lead acid battery: An electrochemical battery that uses lead and lead oxide for electrodes and sulfuric acid for the electrolyte.

Leaded gasoline: A fuel that contains more than 0.05 gram of lead per gallon or more than 0.005 gram of phosphorus per gallon.

Leaded premium gasoline: Gasoline having an antiknock index (R+M/2) greater than 90 and containing more than 0.05 grams of lead or 0.005 grams of phosphorus per gallon.

Leaded regular gasoline: Gasoline having an antiknock index (R+M/2) greater than or equal to 87 and less than or equal to 90 and containing more than 0.05 grams of lead or 0.005 grams of phosphorus per gallon.

Leading edge: In reference to a wind energy conversion system, the area of a turbine blade surface that first comes into contact with the wind.

Lease and plant fuel: Natural gas used in well, field, and lease operations (such as gas used in drilling operations, heaters, dehydrators, and field compressors) and as fuel in natural gas processing plants.

Lease condensate: A mixture consisting primarily of pentanes and heavier hydrocarbons which is recovered as a liquid from natural gas in lease separation facilities. This category excludes natural gas plant liquids, such as butane and propane, which are recovered at downstream natural gas processing plants or facilities.

Lease equipment: All equipment located on the lease except the well to the point of the "Christmas tree."

Lease fuel: Natural gas used in well, field, and lease operations, such as gas used in drilling operations, heaters, dehydrators, and field compressors.

Lease operations: Any well, lease, or field operations related to the exploration for or production of natural gas prior to delivery for processing or transportation out of the field. Gas used in lease operations includes usage such as for drilling operations, heaters, dehydraters, field compressors, and net used for gas lift.

Lease separation facility (lease separator): A facility installed at the surface for the purpose of (a) separating gases from produced crude oil and water at the temperature and pressure conditions set by the separator and/or (b) separating gases from that portion of the produced natural gas stream that liquefies at the temperature and pressure conditions set by the separator.

Leasehold reserves: Natural gas liquid reserves corresponding to the leasehold production defined above.

Lessee: An independent marketer who leases the station and land and has use of tanks, pumps, signs, etc. A lessee dealer typically has a supply agreement with a refiner or distributor and purchases products at dealer tank-wagon prices. The term "lessee dealer" is limited to those dealers who are supplied directly by a refiner or any affiliate or subsidiary of the reporting company. "Direct supply" includes use of commission agent or common carrier delivery.

Levelized cost: The present value of the total cost of building and operating a generating plant over its economic life, converted to equal annual payments. Costs are levelized in real dollars (*i.e.*, adjusted to remove the impact of inflation).

Leverage ratio: A measure that indicates the financial ability to meet debt service requirements and increase the value of the investment to the stockholders. (*i.e.* the ratio of total debt to total assets).

Liability: An amount payable in dollars or by future services to be rendered.

Licensed site capacity: Capacity (number of assemblies) for which the site is currently licensed.

Licensees: Entity that has been granted permission to engage in an activity otherwise unlawful (*i.e.*, hydropower project).

Life extension: Restoration or refurbishment of a plant to its original performance without the installation of new combustion technologies. Life extension results in 10 to 20 years of plant life beyond the anticipated retirement date, but usually does not result in larger capacity.

Lift: The force that pulls a wind turbine blade, as opposed to drag.

Lifting costs: The costs associated with the extraction of a mineral reserve from a producing property.

Light bulbs: A term generally used to describe a manmade source of light. The term is often used when referring to a "bulb" or "tube".

Light-duty vehicles: Vehicles weighing less than 8,500 lbs (include automobiles, motorcycles, and light trucks).

Light gas oils: Liquid petroleum distillates heavier than naphtha, with an approximate boiling range from 401 degrees to 650 degrees Fahrenheit.

Light oil: Lighter fuel oils distilled off during the refining process. Virtually all petroleum used in internal combustion and gas-turbine engines is light oil. Includes fuel oil numbers 1 and 2, kerosene, and jet fuel.

Light rail: An electric railway with a "light volume" traffic capacity compared to "heavy rail." Light rail may use exclusive or shared rights-of-way, high or low platform loading, and multi-car trains or single cars. Also known as "street car," "trolley car," and "tramway."

Light trucks: All single unit two-axle, four-tire trucks, including pickup trucks, sports utility vehicles, vans, motor homes, etc. This is the Department of Transportation definition. The Energy Information defined light truck as all trucks weighing 8,500 pounds or less.

Light water: Ordinary water (H_2O), as distinguished from heavy water or deuterium oxide (D_2O).

Light water reactor (LWR): A nuclear reactor that uses water as the primary coolant and moderator, with slightly enriched uranium as fuel.

Lighting conservation feature: A building feature or practice designed to reduce the amount of energy consumed by the lighting system.

Lighting Demand-Side Management (DSM) program: A DSM program designed to promote efficient lighting systems in new construction or existing facilities. Lighting DSM programs can include: certain types of high-efficiency fluorescent fixtures including T-8 lamp technology, solid state electronic ballasts, specular reflectors, compact fluorescent fixtures, LED and electro-luminescent Emergency Exist Signs, High Pressure Sodium with switchable ballasts, Compact Metal Halide, occupancy sensors, and daylighting controllers.

Lighting equipment: These are light bulbs used to light the building's interior, such as incandescent light bulbs, fluorescent light bulbs, compact fluorescent light bulbs, and high-intensity discharge (HID) lights.

Lights: All of the light bulbs controlled by one switch are counted as one light. For example, a chandelier with multiple lights controlled by one switch is counted as one light. A floor lamp with two separate globes or bulbs controlled by two separate switches would be counted as two lights. Indoor and outdoor lights were counted if they were under the control of the householder. This would exclude lights in the hallway of multifamily buildings.

Lignite: The lowest rank of coal, often referred to as brown coal, used almost exclusively as fuel for steam-electric power generation. It is brownish-black and has a high inherent moisture content, sometimes as high as 45 percent The heat content of lignite ranges from 9 to 17 million Btu per ton on a moist, mineral-matter-free basis. The heat content of lignite consumed in the United States averages 13 million Btu per ton, on the as-received basis (*i.e.*, containing both inherent moisture and mineral matter).

Line loss: Electric energy lost because of the transmission of electricity. Much of the loss is thermal in nature.

Line-miles of seismic exploration: The distance along the Earth's surface that is covered by seismic surveying.

Liquefied natural gas (LNG): Natural gas (primarily methane) that has been liquefied by reducing its temperature to -260 degrees Fahrenheit at atmospheric pressure.

Liquefied petroleum gases: A group of hydrocarbon-based gases derived from crude oil refining or natural gas fractionation. They include ethane, ethylene, propane, propylene, normal butane, butylene, isobutane, and isobutylene. For convenience of transportation, these gases are liquefied through pressurization.

Liquefied refinery gases (LRG): Liquefied petroleum gases fractionated from refinery or still gases. Through compression and/or refrigeration, they are retained in the liquid state. The reported categories are ethane/ethylene, propane/propylene, normal butane/butylene, and isobutane/isobutylene.

Liquid collector: A medium-temperature solar thermal collector, employed predominantly in water heating, which uses pumped liquid as the heat-transfer medium.

Liquid metal fast breeder reactor: A nuclear breeder reactor, cooled by molten sodium, in which fission is caused by fast neutrons.

Load (electric): The amount of electric power delivered or required at any specific point or points on a system. The requirement originates at the energy-consuming equipment of the consumers.

Load control program: A program in which the utility company offers a lower rate in return for having permission to turn off the air conditioner or water heater for short periods of time by remote control. This control allows the utility to reduce peak demand.

Load curve: The relationship of power supplied to the time of occurrence. Illustrates the varying magnitude of the load during the period covered.

Load diversity: The difference between the peak of coincident and noncoincident demands of two or more individual loads.

Load factor: The ratio of the average load to peak load during a specified time interval.

Load following: Regulation of the power output of electric generators within a prescribed area in response to changes in system frequency, tieline loading, or the relation of these to each other, so as to maintain the scheduled system frequency and/or established interchange with other areas within predetermined limits.

Load leveling: Any load control technique that dampens the cyclical daily load flows and increases baseload generation. Peak load pricing and time-of-day charges are two techniques that electric utilities use to reduce peak load and to maximize efficient generation of electricity.

Load loss (3 hours): Any significant incident on an electric utility system that results in a continuous outage of 3 hours or longer to more than 50,000 customers or more than one half of the total customers being served immediately prior to the incident, whichever is less.

Load management technique: Utility demand management practices directed at reducing the maximum kilowatt demand on an electric system and/or modifying the coincident peak demand of one or more classes of service to better meet the utility system capability for a given hour, day, week, season, or year.

Load on equipment: One hundred percent load is the maximum continuous net output of the unit at normal operating conditions during the annual peak load month. For example, if the equipment is capable of operating at 5% overpressure continuously, use this condition for 100% load.

Load reduction request: The issuance of any public or private request to any customer or the general public to reduce the use of electricity for the reasons of maintaining the continuity of service of the reporting entity's bulk electric power supply system. Requests to a customer(s) served under provisions of an interruptible contract are not a reportable action unless the request is made for reasons of maintaining the continuity of service of the reporting entity's bulk electric power supply.

Load shape: A method of describing peak load demand and the relationship of power supplied to the time of occurrence.

Load shedding: Intentional action by a utility that results in the reduction of more than 100 megawatts (MW) of firm customer load for reasons of maintaining the continuity of service of the reporting entity's bulk electric power supply system. The routine use of load control equipment that reduces firm customer load is not considered to be a reportable action.

Local distribution company (LDC): A legal entity engaged primarily in the retail sale and/or delivery of natural gas through a distribution system that includes mainlines (that is, pipelines designed to carry large volumes of gas, usually located under roads or other major right-of-ways) and laterals (that is, pipelines of smaller diameter that connect the end user to the mainline). Since the restructuring of the gas industry, the sale of gas and/or delivery arrangements may be handled by other agents, such as producers, brokers, and marketers that are referred to as "non-LDC."

Long-term debt: Debt securities or borrowings having a maturity of more than one year.

Long-term purchase: A purchase contract under which at least one delivery of material is scheduled to occur during the second calendar year after the contract-signing year. Deliveries also can occur during the contract-signing year, during the first calendar year thereafter, or during any subsequent calendar year.

Long ton: A unit that equals 20 long hundredweight or 2,240 pounds. Used mainly in England.

Longwall mining: An automated form of underground coal mining characterized by high recovery and extraction rates, feasible only in relatively flat-lying, thick, and uniform coalbeds. A high-powered cutting machine is passed across the exposed face of coal, shearing away broken coal, which is continuously hauled away by a floor-level conveyor system. Longwall mining extracts all machine-minable coal between the floor and ceiling within a contiguous block of coal, known as a panel, leaving no support pillars within the panel area. Panel dimensions vary over time and with mining conditions but currently average about 900 feet wide (coal face width) and more than 8,000 feet long (the minable extent of the panel, measured in direction of mining). Longwall mining is done under movable roof supports that are advanced as the bed is cut. The roof in the mined-out area is allowed to fall as the mining advances.

Loop flow: The movement of electric power from generator to load by dividing along multiple parallel paths; it especially refers to power flow along an unintended path that loops away from the most direct geographic path or contract path.

Loss of service (15 minutes): Any loss in service for greater than 15 minutes by an electric utility of firm loads totaling more than 200 MW, or 50 percent of the total load being supplied immediately prior to the incident, whichever is less. However, utilities with a peak load in the prior year of more than 3000 MW are only to report losses of service to firm loads totaling more than 300 MW for greater than 15 minutes. (The DOE shall be notified with service restoration and in any event, within three hours after the beginning of the interruption.)

Low Btu gas: A fuel gas with a heating value between 90 and 200 Btu per cubic foot.

Low E glass: Low-emission glass reflects up to 90% of long-wave radiation, which is heat, but lets in short-wave radiation, which is light. Windows are glazed with a coating that bonds a microscopic, transparent, metallic substance to the inside surface of the double-pane or triple-pane windows.

Low flush toilet: A toilet that uses less water than a standard one during flushing, for the purpose of conserving water resources.

Low flow showerheads: Reduce the amount of water flow through the showerhead from 5 to 6 gallons a minute to 3 gallons a minute.

Low head: Vertical difference of 100 feet or less in the upstream surface water elevation (headwater) and the downstream surface water elevation (tailwater) at a dam.

Low Income Home Energy Assistance Program (LIHEAP): The purpose of LIHEAP is to assist eligible households to meet the cost of heating or cooling in residential dwellings. The Federal government provides the funds to the States that administer the program.

Low power testing: The period of time between a plant's nuclear generating unit's initial fuel loading date and the issuance of its operating (full-power) license. The maximum level of operation during this period is 5 percent of the unit's thermal rating.

Low-pressure sodium lamp: A type of lamp that produces light from sodium gas contained in a bulb operating at a partial pressure of 0.13 to 1.3 Pascal. The yellow light and large size make them applicable to lighting streets and parking lots.

Low temperature collectors: Metallic or nonmetallic collectors that generally operate at temperatures below 110 degrees Fahrenheit and use pumped liquid or air as the heat transfer medium. They usually contain no glazing and no insulation, and they are often made of plastic or rubber, although some are made of metal.

Lubricants: Substances used to reduce friction between bearing surfaces, or incorporated into other materials used as processing aids in the manufacture of other products, or used as carriers of other materials. Petroleum lubricants may be produced either from distillates or residues. Lubricants include all grades of lubricating oils, from spindle oil to cylinder oil to those used in greases.

Lumen: An empirical measure of the quantity of light. It is based upon the spectral sensitivity of the photosensors in the human eye under high (daytime) light levels. Photometrically it is the luminous flux emitted with a solid angle (1 steradian) by a point source having a uniform luminous intensity of 1 candela.

Lumens/Watt (lpw): A measure of the efficacy (efficiency) of lamps. It indicates the amount of light (lumens) emitted by the lamp for each unit of electrical power (Watts) used.

Machine drive (motors): The direct process end use in which thermal or electric energy is converted into mechanical energy. Motors are found in almost every process in manufacturing. Therefore, when motors are found in equipment that is wholly contained in another end use (such as process cooling and refrigeration), the energy is classified there rather than in machine drive.

Made available (vehicle): A vehicle is considered "Made available" if it is available for delivery to dealers or users, whether or not it was actually delivered to them. To be "Made available", the vehicle must be completed and available for delivery; thus, any conversion to be performed by an original equipment manufacturer (OEM) Vehicle Converter or Aftermarket Vehicle Converter must have been completed.

Magma: Naturally occurring molten rock, generated within the earth and capable of intrusion and extrusion, from which igneous rocks are thought to have been derived through solidification and related processes. It may or may not contain suspended solids (such as crystals and rock fragments) and/or gas phases.

Main heating equipment: Equipment primarily used for heating ambient air in the housing unit.

Main heating fuel: The form of energy used most frequently to heat the largest portion of the floorspace of a structure. The energy source designated as the main heating fuel is the source delivered to the site for that purpose, not any subsequent form into which it is transformed on site to deliver the heat energy (*e.g.*, for buildings heated by a steam boiler, the main heating fuel is the main input fuel to the boiler, not the steam or hot water circulated through the building.)

Mains: A system of pipes for transporting gas within a distributing gas utility's retail service area to points of connection with consumer service pipes.

Maintenance expenses: That portion of operating expenses consisting of labor, materials, and other direct and indirect expenses incurred for preserving the operating efficiency and/or physical condition of utility plants used for power production, transmission, and distribution of energy.

Maintenance of boiler plant (expenses): The cost of labor, material, and expenses incurred in the maintenance of a steam plant. Includes furnaces; boilers; coal, ash-handling, and coal-preparation equipment; steam and feed water piping; and boiler apparatus and accessories used in the production of steam, mercury, or other vapor to be used primarily for generating electricity. The point at which an electric steam plant is distinguished from an electric plant is defined as follows:

1. Inlet flange of throttle valve on prime mover.
2. Flange of all steam extraction lines on prime mover.
3. Hotwell pump outlet on condensate lines.
4. Inlet flange of all turbine-room auxiliaries.
5. Connection to line side of motor starter for all boiler-plant equipment.

Maintenance of structures (expenses): The cost of labor, materials, and expenses incurred in maintenance of power production structures. Structures include all buildings and facilities to house, support, or safeguard property or persons.

Maintenance supervision and engineering expenses: The cost of labor and expenses incurred in the general supervision and direction of the maintenance of power generation stations. The supervision and engineering included consists of the pay and expenses of superintendents, engineers, clerks, other employees, and consultants engaged in supervising and directing the maintenance of each utility function. Direct supervision and engineering of specific activities, such as fuel handling, boiler room operations, generator operations, etc., are charged to the appropriate accounts.

Major electric utility: A utility that, in the last 3 consecutive calendar years, had sales or transmission services exceeding one of the following: (1) 1 million megawatthours of total annual sales; (2) 100 megawatthours of annual sales for resale; (3) 500 megawatthours of annual gross interchange out; or (4) 500 megawatthours of wheeling (deliveries plus losses) for others.

Major energy sources: Fuels or energy sources such as electricity, fuel oil, natural gas, district steam, district hot water, and district chilled water. District chilled water is not included in any totals for the sum of major energy sources or fuels; all other major fuels are included in these totals.

Major fuels: Fuels or energy sources such as: electricity, fuel oil, liquefied petroleum gases, natural gas, district steam, district hot water, and district chilled water.

Major interstate pipeline company: A company whose combined sales for resale, including gas transported interstate or stored for a fee, exceeded 50 million thousand cubic feet in the previous year.

Make-up air: Air brought into a building from outside to replace exhaust air.

Manhattan Project: The U.S. Government project that produced the first nuclear weapons during World War II. Started in 1942, the Manhattan Project formally ended in 1946. The Hanford Site, Oak Ridge Reservation, and Los Alamos National Laboratory were created for this effort.

Manual dimmer switches: These are like residential-style dimmer switches. They are not generally used with fluorescent and high-intensity discharge (HID) lamps.

Manufactured gas: A gas obtained by destructive distillation of coal or by the thermal decomposition of oil, or by the reaction of steam passing through a bed of heated coal or coke. Examples are coal gases, coke oven gases, producer gas, blast furnace gas, blue (water) gas, carbureted water gas. Btu content varies widely.

Manufacturing: An energy-consuming subsector of the industrial sector that consists of all facilities and equipment engaged in the mechanical, physical, chemical, or electronic transformation of materials, substances, or components into new products. Assembly of component parts of products is included, except for that which is included in construction.

Manufacturing division: One of 10 fields of economic activity defined by the Standard Industrial Classification Manual. The manufacturing division includes all establishments engaged in the mechanical or chemical transformation of materials or substances into new products. The other divisions of the U.S. economy are agriculture, forestry, fishing, hunting, and trapping; mining; construction; transportation, communications, electric, gas, and sanitary services; wholesale trade; retail trade; finance, insurance, and real estate; personal, business, professional, repair, recreation, and other services; and public administration. The establishments in the manufacturing division constitute the universe for the MECS.

Manufacturing establishment: An economic unit at a single physical location where mechanical or chemical transformation of materials or substances into new products are performed.

Marginal cost: The change in cost associated with a unit change in quantity supplied or produced.

Marine freight: Freight transported over rivers, canals, the Great Lakes, and domestic ocean waterways.

Market clearing price: The price at which supply equals demand for the Day-ahead or hour-ahead markets.

Market-based pricing: Prices of electric power or other forms of energy determined in an open market system of supply and demand under which prices are set solely by agreement as to what buyers will pay and sellers will accept. Such prices could recover less or more than full costs, depending upon what the buyers and sellers see as their relevant opportunities and risks.

Market price contract: A contract in which the price of uranium is not specifically determined at the time the contract is signed but is based instead on the prevailing market price at the time of delivery. A market price contract may include a floor price, that is, a lower limit on the eventual settled price. The floor price and the method of price escalation generally are determined when the contract is signed. The contract may also include a price ceiling or a discount from the agreed-upon market price reference.

Market price settlement (uranium): The price paid for uranium delivery under a market-price contract. The price is commonly (but not always) determined at or sometime before delivery and may be related to a floor price, ceiling price, or discount.

Marketable coke: Those grades of coke produced in delayed or fluid cokers that may be recovered as relatively pure carbon. This "green" coke may be sold as is or further purified by calcining.

Marketed production: Gross withdrawals less gas used for repressuring, quantities vented and flared, and nonhydrocarbon gases removed in treating or processing operations. Includes all quantities of gas used in field and processing plant operations.

Masonry: A general term covering wall construction using masonry materials such as brick, concrete block, stone, and tile that are set in mortar; also included is stucco. The category does not include concrete panels because concrete panels represent a different method of constructing buildings. Concrete panels are reported separately.

Masonry stove: A type of heating appliance similar to a fireplace, but much more efficient and clean burning. They are made of masonry and have long channels through which combustion gases give up their heat to the heavy mass of the stove, which releases the heat slowly into a room. Often called Russian or Finnish fireplaces.

Mass burn facility: A type of municipal solid waste (MSW) incineration facility in which MSW is burned with only minor presorting to remove oversize, hazardous, or explosive materials.

Master-metering: Measurement of electricity or natural gas consumption of several tenants or housing units using a single meter. That is, one meter measures the energy usage for several households collectively.

Maximum deliverability: The maximum deliverability rate (Mcf/d) estimated at the present developed maximum operating capacity.

Maximum demand: The greatest of all demands of the load that has occurred within a specified period of time.

Maximum dependable capacity, net: The gross electrical output measured at the output terminals of the turbine generator(s) during the most restrictive seasonal conditions, less the station service load.

Maximum established site capacity (reactors): The maximum established spent fuel capacity for the site is defined by DOE as the maximum number of intact assemblies that will be able to be stored at some point in the future (between the reporting date and the reactor's end of life) taking into account any established or current studies or engineering evaluations at the time of submittal for licensing approval from the NRC.

Maximum hourly load: This is determined by the interval in which the 60-minute integrated demand is the greatest.

Maximum streamflow: The maximum rate of water flow past a given point during a specified period.

Mcf: One thousand cubic feet.

Mean indoor temperature: The "usual" temperature. If different sections of the house are kept at different temperatures, the reported temperature is for the section where the people are. A thermostat setting is accepted if the temperature is not known.

Mean operating hours: The arithmetic average number of operating hours per building is the weighted sum of the number of operating hours divided by the weighted sum of the number of buildings.

Mean power output (of a wind turbine): The average power output of a wind energy conversion system at a given mean wind speed based on a Raleigh frequency distribution.

Mean square feet per building: The arithmetic average square feet per building is the weighted sum of the total square feet divided by the weighted sum of the number of buildings.

Measured heated area of residence: The floor area of the housing unit that is enclosed from the weather and heated. Basements are included whether or not they contain finished space. Garages are included if they have a wall in common with the house. Attics that have finished space and attics that have some heated space are included. Crawl spaces are not included even if they are enclosed from the weather. Sheds and other buildings that are not attached to the house are not included. "Measured" area means the measurement of the dimensions of the home, using a metallic, retractable, 50-foot tape measure. "Heated area" is that portion of the measured area that is heated during most of the season. Rooms that are shut off during the heating season to save on fuel are not counted. Attached garages that are unheated and unheated areas in the attics and basements are also not counted.

Measured resources, coal: Coal resources for which estimates of the rank, quality, and quantity have been computed, within a margin of error of less than 20 percent, from sample analyses and measurements from closely spaced and geologically well known sample sites. Measured resources are computed from dimensions revealed in outcrops, trenches, mine workings, and drill holes. The points of observation and measurement are so closely spaced and the thickness and extent of coals are so well defined that the tonnage is judged to be accurate within 20 percent. Although the spacing of the points of observation necessary to demonstrate continuity of the coal differs from region to region, according to the character of the coalbeds, the point of observation are no greater than 1/2 mile apart. Measured coal is projected to extend as a belt 1/4 mile wide from the outcrop or points of observation or measurement.

Median: The middle number of a data set when the measurements are arranged in ascending (or descending) order.

Median streamflow: The middle rate of flow of water past a given point for which there have been several greater and lesser rates of flow occurring during a specified period.

Median water condition: The middle precipitation and run-off condition for a distribution of water conditions that have happened over a long period of time. Usually determined by examining the water supply record of the period in question.

Medium pressure: For valves and fittings, implies that they are suitable for working pressures between 125 to 175 pounds per square inch.

Medium-temperature collector: A collector designed to operate in the temperature range of 140 degrees to 180 degrees Fahrenheit, but that can also operate at a temperature as low as 110 degrees Fahrenheit. The collector typically consists of a metal frame, metal absorption panels with integral flow channels (attached tubing for liquid collectors or integral ducting for air collectors), and glazing and insulation on the sides and back.

Megawatt (MW): One million watts of electricity.

Megawatt electric (MWe): One million watts of electric capacity.

Megawatthour (MWh): One thousand kilowatt-hours or 1 million watt-hours.

Mercaptan: An organic chemical compound that has a sulfur like odor that is added to natural gas before distribution to the consumer, to give it a distinct, unpleasant odor (smells like rotten eggs). This serves as a safety device by allowing it to be detected in the atmosphere, in cases where leaks occur.

Merchant coke plant: A coke plant where coke is produced primarily for sale on the commercial (open) market.

Merchant facilities: High-risk, high-profit facilities that operate, at least partially, at the whims of the market, as opposed to those facilities that are constructed with close cooperation of municipalities and have significant amounts of waste supply guaranteed.

Merchant MTBE plants: MTBE (methyl tertiary butyl ether) production facilities primarily located within petrochemical plants rather than refineries. Production from these units is sold under contract or on the spot market to refiners or other gasoline blenders.

Merchant oxygenate plants: Oxygenate production facilities that are not associated with a petroleum refinery. Production from these facilities is sold under contract or on the spot market to refiners or other gasoline blenders.

Mercury vapor lamp: A high-intensity discharge lamp that uses mercury as the primary light-producing element. Includes clear, phosphor coated, and self-ballasted lamps.

Merger: A combining of companies or corporations into one, often by issuing stock of the controlling corporation to replace the greater part of that of the other.

Met: An approximate unit of heat produced by a resting person, equal to about 18.5 Btu per square foot per hour.

Metal halide lamp: A high-intensity discharge lamp type that uses mercury and several halide additives as light-producing elements. These lights have the best Color Rendition Index (CRI) of the high-intensity discharge lamps. They can be used for commercial interior lighting or for stadium lights.

Metallic: The metallic material composition of the collector's absorber system.

Metallurgical coal: Coking coal and pulverized coal consumed in making steel.

Metered data: End-use data obtained through the direct measurement of the total energy consumed for specific uses within the individual household. Individual appliances can be submetered by connecting the recording meters directly to individual appliances.

Metered peak demand: The presence of a device to measure the maximum rate of electricity consumption per unit of time. This device allows electric utility companies to bill their customers for maximum consumption, as well as for total consumption.

Methane: A colorless, flammable, odorless hydrocarbon gas (CH_4) which is the major component of natural gas. It is also an important source of hydrogen in various industrial processes. Methane is a greenhouse gas.

Methanogens: Bacteria that synthesize methane, requiring completely anaerobic conditions for growth.

Methanol (CH_3OH): A light, volatile alcohol eligible for gasoline blending.

Methanol blend: Mixtures containing 85 percent or more (or such other percentage, but not less than 70 percent) by volume of methanol with gasoline. Pure methanol is considered an "other alternative fuel."

Methanotrophs: Bacteria that use methane as food and oxidize it into carbon dioxide.

Methyl chloroform (trichloroethane): An industrial chemical (CH_3CCl_3) used as a solvent, aerosol propellant, and pesticide and for metal degreasing.

Methylene chloride: A colorless liquid, nonexplosive and practically nonflammable. Used as a refrigerant in centrifugal compressors, a solvent for organic materials, and a component in nonflammable paint removers.

Metric conversion factors (for floorspace): Floorspace estimates may be converted to metric units by using the relationship, 1 square foot is approximately equal to .0929 square meters. Energy estimates may be converted to metric units by using the relationship, 1 Btu is approximately equal to 1,055 joules. One kilowatthour is exactly 3,600,000 joules. One gigajoule is approximately 278 kilowatthours (kWh).

Metric ton: A unit of weight equal to 2,204.6 pounds.

Metropolitan: Located within the boundaries of a metropolitan area.

Metropolitan area: A geographic area that is a metropolitan statistical area or a consolidated metropolitan statistical area as defined by the U.S. Office of Management and Budget.

Metropolitan statistical area (MSA): A county or group of contiguous counties (towns and cities in New England) that has (1) at least one city with 50,000 or more inhabitants; or (2) an urbanized area of 50,000 inhabitants and a total population of 100,000 or more inhabitants (75,000 in New England). These areas are defined by the U.S. Office of Management and Budget. The contiguous counties or other jurisdictions to be included in an MSA are those that, according to certain criteria, are essentially metropolitan in character and are socially and economically integrated with the central city or urbanized area.

Microcrystalline wax: Wax extracted from certain petroleum residues having a finer and less apparent crystalline structure than paraffin wax and having the following physical characteristics: penetration at 77 degrees Fahrenheit (D1321)-60 maximum; viscosity at 210 degrees Fahrenheit in Saybolt Universal Seconds (SUS); (D88)-60 SUS (10.22 centistokes) minimum to 150 SUS (31.8 centistokes) maximum; oil content (D721)-5 percent minimum.

Microgroove: A small groove scribed into the surface of a solar photovoltaic cell which is filled with metal for contacts.

Micrometer (or Micron): One-millionth of a meter. It can also be expressed as 10^{-6} meter.

Microwave oven: A household cooking appliance consisting of a compartment designed to cook or heat food by means of microwave energy. It may also have a browning coil and convection heating as additional features.

Midgrade gasoline: Gasoline having an antiknock index, *i.e.*, octane rating, greater than or equal to 88 and less than or equal to 90.

Mid-size passenger car: A passenger car with between 110 and 119 cubic feet of interior passenger and luggage volume.

Middle distillates: A general classification of refined petroleum products that includes distillate fuel oil and kerosene.

Middlings: In coal preparation, this material called mid-coal is neither clean nor refuse; due to their intermediate specific gravity, middlings sink only partway in the washing vessels and are removed by auxiliary means.

Miles per gallon (MPG): A measure of vehicle fuel efficiency. Miles per gallon or MPG represents "Fleet Miles per Gallon." For each subgroup or "table cell," MPG is computed as the ratio of the total number of miles traveled by all vehicles in the subgroup to the total number of gallons consumed. MPGs are assigned to each vehicle using the EPA certification files and adjusted for on-road driving.

Military use: Includes sales to the Armed Forces, including volumes sold to the Defense Fuel Supply Center (DFSC) for use by all branches of the Department of Defense (DOD).

Mill: A monetary cost and billing unit used by utilities; it is equal to 1/1000 of the U.S. dollar (equivalent to 1/10 of 1 cent).

Mill capital: Cost for transportation and equipping a plant for processing ore or other feed materials.

Mill feed: Uranium ore supplied to a crusher or grinding mill in an ore-dressing process.

Milling: The grinding or crushing of ore, concentration, and other benefication, including the removal of valueless or harmful constituents and preparation for market.

Milling capacity: The maximum rate at which a mill is capable of treating ore or producing concentrate.

Milling of uranium: The processing of uranium from ore mined by conventional methods, such as underground or openpit methods, to separate the uranium from the undesired material in the ore.

Minable: Capable of being mined under current mining technology and environmental and legal restrictions, rules, and regulations.

Mine capital: Cost for exploration and development, pre-mining stripping, shaft sinking, and mine development (including in situ leaching), as well as the mine plant and its equipment.

Mine count: The number of mines, or mines collocated with preparation plants or tipples, located in a particular geographic area (state or region). If a mine is mining coal across two counties within a state, or across two states, then it is counted as two operations.

Mineral: Any of the various naturally occurring inorganic substances, such as metals, salt, sand, stone, sulfur, and water, usually obtained from the earth.

Mineral lease: An agreement wherein a mineral interest owner (lessor) conveys to another party (lessee) the rights to explore for, develop, and produce specified minerals. The lessee acquires a working interest and the lessor retains a non-operating interest in the property, referred to as the royalty interest, each in proportions agreed upon.

Mineral-matter-free basis: Mineral matter in coal is the parent material in coal from which ash is derived and which comes from minerals present in the original plant materials that formed the coal, or from extraneous sources such as sediments and precipitates from mineralized water. Mineral matter in coal cannot be analytically determined and is commonly calculated using data on ash and ash-forming constituents. Coal analyses are calculated to the mineral matter free basis by adjusting formulas used in calculations in order to deduct the weight of mineral matter from the total coal.

Mineral rights: The ownership of the minerals beneath the earth's surface with the right to remove them. Mineral rights may be conveyed separately from surface rights.

Mining: An energy-consuming subsector of the industrial sector that consists of all facilities and equipment used to extract energy and mineral resources.

Mini van: Small van that first appeared with that designation in 1984. Any of the smaller vans built on an automobile-type frame. Earlier models such as the Volkswagen van are now included in this category.

Minimum streamflow: The lowest rate of flow of water past a given point during a specified period.

Mining operation: One mine and/or tipple at a single physical location.

Minority carrier: A current carrier, either an electron or a hole, that is in the minority in a specific layer of a semiconductor material; the diffusion of minority carriers under the action of the cell junction voltage is the current in a photovoltaic device.

Miscellaneous petroleum products: Includes all finished products not classified elsewhere (*e.g.*, petrolatum lube refining byproducts (aromatic extracts and tars), absorption oils, ram-jet fuel, petroleum rocket fuels, synthetic natural gas feedstocks, and specialty oils).

Miscellaneous reserves: A supply source having not more than 50 billion cubic feet of dedicated recoverable salable reserves and that falls within the definition of **Supply Source**.

Mixed waste: Waste containing both radioactive and hazardous constituents.

MMbbl/d: One million barrels of oil per day.

MMBtu: One million British thermal units.

MMcf: One million cubic feet.

MMgal/d: One million gallons per day.

MMst: One million short tons.

Mobile home: A housing unit built on a movable chassis and moved to the site. It may be placed on a permanent or temporary foundation and may contain one room or more. If rooms are added to the structure, it is considered a single-family housing unit. A manufactured house assembled on site is a single-family housing unit, not a mobile home.

Moderator: A material, such as ordinary water, heavy water, or graphite, used in a reactor to slow down high-velocity neutrons, thus increasing the likelihood of further fission.

Modules: Photovoltaic cells or an assembly of cells into panels (modules) intended for and shipped for final consumption or to another organization for resale. When exported, incomplete modules and unencapsulated cells are also included. Modules used for space applications are not included.

Moist (coal) basis: "Moist" coal contains its natural inherent or bed moisture, but does not include water adhering to the surface. Coal analyses expressed on a moist basis are performed or adjusted so as to describe the data when the coal contains only that moisture that exists in the bed in its natural state of deposition and when the coal has not lost any moisture due to drying.

Moisture content: The water content of a substance (a solid fuel) as measured under specified conditions being the "dry basis," which equals the weight of the wet sample minus the weight of a (bone) dry sample divided by the weight of the dry sample times 100 (to get percent); "wet basis," which is equal to the weight of the wet sample minus the weight of the dry sample divided by the weight of the wet sample times 100.

Mole: The quantity of a compound or element that has a weight in grams numerically equal to its molecular weight.

Montreal Protocol: The Montreal Protocol on Substances that Deplete the Ozone Layer (1987). An international agreement, signed by most of the industrialized nations, to substantially reduce the use of chlorofluorocarbons (CFCs). Signed in January 1989, the original document called for a 50-percent reduction in CFC use by 1992 relative to 1986 levels. The subsequent London Agreement called for a complete elimination of CFC use by 2000.

Motor gasoline blending: Mechanical mixing of motor gasoline blending components, and oxygenates when required, to produce finished motor gasoline. Finished motor gasoline may be further mixed with other motor gasoline blending components or oxygenates, resulting in increased volumes of finished motor gasoline and/or changes in the formulation of finished motor gasoline (*e.g.*, conventional motor gasoline mixed with MTBE to produce oxygenated motor gasoline).

Motor gasoline blending components: Naphthas (*e.g.*, straight-run gasoline, alkylate, reformate, benzene, toluene, xylene) used for blending or compounding into finished motor gasoline. These components include reformulated gasoline blendstock for oxygenate blending (RBOB) but exclude oxygenates (alcohols, ethers), butane, and pentanes plus.

Motor gasoline (finished): A complex mixture of relatively volatile hydrocarbons with or without small quantities of additives, blended to form a fuel suitable for use in spark-ignition engines. Motor gasoline, as defined in ASTM Specification D 4814 or Federal Specification VV-G-1690C, is characterized as having a boiling range of 122 to 158 degrees Fahrenheit at the 10 percent recovery point to 365 to 374 degrees Fahrenheit at the 90 percent recovery point. Motor Gasoline includes conventional gasoline; all types of oxygenated gasoline, including gasohol; and reformulated gasoline, but excludes aviation gasoline.

Motor speed: The number of revolutions that the motor turns in a given time period (*i.e.* revolutions per minute, rpm).

MPG: Miles per gallon.

MPG shortfall: The difference between actual on-road MPG and EPA laboratory test MPG. MPG shortfall is expressed as gallons per mile ratio (GPMR).

MSHA ID number: Seven (7)-digit code assigned to a mining operation by the Mine Safety and Health Administration.

MTBE (methyl tertiary butyl ether) $(CH_3)_3COCH_3$: An ether intended for gasoline blending as described in "**Oxygenates.**"

Multiple completion well: A well equipped to produce oil and/or gas separately from more than one reservoir. Such wells contain multiple strings of tubing or other equipment that permit production from the various completions to be measured and accounted for separately. For statistical purposes, a multiple completion well is reported as one well and classified as either an oil well or a gas well. If one of the several completions in a given well is an oil completion, the well is classified as an oil well. If all of the completions in a given well are gas completions, the well is classified as a gas well.

Multiple cropping: A system of growing several crops on the same field in one year.

Multiple purpose project: The development of hydroelectric facilities to serve more than one function. Some of the uses include hydroelectric power, irrigation, water supply, water quality control, and/or fish and wildlife enhancement.

Multiple purpose reservoir: Stored water and its usage governed by advanced water resource conservation practices to achieve more than one water control objective. Some of the objectives include flood control, hydroelectric power development, irrigation, recreation usage, and wilderness protection.

Municipal solid waste: Residential solid waste and some nonhazardous commercial, institutional, and industrial wastes.

Municipal waste: As defined in the Energy Security Act (P.L. 96-294; 1980) as "any organic matter, including sewage, sewage sludge, and industrial or commercial waste, and mixtures of such matter and inorganic refuse from any publicly or privately operated municipal waste collection or similar disposal system, or from similar waste flows (other than such flows which constitute agricultural wastes or residues, or wood wastes or residues from wood harvesting activities or production of forest products)."

Municipal waste to energy project or plant: A facility that produces fuel or energy from municipal solid waste.

Municipality: A village town, city, county, or other political subdivision of a State.

NAICS (North American Industry Classification System: A coding system developed jointly by the United States, Canada, and Mexico to classify businesses and industries according to the type of economic activity in which they are engaged. NAICS replaces the Standard Industrial Classification (SIC) codes.

Name plate: A metal tag attached to a machine or appliance that contains information such as brand name, serial number, voltage, power ratings under specified conditions, and other manufacturer supplied data.

Naphtha less than 401 degrees Fahrenheit: A naphtha with a boiling range of less than 401 degrees Fahrenheit that is intended for use as a petrochemical feedstock.

Naphthas: Refined or partly refined light distillates with an approximate boiling point range of 27 degrees to 221 degrees Centigrade. Blended further or mixed with other materials, they make high-grade motor gasoline or jet fuel. Also, used as solvents, petrochemical feedstocks, or as raw materials for the production of town gas.

Naphtha-type jet fuel: A fuel in the heavy naphtha boiling range having an average gravity of 52.8 degrees API, 20 to 90 percent distillation temperatures of 290 degrees to 470 degrees Fahrenheit, and meeting Military Specification MIL-T-5624L (Grade JP-4). It is used primarily for military turbojet and turboprop aircraft engines because it has a lower freeze point than other aviation fuels and meets engine requirements at high altitudes and speeds.

National Association of Regulatory Utility Commissioners (NARUC): An affiliation of the public service commissioners to promote the uniform treatment of members of the railroad, public utilities, and public service commissions of the 50 states, the District of Columbia, the Commonwealth of Puerto Rico, and the territory of the Virgin Islands.

National Defense Authorization Act: The federal law, enacted in 1994 and amended in 1995, that required the Secretary of Energy to prepare the Baseline Report.

National priorities list: The Environmental Protection Agency's list of the most serious uncontrolled or abandoned hazardous waste sites identified for possible long-term remedial action under the Comprehensive Environmental Response, Compensation, and Liability Act (CERCLA). The list is based primarily on the score a site receives from the Environmental Protection Agency Hazard Ranking System. The Environmental Protection Agency is required to update the National Priorities List at least once a year.

National Rural Electric Cooperative Association (NRECA): A national organization dedicated to representing the interests of cooperative electric utilities and the consumers they serve. Members come from the 46 states that have an electric distribution cooperative.

National uranium resource evaluation (NURE): A program begun by the U.S. Atomic Energy Commission (AEC) in 1974 to make a comprehensive evaluation of U.S. uranium resources and continued through 1983 by the AEC's successor agencies, the Energy Research and Development Administration (ERDA), and the Department of Energy (DOE). The NURE program included aerial radiometric and magnetic surveys, hydrogeochemical and stream sediment surveys, geologic drilling in selected areas, geophysical logging of selected boreholes, and geologic studies to identify and evaluate geologic environments favorable for uranium.

Native gas: Gas in place at the time that a reservoir was converted to use as an underground storage reservoir in contrast to injected gas volumes.

Natural gas: A gaseous mixture of hydrocarbon compounds, the primary one being **methane**.

1. **Wet natural gas:** A mixture of hydrocarbon compounds and small quantities of various nonhydrocarbons existing in the gaseous phase or in solution with crude oil in porous rock formations at reservoir conditions. The principal hydrocarbons normally contained in the mixture are methane, ethane, propane, butane, and pentane. Typical nonhydrocarbon gases that may be present in reservoir natural gas are water vapor, carbon dioxide, hydrogen sulfide, nitrogen and trace amounts of helium. Under reservoir conditions, natural gas and its associated liquefiable portions occur either in a single gaseous phase in the reservoir or in solution with crude oil and are not distinguishable at the time as separate substances. **Associated-dissolved natural gas:** Natural gas that occurs in crude oil reservoirs either as free gas (associated) or as gas in solution with crude oil (dissolved gas).
2. **Nonassociated natural gas:** Natural gas that is not in contact with significant quantities of crude oil in the reservoir.
3. **Dry natural gas:** Natural gas which remains after: 1) the liquefiable hydrocarbon portion has been removed from the gas stream (*i.e.*, gas after lease, field, and/or plant separation); and 2) any volumes of nonhydrocarbon gases have been removed where they occur in sufficient quantity to render the gas unmarketable.

Natural gas field facility: A field facility designed to process natural gas produced from more than one lease for the purpose of recovering condensate from a stream of natural gas; however, some field facilities are designed to recover propane, normal butane, and pentanes plus, etc., and to control the quality of natural gas to be marketed.

Natural gas gross withdrawals: Full well-stream volume of produced natural gas, excluding condensate separated at the lease.

Natural gas hydrates: Solid, crystalline, wax-like substances composed of water, methane, and usually a small amount of other gases, with the gases being trapped in the interstices of a water-ice lattice.

They form beneath permafrost and on the ocean floor under conditions of moderately high pressure and at temperatures near the freezing point of water.

Natural gas liquids (NGL): A general term for all liquid products separated from natural gas in gas processing or cycling plants. They include natural gas plant liquids and lease condensate.

Natural gas liquids production: The volume of natural gas liquids removed from natural gas in lease separators, field facilities, gas processing plants, or cycling plants during the report year.

Natural gas marketed production: Gross withdrawals of natural gas from production reservoirs, less gas used for reservoir repressuring, nonhydrocarbon gases removed in treating and processing operations, and quantities vented and flared.

Natural gas plant liquids: Those hydrocarbons in natural gas that are separated as liquids at downstream gas processing plants or at fractionating and cycling plants. Data on lease condensate are excluded. Products obtained include liquefied petroleum gases and pentanes plus.

Natural Gas Policy Act of 1978 (NGPA): Signed into law on November 9, 1978, the NGPA is a framework for the regulation of most facets of the natural gas industry.

Natural gas processing plants: Facilities designed to recover natural gas liquids from a stream of natural gas that may or may not have passed through lease separators and/or field separation facilities. These facilities also control the quality of the natural gas to be marketed. **Cycling** plants are classified as natural gas processing plants.

Natural gas utility demand-side management (DSM) program sponsor: A DSM (demand-side management) program sponsored by a natural gas utility that suggests ways to increase the energy efficiency of buildings, to reduce energy costs, to change the usage patterns, or to promote the use of a different energy source.

Natural gasoline: A term used in the gas processing industry to refer to a mixture of liquid hydrocarbons (mostly pentanes and heavier hydrocarbons) extracted from natural gas. It includes isopentane.

Natural reservoir pressure: The energy within an oil or gas reservoir that causes the oil or gas to rise (unassisted by other forces) to the earth's surface when the reservoir is penetrated by an oil or gas well. The energy may be the result of "dissolved gas drive," "gas cap drive," or "water drive." Regardless of the type of drive, the principle is the same: the energy of the gas or water, creating a natural pressure, forces the oil or gas to the well bore.

Natural streamflow: The rate of flow of water past a given point of an uncontrolled stream or regulated streamflow adjusted to eliminate the effects of reservoir storage or upstream diversions at a set time interval.

Natural uranium: Uranium with the U-235 isotope present at a concentration of 0.711 percent (by weight), that is, uranium with its isotopic content exactly as it is found in nature.

Net cell shipments: Represents the difference between cell shipments and cell purchases.

Net electricity consumption: Consumption of electricity computed as generation, plus imports, minus exports, minus transmission and distribution losses.

Net energy for load: Net generation of main generating units that are system-owned or system-operated, plus energy receipts minus energy deliveries.

Net energy for system: The sum of energy an electric utility needs to satisfy their service areas, including full and partial requirements consumers.

Net generation: The amount of gross generation less the electrical energy consumed at the generating station(s) for station service or auxiliaries.

Net head: The gross head minus all hydraulic losses except those chargeable to the turbine.

Net income: Operating income plus other income and extraordinary income less operating expenses, taxes, interest charges, other deductions, and extraordinary deductions.

Net interstate flow of electricity: The difference between the sum of electricity sales and losses within a state and the total amount of electricity generated within that state. A positive number indicates that more electricity (including associated losses) came into the state than went out of the state during the year; conversely, a negative number indicates that more electricity (including associated losses) went out of the state than came into the state.

Net module shipments: Represents the difference between module shipments and module purchases. When exported, incomplete modules and unencapsulated cells are also included.

Net photovoltaic module shipment: The difference between photovoltaic module shipments and photovoltaic module purchases.

Net profits interest: A contractual arrangement under which the beneficiary, in exchange for consideration paid, receives a stated percentage of the net profits. That type of arrangement is considered a nonoperating interest, as distinguished from a working interest, because it does not involve the rights and obligations of operating a mineral property (costs of exploration, development, and operation). The net profits interest does not bear any part of net losses.

Net summer capacity: The maximum output, commonly expressed in megawatts (MW), that generating equipment can supply to system load, as demonstrated by a multi-hour test, at the time of summer peak demand (period of May 1 through October 31). This output reflects a reduction in capacity due to electricity use for station service or auxiliaries.

Net winter capacity: The maximum output, commonly expressed in megawatts (MW), that generating equipment can supply to system load, as demonstrated by a multi-hour test, at the time of peak winter demand (period of November 1 though April 30). This output reflects a reduction in capacity due to electricity use for station service or auxiliaries.

Netback purchase: Refers to a crude oil purchase agreement wherein the price paid for the crude is determined by sales prices of the types of products that are derivable from that crude as well as other considerations (*e.g.*, transportation and processing costs). Typically, the price is calculated based on product prices extant on or near the cargo's date of importation.

New field: A field discovered during the report year.

New field discoveries: The volumes of proved reserves of crude oil, natural gas, and/or natural gas liquids discovered in new fields during the report year.

New reservoir: A reservoir discovered during the report year.

Nitrogen dioxide: A compound of nitrogen and oxygen formed by the oxidation of nitric oxide (NO) which is produced by the combustion of solid fuels.

Nitrogen oxides (NO_x): Compounds of nitrogen and oxygen produced by the burning of fossil fuels.

Nitrous oxide (N_2O): A colorless gas, naturally occurring in the atmosphere. Nitrous oxide has a 100-year Global Warming Potential of 310.

No. 1 diesel fuel: A light distillate fuel oil that has distillation temperatures of 550 degrees Fahrenheit at the 90-percent point and meets the specifications defined in ASTM Specification D 975. It is used in high-speed diesel engines, such as those in city buses and similar vehicles.

No. 1 distillate: A light petroleum distillate that can be used as either a diesel fuel or a fuel oil (see **No. 1 fuel oil** (below).

No. 1 fuel oil: A light distillate fuel oil that has distillation temperatures of 400 degrees Fahrenheit at the 10-percent recovery point and 550 degrees Fahrenheit at the 90-percent point and meets the specifications defined in ASTM Specification D 396. It is used primarily as fuel for portable outdoor stoves and portable outdoor heaters.

No. 2 diesel fuel: A fuel that has distillation temperatures of 500 degrees Fahrenheit at the 10-percent recovery point and 640 degrees Fahrenheit at the 90-percent recovery point and meets the specifications defined in ASTM Specification D 975. It is used in high-speed diesel engines, such as those in railroad locomotives, trucks, and automobiles.

No. 2 distillate: A petroleum distillate that can be used as either a diesel fuel (see **No. 2 diesel fuel** above) or a fuel oil (see **No. 2 fuel oil** below).

No. 2 fuel oil (heating oil): A distillate fuel oil that has distillation temperatures of 400 degrees Fahrenheit at the 10-percent recovery point and 640 degrees Fahrenheit at the 90-percent recovery point and meets the specifications defined in ASTM Specification D 396. It is used in atomizing type burners for domestic heating or for moderate capacity commercial/industrial burner units.

No. 2 fuel oil and No. 2 diesel sold to consumers for all other end uses: Those consumers who purchase fuel oil or diesel fuel for their own use including: commercial/institutional buildings (including apartment buildings), manufacturing and non-manufacturing establishments, farms (including farm houses), motor vehicles, commercial or private boats, military, governments, electric utilities, railroads, construction, logging or any other nonresidential end-use purpose.

No. 2 fuel oil sold to private homes for heating: Private household customers who purchase fuel oil for the specific purpose of heating their home, water heating, cooking, etc., excluding farm houses, farming and apartment buildings.

No. 4 fuel oil: A distillate fuel oil made by blending distillate fuel oil and residual fuel oil stocks. It conforms with ASTM Specification D 396 or Federal Specification VV-F-815C and is used extensively in industrial plants and in commercial burner installations that are not equipped with preheating facilities. It also includes No. 4 diesel fuel used for low- and medium-speed diesel engines and conforms to ASTM Specification D 975.

No. 5 and no. 6 fuel oil sold directly to the ultimate consumer: Includes ships, mines, smelters, manufacturing plants, electric utilities, drilling, railroad.

No. 5 and no. 6 fuel oil sold to refiners or other dealers who will resale the product: Includes all volumes of No. 5 and No. 6 fuel oil purchased by a trade or business with the intent of reselling the product to the ultimate consumers.

NOAA: National Oceanic and Atmospheric Administration.

NOAA division: One of the 345 weather divisions designated by the National Oceanic and Atmospheric Administration (NOAA) encompassing the 48 contiguous states. These divisions usually follow county borders to encompass counties with similar weather conditions. The NOAA division does not follow county borders when weather conditions vary considerably within a county;

such is likely to happen when the county borders the ocean or contains high mountains. A state contains an average of seven NOAA divisions; a NOAA division contains an average of nine counties.

Noload loss: Power and energy lost by an electric system when not operating under demand.

Nominal dollars: A measure used to express nominal price.

Nominal price: The price paid for a product or service at the time of the transaction. Nominal prices are those that have not been adjusted to remove the effect of changes in the purchasing power of the dollar; they reflect buying power in the year in which the transaction occurred.

Nonassociated natural gas: Natural gas that is not in contact with significant quantities of crude oil in the reservoir.

Nonattainment area: Any area that does not meet the national primary or secondary ambient air quality standard established by the Environmental Protection Agency for designated pollutants, such as carbon monoxide and ozone.

Nonbranded product: Any refined petroleum product that is not a branded product.

Noncoincident demand: Sum of two or more demands on individual systems that do not occur in the same demand interval.

Noncoincidental peak load: The sum of two or more peak loads on individual systems that do not occur in the same time interval. Meaningful only when considering loads within a limited period of time, such as a day, week, month, a heating or cooling season, and usually for not more than 1 year.

Nonconventional plant (uranium): A facility engineered and built principally for processing of uraniferous solutions that are produced during in situ leach mining, from heap leaching, or in the manufacture of other commodities, and the recovery, by chemical treatment in the plant's circuits, of uranium from the processing solutions.

Nondedicated vehicle: A motor vehicle capable of operating on an alternative fuel and /or on either gasoline or diesel.

Nonfirm power: Power or power-producing capacity supplied or available under a commitment having limited or no assured availability.

Nonfuel components: Components that are not associated with a particular fuel. These include, but are not limited to, control spiders, burnable poison rod assemblies, control rod elements, thimble plugs, fission chambers, primary and secondary neutron sources, and BWR (boiling water reactor) channels.

Nonfuel use (of energy): Use of energy as feedstock or raw material input.

Nonfungible product: A gasoline blend or blendstock that cannot be shipped via existing petroleum product distribution systems because of incompatibility problems. Gasoline/ethanol blends, for example, are contaminated by water that is typically present in petroleum product distribution systems.

Nonhydrocarbon gases: Typical nonhydrocarbon gases that may be present in reservoir natural gas, such as carbon dioxide, helium, hydrogen sulfide, and nitrogen.

Nonmethane volatile organic compounds (NMVOC): Organic compounds, other than methane, that participate in atmospheric photochemical reactions.

Nonoperating interest: Any mineral lease interest (*e.g.*, royalty, production payment, net profits interest) that does not involve the rights and obligations of operating a mineral property.

Nonproducing reservoir: Reservoir in which oil and / or gas proved reserves have been identified, but which did not produce during the report year to the owned or contracted interest of the reporting company regardless of the availability and / or operation of production, gathering, or transportation facilities.

Nonrenewable fuels: Fuels that cannot be easily made or "renewed," such as oil, natural gas, and coal.

Nonrequirements consumer: A wholesale consumer (unlike a full or partial requirements consumer) that purchases economic or coordination power to supplement their own or another system's energy needs.

Nonresidential building: A building used for some purpose other than residential. Nonresidential buildings comprise three groups: commercial, manufacturing / industrial, and agricultural.

Nonroad alternative fuel vehicle (nonroad AFV): An alternative fuel vehicle designed for off-road operation and use for surface / air transportation, industrial, or commercial purposes. Nonroad AFVs include forklifts and other industrial vehicles, rail locomotives, self-propelled electric rail cars, aircraft, airport service vehicles, construction vehicles, agricultural vehicles, and marine vessels. Recreational AFVs (golf carts, snowmobiles, pleasure watercraft, etc.) are excluded from the definition.

Nonspinning reserve: The generating capacity not currently running but capable of being connected to the bus and load within a specified time.

Nonutility generation: Electric generation by end-users, or small power producers under the Public Utility Regulatory Policies Act, to supply electric power for industrial, commercial, and military operations, or sales to electric utilities.

Nonutility power producer: A corporation, person, agency, authority, or other legal entity or instrumentality that owns or operates facilities for electric generation and is not an electric utility. Nonutility power producers include qualifying **cogenerators,** qualifying small power producers, and other nonutility generators (including independent power producers). Nonutility power producers are without a designated franchised service area and do not file forms listed in the Code of Federal Regulations, Title 18, Part 141.

Normal butane (C_4H_{10}): A normally gaseous straight-chain hydrocarbon. It is a colorless paraffinic gas that boils at a temperature of 31.1 degrees Fahrenheit. It is extracted from natural gas or refinery gas streams.

North American Electric Reliability Council (NERC): A council formed in 1968 by the electric utility industry to promote the reliability and adequacy of bulk power supply in the electric utility systems of North America. NERC consists of regional reliability councils and encompasses essentially all the power regions of the contiguous United States, Canada, and Mexico.

North American Industrial Classification System (NAICS): A new classification scheme, developed by the Office of Management and Budget to replace the Standard Industrial Classification (SIC) System, that categorizes establishments according to the types of production processes they primarily use.

Nuclear electric power (nuclear power): Electricity generated by the use of the thermal energy released from the fission of nuclear fuel in a reactor.

Nuclear fuel: Fissionable materials that have been enriched to such a composition that, when placed in a nuclear reactor, will support a self-sustaining fission chain reaction, producing heat in a controlled manner for process use.

Nuclear reactor: An apparatus in which a nuclear fission chain reaction can be initiated, controlled, and sustained at a specific rate. A reactor includes fuel (fissionable material), moderating material to control the rate of fission, a heavy-walled pressure vessel to house reactor components, shielding to protect personnel, a system to conduct heat away from the reactor, and instrumentation for monitoring and controlling the reactor's systems.

Number of mines: The number of mines, or mines collocated with preparation plants or tipples, located in a particular geographic area (State or region). If a mine is mining coal across two counties within a State, or across two States, then it is counted as two operations.

Number of mining operations: The number of mining operations includes preparation plants with greater than 5,000 total direct labor hours. Mining operations that consist of a mine and preparation plant, or a preparation plant only, will be counted as two operations if the preparation plant processes both underground and surface coal. Excluded are silt, culm, refuse bank, slurry dam, and dredge operations except for Pennsylvania anthracite. Excludes mines producing less than 10,000 short tons of coal during the year.

Occupancy sensors: These are also known as "ultrasonic switchers." When movement is detected, the lights are turned on and remain on as long as there is movement in the room.

Ocean energy systems: Energy conversion technologies that harness the energy in tides, waves, and thermal gradients in the oceans.

Ocean thermal energy conversion (OTEC): The process or technologies for producing energy by harnessing the temperature differences (thermal gradients) between ocean surface waters and that of ocean depths. Warm surface water is pumped through an evaporator containing a working fluid in a closed Rankine-cycle system. The vaporized fluid drives a turbine/generator.

Octane: A flammable liquid hydrocarbon found in petroleum. Used as a standard to measure the anti-knock properties of motor fuel.

Octane rating: A number used to indicate gasoline's antiknock performance in motor vehicle engines. The two recognized laboratory engine test methods for determining the antiknock rating, *i.e.*, octane rating, of gasolines are the Research method and the Motor method. To provide a single number as guidance to the consumer, the antiknock index (R + M)/2, which is the average of the Research and Motor octane numbers, was developed.

OEM: Original Equipment Manufacturer.

Off-highway use: Includes petroleum products sales for use in:

1. **Construction.** Construction equipment including earthmoving equipment, cranes, stationary generators, air compressors, etc.
2. **Other.** Sales for off-highway uses other than construction. Sales for logging are included in this category. Volumes for off-highway use by the agriculture industry are reported under "Farm Use" (which includes sales for use in tractors, irrigation pumps, other agricultural machinery, etc.)

Off-hours equipment reduction: A conservation feature where there is a change in the temperature setting or reduction in the use of heating, cooling, domestic hot water heating, lighting or any other equipment either manually or automatically.

Off peak: Period of relatively low system demand. These periods often occur in daily, weekly, and seasonal patterns; these off-peak periods differ for each individual electric utility.

Off peak gas: Gas that is to be delivered and taken on demand when demand is not at its peak.

Off-system: Any point not on, or directly interconnected with, a transportation, storage, and/or distribution system operated by a natural gas company within a state.

Offshore: That geographic area that lies seaward of the coastline. In general, the coastline is the line of ordinary low water along with that portion of the coast that is in direct contact with the open sea or the line marking the seaward limit of inland water.

If a state agency uses a different basis for classifying onshore and offshore areas, the state classification should be used (*e.g.*, Cook Inlet in Alaska is classified as offshore; for Louisiana, the coastline is defined as the Chapman Line, as modified by subsequent adjudication).

Offshore reserves and production: Unless otherwise dedicated, reserves and production that are in either state or Federal domains, located seaward of the coastline.

Offsite-produced energy for heat, power, and electricity generation: This measure of energy consumption, which is equivalent to purchased energy includes energy produced off-site and consumed onsite. It excludes energy produced and consumed onsite, energy used as raw material input, and electricity losses.

Offsystem (natural gas): Natural gas that is transported to the end user by the company making final delivery of the gas to the end user. The end user purchases the gas from another company, such as a producer or marketer, not from the delivering company (typically a local distribution company or a pipeline company).

Ohm: A measure of the electrical resistance of a material equal to the resistance of a circuit in which the potential difference of 1 volt produces a current of 1 ampere.

Ohm's Law: In a given electrical circuit, the amount of current in amperes is equal to the pressure in volts divided by the resistance, in ohms. The principle is named after the German scientist Georg Simon Ohm.

Oil: A mixture of hydrocarbons usually existing in the liquid state in natural underground pools or reservoirs. Gas is often found in association with oil.

Oil company use: Includes sales to drilling companies, pipelines or other related oil companies not engaged in the selling of petroleum products. Includes fuel oil that was purchased or produced and used by company facilities for the operation of drilling equipment, other field or refinery operations, and space heating at petroleum refineries, pipeline companies, and oil-drilling companies. Oil used to bunker vessels is counted under vessel bunkering. Sales to other oil companies for field use are included, but sales for use as refinery charging stocks are excluded.

Oil reservoir: An underground pool of liquid consisting of hydrocarbons, sulfur, oxygen, and nitrogen trapped within a geological formation and protected from evaporation by the overlying mineral strata.

Oil shale: A sedimentary rock containing kerogen, a solid organic material.

Oil stocks: Oil stocks include crude oil (including strategic reserves), unfinished oils, natural gas plant liquids, and refined petroleum products.

Oil well: A well completed for the production of crude oil from at least one oil zone or reservoir.

Oil well (casinghead) gas: Associated and dissolved gas produced along with crude oil from oil completions.

Old field: A field discovered prior to the report year.

Old reservoir: A reservoir discovered prior to the report year.

On-highway use (diesel): Includes sales for use in motor vehicles. Volumes used by companies in the marketing and distribution of petroleum products are also included.

On peak: Periods of relatively high system demand. These periods often occur in daily, weekly, and seasonal patterns; these on-peak periods differ for each individual electric utility.

On-system: Any point on or directly interconnected with a transportation, storage, or distribution system operated by a natural gas company.

On-system sales: Sales to customers where the delivery point is a point on, or directly interconnected with, a transportation, storage, and/or distribution system operated by the reporting company.

One sun: Natural solar insulation falling on an object without concentration or diffusion of the solar rays.

One-time fee: The fee assessed a nuclear utility for spent nuclear fuel (SNF) or solidified high-level radioactive waste derived from SNF, which fuel was used to generate electricity in a civilian nuclear power reactor prior to April 7, 1983, and which is assessed by applying industry-wide average dollar-per-kilogram charges to four distinct ranges of fuel burnup so that equivalent to an industry-wide average charge of 1.0 mill per kilowatthour.

Onsite transportation: The direct nonprocess end use that includes energy used in vehicles and transportation equipment that primarily consume energy within the boundaries of the establishment. Energy used in vehicles that are found primarily offsite, such as delivery trucks, is not measured by the MECS.

Onsystem (natural gas): Natural gas that is sold (and transported) to the end user by the company making final delivery of the gas to the end user. Companies that make final delivery of natural gas are typically local distribution companies or pipeline companies.

OPEC (Organization of Petroleum Exporting Countries): The acronym for the Organization of Petroleum Exporting Countries that have organized for the purpose of negotiating with oil companies on matters of oil production, prices, and future concession rights. Current members (as of the date of writing this definition) are Algeria, Indonesia, Iran, Iraq, Kuwait, Libya, Nigeria, Qatar, Saudi Arabia, the United Arab Emirates, and Venezuela.

Open access: A regulatory mandate to allow others to use a utility's transmission and distribution facilities to move bulk power from one point to another on a nondiscriminatory basis for a cost-based fee.

Open market coal: Coal is sold in the open market, *i.e.*, coal sold to companies other than the reporting company's parent company or an operating subsidiary of the parent company.

Open refrigeration unit: Refrigeration in cabinets (units) without covers or with flexible covers made of plastic or some other material, hung in strips or curtains (fringed material, usually plastic, that push aside like a bead curtain). Flexible covers stop the flow of warm air into the refrigerated space.

Operable capacity: The amount of capacity that, at the beginning of the period, is in operation; not in operation and not under active repair, but capable of being placed in operation within 30 days; or not in operation but under active repair that can be completed within 90 days. Operable capacity is the sum of the operating and idle capacity and is measured in barrels per calendar day or barrels per stream day.

Operable generators/units: Electric generators or generating units that are available to provide power to the grid or generating units that have been providing power to the grid but are temporarily shut down. This includes units in standby status, units out of service for an indefinite period, and new units that have their construction complete and are ready to provide test generation. A nuclear unit is operable once it receives its Full Power Operating License.

Operable nuclear unit (U.S.): A U.S. nuclear generating unit that has completed low-power testing and is in possession of a full-power operating license issued by the Nuclear Regulatory Commission.

Operable nuclear unit (foreign): A nuclear generating unit outside the United States that generates electricity for a grid.

Operable refineries: Refineries that were in one of the following three categories at the beginning of a given year: in operation; not in operation and not under active repair, but capable of being placed into operation within 30 days; or not in operation, but under active repair that could be completed within 90 days.

Operable unit: A unit available to provide electric power to the grid.

Operable utilization rate: Represents the use of the atmospheric crude oil distillation units. The rate is calculated by dividing the gross input to these units by the operable refining capacity of the units.

Operated: Exercised management responsibility for the day-to-day operations of natural gas production, gathering, treating, processing, transportation, storage, and/or distribution facilities and/or a synthetic natural gas plant.

Operating capacity: The component of operable capacity that is in operation at the beginning of the period.

Operating day: A normal business day. Days when a company conducts business due to emergencies or other unexpected events are not included.

Operating expenses: Segment expenses related both to revenue from sales to unaffiliated customers and revenue from intersegment sales or transfers, excluding loss on disposition of property, plant, and equipment; interest expenses and financial charges; foreign currency translation effects; minority interest; and income taxes.

Operating income: Operating revenues less operating expenses. Excludes items of other revenue and expense, such as equity in earnings of unconsolidated affiliates, dividends, interest income and expense, income taxes, extraordinary items, and cumulative effects of accounting changes.

Operating revenues: Segment revenues both from sales to unaffiliated customers (*i.e.*, revenue from customers outside the enterprise as reported in the company's consolidated income statement) and from intersegment sales or transfers, if any, of product and services similar to those sold to unaffiliated customers, excluding equity in earnings of unconsolidated affiliates; dividend and interest income; gain on disposition of property, plant, and equipment; and foreign currency translation effects.

Operating subsidiary: Company that operates a coal mining operation and is owned by another company (*i.e.*, the parent company).

Operating unit: A unit that is in operation at the beginning of the reporting period.

Operating utilization rate: Represents the use of the atmospheric crude oil distillation units. The rate is calculated by dividing the gross input to these units by the operating refining capacity of the units.

Operator, gas plant: The person responsible for the management and day-to-day operation of one or more natural gas processing plants as of December 31 of the report year. The operator is generally a working-interest owner or a company under contract to the working-interest owner(s). Plants shut down during the report year are also to be considered "operated" as of December 31.

Operator, oil and/or gas well: The person responsible for the management and day-to-day operation of one or more crude oil and/or natural gas wells as of December 31 of the report year. The operator is generally a working-interest owner or a company under contract to the working-interest owner(s). Wells included are those that have proved reserves of crude oil, natural gas, and/or lease condensate in the reservoirs associated with them, whether or not they are producing. Wells abandoned during the report year are also to be considered "operated" as of December 31.

Optional delivery commitment: A provision to allow the conditional purchase or sale of a specific quantity of material in addition to the firm quantity in the contract.

Order: A ruling issued by a utility commission granting or denying an application in whole or in part. The order explains the basis for the decision, noting any dispute with the factual assertions of the applicant. Also applied to a final regulation of a utility commission.

Organic content: The share of a substance that is of animal or plant origin.

Organic waste: Waste material of animal or plant origin.

Organization for Economic Cooperation and Development (OECD): An international organization helping governments tackle the economic, social and governance challenges of a globalized economy. Its membership comprises about 30 member countries. With active relationships with some 70 other countries, NGOs and civil society, it has a global reach.

Organization of Petroleum Exporting Countries (OPEC): Countries that have organized for the purpose of negotiating with oil companies on matters of oil production, prices, and future concession rights. Current members (as of the date of writing this definition) are Algeria, Indonesia, Iran, Iraq, Kuwait, Libya, Nigeria, Qatar, Saudi Arabia, the United Arab Emirates, and Venezuela.

Original cost: The initial amount of money spent to acquire an asset. It is equal to the price paid, or present value of the liability incurred, or fair value of stock issued, plus normal incidental costs necessary to put the asset into its initial use.

Original equipment manufacturer(OEM): A company that provides the original design and materials for manufacture and engages in the assembly of vehicles. The OEM is directly responsible for manufacturing, marketing, and providing warranties for the finished product.

Original equipment manufacturer vehicle: A vehicle produced and marketed by an original equipment manufacturer (OEM), including gasoline and diesel vehicles as well as alternative-fuel vehicles. A vehicle manufactured by an OEM but converted to an alternative-fuel vehicle before its initial delivery to an end-user (for example, through a contract between a conversion company and the OEM) is considered to be an OEM vehicle as long as that vehicle is still covered under the OEM's warranty.

Other: The "other" category is defined as representing electricity consumers not elsewhere classified. This category includes public street and highway lighting service, public authority service to public authorities, railroad and railway service, and interdepartmental services.

Other capital costs: Costs for items or activities not included elsewhere under capital-cost tabulations, such as for and decommissioning, dismantling, and reclamation.

Other demand-side management (DSM) assistance programs: A DSM program assistance that includes alternative-rate, fuel-switching, and any other DSM assistance programs that are offered to consumers to encourage their participation in DSM programs.

Other end users: For motor gasoline, all direct sales to end users other than those made through company outlets. For No. 2 distillate, all direct sales to end users other than residential, commercial/institutional, industrial sales, and sales through company outlets. Included in the "other end users" category are sales to utilities and agricultural users.

Other energy operations: Energy operations not included under Petroleum or Coal. "Other energy" includes nuclear, oil shale, tar sands, coal liquefaction and gasification, geothermal, solar, and other forms of non-conventional energy.

Other finished: Motor gasoline not included in the oxygenated or reformulated gasoline categories.

Other gas: Includes manufactured gas, coke-oven gas, blast-furnace gas, and refinery gas. Manufactured gas is obtained by distillation of coal, by the thermal decomposition of oil, or by the reaction of steam passing through a bed of heated coal or coke.

Other generation: Electricity originating from these sources: biomass, fuel cells, geothermal heat, solar power, waste, wind, and wood.

Other hydrocarbons: Materials received by a refinery and consumed as a raw material. Includes hydrogen, coal tar derivatives, gilsonite, and natural gas received by the refinery for reforming into hydrogen. Natural gas to be used as fuel is excluded.

Other industrial plant: Industrial users, not including coke plants, engaged in the mechanical or chemical transformation of materials or substances into new products (manufacturing); and companies engaged in the agriculture, mining, or construction industries.

Other load management: Demand-Side Management (DSM) program other than Direct Load Control and Interruptible Load that limits or shifts peak load from on-peak to off-peak time periods. It includes technologies that primarily shift all or part of a load from one time-of-day to another and secondarily may have an impact on energy consumption. Examples include space heating and water heating storage systems, cool storage systems, and load limiting devices in energy management systems. This category also includes programs that aggressively promote time-of-use rates and other innovative rates such as real time pricing. These rates are intended to reduce consumer bills and shift hours of operation of equipment from on-peak to off-peak periods through the application of time-differentiated rates.

Other oils equal to or greater than 401 degrees Fahrenheit: Oils with a boiling range equal to or greater than 401 degrees Fahrenheit that are intended for use as a petrochemical feedstock.

Other operating costs: Costs for other items or activities not included elsewhere in operating-cost tabulations, but required to support the calculation of a cutoff grade for ore reserves estimation.

Other oxygenates: Other aliphatic alcohols and aliphatic ethers intended for motor gasoline blending (*e.g.*, isopropyl ether (IPE) or n-propanol).

Other power producers: Independent power producers that generate electricity and cogeneration plants that are not included in the other industrial, coke and commercial sectors.

Other refiners: Refiners with a total refinery capacity in the United States and its possessions of less than 275,000 barrels per day as of January 1, 1982.

Other service to public authorities: Electricity supplied to municipalities, divisions or agencies of state or Federal governments, under special contracts or agreements or service classifications applicable only to public authorities.

Other single-unit truck: A motor vehicle consisting primarily of a single motorized device with more than two axles or more than four tires.

Other supply contracts: Any contracted gas supply other than owned reserves, producer-contracted reserves, and interstate pipeline purchases that are used for acts and services for which the company has received certificate authorization from FERC. Purchases from intrastate pipelines pursuant to Section 311(b) of the NGPA of 1978 are included with other supply contracts.

Other trucks/vans: Those trucks and vans that weigh more than 8,500 lbs GVW.

Other unavailable capability: Net capability of main generating units that are unavailable for load for easons other than full-forced outage or scheduled maintenance. Legal restrictions or other causes make these units unavailable.

Outage: The period during which a generating unit, transmission line, or other facility is out of service.

Outer Continental Shelf: Offshore Federal domain.

Output: The amount of power or energy produced by a generating unit, station, or system.

Oven: An appliance that is an enclosed compartment supplied with heat and used for cooking food. Toaster ovens are not considered ovens. The range stove top or burners and the oven are considered two separate appliances, although they are often purchased as one appliance.

Overburden: Any material, consolidated or unconsolidated, that overlies a coal deposit.

Overburden ratio: Overburden ratio refers to the amount of overburden that must be removed to excavate a given quantity of coal. It is commonly expressed in cubic yards per ton of coal, but is sometimes expressed as a ratio comparing the thickness of the overburden with the thickness of the coalbed.

Owned reserves: Any reserve of natural gas that the reporting company owns as a result of oil and gas leases, fee-mineral ownership, royalty reservations, or lease or royalty reservations and assignments committed to services under certificate authorizations by FERC. Company-owned recoverable natural gas in underground storage is classified as owned reserves.

Owned/rented: The relationship of a housing unit's occupants to the structure itself, not the land on which the structure is located. "Owned" means the owner or co-owner is a member of the household and the housing unit is either fully paid for or mortgaged. A household is classified "rented" even if the rent is paid by someone not living in the unit. "Rent-free" means the unit is not owned or being bought and no money is paid or contracted for rent. Such units are usually provided in exchange for services rendered or as an allowance or favor from a relative or friend not living in the unit. Unless shown separately, rent-free households are grouped with rented households.

Owner occupied: Having the owner or the owner's business represented at the site. A building is considered owner occupied if an employee or representative of the owner (such as a building engineer or building manager) maintains office space in the building. Similarly, a chain store is considered owner occupied even though the actual owner may not be in the building but headquartered elsewhere. Other examples of the owner's business occupying a building include State-owned university buildings, elementary and secondary schools owned by a public school district, and a post office where the building is owned by the U.S. Postal Service.

Owners equity: Interest of the owners in the assets of the business represented by capital contributions and retained earnings.

Ownership of building: The individual, agency, or organization that owns the building. For consumption surveys, building ownership is grouped into the following categories: Federal, State, or local government agency; a privately owned utility company; a church, synagogue, or other religious group; or any other type of individual or group.

Overriding royalty: A royalty interest, in addition to the basic royalty, created out of the working interest; it is, therefore, limited in its duration to the life of the lease under which it is created.

Oxidize: To chemically transform a substance by combining it with oxygen.

Oxygenated gasoline: Finished motor gasoline, other than reformulated gasoline, having an oxygen content of 2.7 percent or higher by weight and required by the U.S. Environmental Protection Agency (EPA) to be sold in areas designated by EPA as carbon monoxide (CO) nonattainment areas.

Oxygenated gasoline (includes Gasohol): Finished motor gasoline, other than reformulated gasoline, having an oxygen content of 2.7 percent or higher by weight. Includes gasohol.

Oxygenates: Substances which, when added to gasoline, increase the amount of oxygen in that gasoline blend. Ethanol, Methyl Tertiary Butyl Ether (MTBE), Ethyl Tertiary Butyl Ether (ETBE), and methanol are common oxygenates.

Ozone: A molecule made up of three atoms of oxygen. Occurs naturally in the stratosphere and provides a protective layer shielding the Earth from harmful ultraviolet radiation. In the troposphere, it is a chemical oxidant, a greenhouse gas, and a major component of photochemical smog.

Ozone precursors: Chemical compounds, such as carbon monoxide, methane, nonmethane hydrocarbons, and nitrogen oxides, which in the presence of solar radiation react with other chemical compounds to form ozone.

Packaged air conditioning units: Usually mounted on the roof or on a slab beside the building. (These are known as self-contained units, or Direct Expansion (DX). They contain air conditioning equipment as well as fans, and may or may not include heating equipment.) These are self-contained units that contain the equipment that generates cool air and the equipment that distributes the cooled air. These units commonly consume natural gas or electricity. The units are mounted on the rooftop, exposed to the elements. They typically blow cool air into the building through duct work, but other types of distribution systems may exist. The units usually serve more than one room. There are often several units on the roof of a single building. Also known as: Packaged Terminal Air Conditioners (PTAC). These packaged units are often constructed as a single unit for heating and for cooling.

Packaged units: Units built and assembled at a factory and installed as a self-contained unit to heat or cool all or portions of a building. Packaged units are in contrast to engineer-specified units built up from individual components for use in a given building. Packaged Units can apply to heating equipment, cooling equipment, or combined heating and cooling equipment. Some types of electric packaged units are also called "Direct Expansion" or DX units.

Parabolic dish: A high-temperature (above 180 degrees Fahrenheit) solar thermal concentrator, generally bowl-shaped, with two-axis tracking.

Parabolic trough: A high-temperature (above 180 degrees Fahrenheit) solar thermal concentrator with the capacity for tracking the sun using one axis of rotation.

Paraffin (oil): A light-colored, wax-free oil obtained by pressing paraffin distillate.

Paraffin (wax): The wax removed from paraffin distillates by chilling and pressing. When separating from solutions, it is a colorless, more or less translucent, crystalline mass, without odor and taste, slightly greasy to touch, and consisting of a mixture of solid hydrocarbons in which the paraffin series predominates.

Paraffinic hydrocarbons: Straight-chain hydrocarbon compounds with the general formula C_nH_{2n+2}.

Parent: A firm that directly or indirectly controls another entity.

Parent company: An affiliated company that exercises ultimate control over a business entity, either directly or indirectly, through one or more intermediaries.

Partial requirements consumer: A wholesale consumer with generating resources insufficient to carry all its load and whose energy seller is a long-term firm power source supplemental to the consumer's own generation or energy received from others. The terms and conditions of sale are similar to those for a full equirements consumer.

Particulate: A small, discrete mass of solid or liquid matter that remains individually dispersed in gas or liquid emissions. Particulates take the form of aerosol, dust, fume, mist, smoke, or spray. Each of these forms has different properties.

Passenger-miles traveled: The total distance traveled by all passengers. It is calculated as the product of the occupancy rate in vehicles and the vehicle miles traveled.

Passive solar heating: A solar heating system that uses no external mechanical power, such as pumps or blowers, to move the collected solar heat.

Payables to municipality: The amounts payable by the utility department to the municipality or its other departments that are subject to current settlement.

Payment method for utilities: The method by which fuel suppliers or utility companies are paid for all electricity, natural gas, fuel oil, kerosene, or liquefied petroleum gas used by a household. Households that pay the utility company directly are classified as "all paid by household." Households that pay directly for at least one but not all of their fuels used and that has at least one fuel charge included in the rent were classified as "some paid, some included in rent." Households for which all fuels used are included in rent were classified as "all included in rent." If the household did not fall into one of these categories, it was classified as "other." Examples of households falling into the "other" category are: (1) households for which fuel bills were paid by a social service agency or a relative, and (2) households that paid for some of their fuels used but paid for other fuels through another arrangement.

Peak day withdrawal: The maximum daily withdrawal rate (Mcf/d) experienced during the reporting period.

Peak demand: The maximum load during a specified period of time.

Peak kilowatt: One thousand peak watts.

Peak load: The maximum load during a specified period of time.

Peak load month: The month of greatest plant electrical generation during the winter heating season (Oct-Mar) and summer cooling season (Apr-Sept), respectively.

Peak load plant: A plant usually housing old, low-efficiency steam units, gas turbines, diesels, or pumped-storage hydroelectric equipment normally used during the peak-load periods.

Peak megawatt: One million peak watts.

Peak watt: A manufacturer's unit indicating the amount of power a photovoltaic cell or module will produce at standard test conditions (normally 1,000 watts per square meter and 25 degrees Celsius).

Peaking capacity: Capacity of generating equipment normally reserved for operation during the hours of highest daily, weekly, or seasonal loads. Some generating equipment may be operated at certain times as peaking capacity and at other times to serve loads on an around-the-clock basis.

Peat: Peat consists of partially decomposed plant debris. It is considered an early stage in the development of coal. Peat is distinguished from lignite by the presence of free cellulose and a high moisture content (exceeding 70 percent). The heat content of air-dried peat (about 50 percent moisture) is about 9 million Btu per ton. Most U.S. peat is used as a soil conditioner. The first U.S. electric power plant fueled by peat began operation in Maine in 1990.

Pentanes plus: A mixture of hydrocarbons, mostly pentanes and heavier, extracted from natural gas. Includes isopentane, natural gasoline, and plant condensate.

Percent difference: The relative change in a quantity over a specified time period. It is calculated as follows: the current value has the previous value subtracted from it; this new number is divided by the absolute value of the previous value; then this new number is multiplied by 100.

Percent utilization: The ratio of total production to productive capacity, times 100.

Perfluorocarbons (PFCs): A group of man-made chemicals composed of one or two carbon atoms and four to six fluorine atoms, containing no chlorine. PFCs have no commercial uses and are emitted as a byproduct of aluminum smelting and semiconductor manufacturing. PFCs have very high 100-year Global Warming Potentials and are very long-lived in the atmosphere.

Perfluoromethane: A compound (CF_4) emitted as a byproduct of aluminum smelting.

Permanently discharged fuel: Spent nuclear fuel for which there are no plans for reinsertion in the reactor core.

Permeability: The ease with which fluid flows through a porous medium.

Persian Gulf: The countries that surround the Persian Gulf are: Bahrain, Iran, Iraq, Kuwait, Qatar, Saudi Arabia, and the United Arab Emirates.

Person: An individual, a corporation, a partnership, an association, a joint-stock company, a business trust, or an unincorporated organization.

Person-year: One whole year, or fraction thereof, worked by an employee, including contracted manpower. Expressed as a quotient (to two decimal places) of the time units worked during a year (hours, weeks, or months) divided by the like total time units in a year. For example: 80 hours worked is 0.04 (rounded) of a person-year; 8 weeks worked is 0.15 (rounded) of a person-year; 12 months worked is 1.0 person-year. Contracted manpower includes survey crews, drilling crews, consultants, and other persons who worked under contract to support a firm's ongoing operations.

Personal computer: A microcomputer for producing written, programmed, or coded material; playing games; or doing calculations.

Petrochemical feedstocks: Chemical feedstocks derived from petroleum principally for the manufacture of chemicals, synthetic rubber, and a variety of plastics.

Petrochemicals: Organic and inorganic compounds and mixtures that include but are not limited to organic chemicals, cyclic intermediates, plastics and resins, synthetic fibers, elastomers, organic dyes, organic pigments, detergents, surface active agents, carbon black, and ammonia.

Petroleum: A broadly defined class of liquid hydrocarbon mixtures. Included are crude oil, lease condensate, unfinished oils, refined products obtained from the processing of crude oil, and natural gas plant liquids.

Petroleum Administration for Defense District (PADD): A geographic aggregation of the 50 States and the District of Columbia into five Districts, with PADD I further split into three subdistricts. The PADDs include the States listed below:

PADD I (East Coast):

PADD IA (New England): Connecticut, Maine, Massachusetts, New Hampshire, Rhode Island, and Vermont.

PADD IB (Central Atlantic): Delaware, District of Columbia, Maryland, New Jersey, New York, and Pennsylvania.

PADD IC (Lower Atlantic): Florida, Georgia, North Carolina, South Carolina, Virginia, and West Virginia.

PADD II (Midwest): Illinois, Indiana, Iowa, Kansas, Kentucky, Michigan, Minnesota, Missouri, Nebraska, North Dakota, Ohio, Oklahoma, South Dakota, Tennessee, and Wisconsin.

PADD III (Gulf Coast): Alabama, Arkansas, Louisiana, Mississippi, New Mexico, and Texas.

PADD IV (Rocky Mountain): Colorado, Idaho, Montana, Utah, and Wyoming.

PADD V (West Coast): Alaska, Arizona, California, Hawaii, Nevada, Oregon, and Washington.

Petroleum coke, catalyst: The carbonaceous residue that is deposited on and deactivates the catalyst used in many catalytic operations (*e.g.*, catalytic cracking). Carbon is deposited on the catalyst, thus deactivating the catalyst. The catalyst is reactivated by burning off the carbon, which is used as a fuel in the refining process. That carbon or coke is not recoverable in a concentrated form.

Petroleum coke, marketable: Those grades of coke produced in delayed or fluid cokers that may be recovered as relatively pure carbon. Marketable petroleum coke may be sold as is or further purified by calcining.

Petroleum consumption: The sum of all refined petroleum products supplied. For each refined petroleum product, the amount supplied is calculated by adding production and imports, then subtracting changes in primary stocks (net withdrawals are a plus quantity and net additions are a minus quantity) and exports.

Petroleum imports: Imports of petroleum into the 50 states and the District of Columbia from foreign countries and from Puerto Rico, the Virgin Islands, and other U.S. territories and possessions. Included are imports for the Strategic Petroleum Reserve and withdrawals from bonded warehouses for onshore consumption, offshore bunker use, and military use. Excluded are receipts of foreign petroleum into bonded warehouses and into U.S. territories and U.S. Foreign Trade Zones.

Petroleum jelly: A semi-solid oily product produced from de-waxing lubricating oil basestocks.

Petroleum products: Petroleum products are obtained from the processing of crude oil (including lease condensate), natural gas, and other hydrocarbon compounds. Petroleum products include unfinished oils, liquefied petroleum gases, pentanes plus, aviation gasoline, motor gasoline, naphtha-type jet fuel, kerosene-type jet fuel, kerosene, distillate fuel oil, residual fuel oil, petrochemical feedstocks, special naphthas, lubricants, waxes, petroleum coke, asphalt, road oil, still gas, and miscellaneous products.

Petroleum refinery: An installation that manufactures finished petroleum products from crude oil, unfinished oils, natural gas liquids, other hydrocarbons, and alcohol.

Petroleum stocks, primary: For individual products, quantities that are held at refineries, in pipelines and at bulk terminals that have a capacity of 50,000 barrels or more, or that are in transit thereto. Stocks held by product retailers and resellers, as well as tertiary stocks held at the point of consumption, are excluded. Stocks of individual products held at gas processing plants are excluded from individual product stimates but are included in other oils estimates and total.

pH: A measure of acidity or alkalinity. A pH of 7 represents neutrality. Acid substances have lower pH. Basic substances have higher pH.

Photosynthesis: The manufacture by plants of carbohydrates and oxygen from carbon dioxide and water in the presence of chlorophyll, with sunlight as the energy source. Carbon is sequestered and oxygen and water vapor are released in the process.

Photovoltaic and solar thermal energy (as used at electric utilities): Energy radiated by the sun as electromagnetic waves (electromagnetic radiation) that is converted at electric utilities into electricity by means of solar (photovoltaic) cells or concentrating (focusing) collectors.

Photovoltaic cell (PVC): An electronic device consisting of layers of semiconductor materials fabricated to form a junction (adjacent layers of materials with different electronic characteristics) and electrical contacts and being capable of converting incident light directly into electricity (direct current).

Photovoltaic module: An integrated assembly of interconnected photovoltaic cells designed to deliver a selected level of working voltage and current at its output terminals, packaged for protection against environmental degradation, and suited for incorporation in photovoltaic power systems.

Pig iron: Crude, high-carbon iron produced by reduction of iron ore in a blast furnace.

Pipeline, distribution: A pipeline that conveys gas from a transmission pipeline to its ultimate consumer.

Pipeline freight: Refers to freight carried through pipelines, including natural gas, crude oil, and petroleum products (excluding water). Energy is consumed by various electrical components of the pipeline, including, valves, other, appurtenances attaches to the pipe, compressor units, metering stations, regulator stations, delivery stations, holders and fabricated assemblies.

Pipeline fuel: Gas consumed in the operation of pipelines, primarily in compressors.

Pipeline, gathering: A pipeline that conveys gas from a production well/field to a gas processing plant or transmission pipeline for eventual delivery to end-use consumers.

Pipeline (natural gas): A continuous pipe conduit, complete with such equipment as valves, compressor stations, communications systems, and meters for transporting natural and/or supplemental gas from one point to another, usually from a point in or beyond the producing field or processing plant to another pipeline or to points of utilization. Also refers to a company operating such facilities.

Pipeline (petroleum): Crude oil and product pipelines used to transport crude oil and petroleum products, respectively (including interstate, intrastate, and intracompany pipelines), within the 50 states and the District of Columbia.

Pipeline purchases: Gas supply contracted from and volumes purchased from other natural gas companies as defined by the Natural Gas Act, as amended (52 Stat. 821), excluding independent producers, as defined in Paragraph 154.91(a), Chapter I, Title 18 of the Code of Federal Regulations.

Pipeline quality natural gas: A mixture of hydrocarbon compounds existing in the gaseous phase with sufficient energy content, generally above 900 British thermal units, and a small enough share of impurities for transport through commercial gas pipelines and sale to end-users.

Pipeline, transmission: A pipeline that conveys gas from a region where it is produced to a region where it is to be distributed.

Pipelines, rate regulated: FRS (Financial Reporting System Survey) establishes three pipeline segments: crude/liquid (raw materials); natural gas; and refined products. The pipelines included in these segments are all federally or State rate-regulated pipeline operations, which are included in the reporting company's consolidated financial statements. However, at the reporting company's option, intrastate pipeline operations may be included in the U.S. Refining/Marketing Segment if: they would comprise less than 5 percent of U.S. Refining/Marketing Segment net PP&E, revenues, and earnings in the aggregate; and if the inclusion of such pipelines in the consolidated financial statements adds less than $100 million to the net PP&E reported for the U.S. Refining/Marketing Segment.

Pitcheblende: Uranium oxide (U_3O_8). It is the main component of high-grade African or domestic uranium ore and also contains other oxides and sulfides, including radium, thorium, and lead components.

Place in service: A vehicle is placed in service if that vehicle is new to the fleet and has not previously been in service for the fleet. These vehicles can be acquired as additional vehicles (increases the size of the company fleet), or as replacement vehicles to replace vehicles that are being retired from service (does not increase the size of the company fleet).

Planetary albedo: The fraction of incident solar radiation that is reflected by the Earth-atmosphere system and returned to space, mostly by backscatter from clouds in the atmosphere.

Planned generator: A proposal by a company to install electric generating equipment at an existing or planned facility or site. The proposal is based on the owner having obtained either (1) all environmental and regulatory approvals, (2) a signed contract for the electric energy, or (3) financial closure for the facility.

Plant: A term commonly used either as a synonym for an industrial establishment or a generating facility or to refer to a particular process within an establishment.

Plant condensate: One of the natural gas liquids, mostly pentanes and heavier hydrocarbons, recovered and separated as liquids at gas inlet separators or scrubbers in processing plants.

Plant hours connected to load: The number of hours the plant is synchronized to load over a time interval usually of 1 year.

Plant liquids: Those volumes of natural gas liquids recovered in natural gas processing plants.

Plant or gas processing plant: A facility designated to achieve the recovery of natural gas liquids from the stream of natural gas, which may or may not have been processed through lease separators and field facilities, and to control the quality of the natural gas to be marketed.

Plant products: Natural gas liquids recovered from natural gas processing plants (and in some cases from field facilities), including ethane, propane, butane, butane-propane mixtures, natural gasoline, plant condensate, and lease condensate.

Plant use: The electric energy used in the operation of a plant. Included is the energy required for pumping at pump-storage plants.

Plant-use electricity: The electric energy used in the operation of a plant. This energy total is subtracted from the gross energy production of the plant.

Plugged-back footage: Under certain conditions, drilling operations may be continued to a greater depth than that at which a potentially productive formation is found. If production is not established at the greater depth, the well may be completed in the shallower formation. Except in special situations, the length of the well bore from the deepest depth at which the well is completed to the maximum depth drilled is defined as "plugged-back footage." Plugged-back footage is included in total footage drilled but is not reported separately.

Plutonium (Pu): A heavy, fissionable, radioactive, metallic element (atomic number 94) that occurs naturally in trace amounts. It can also result as a byproduct of the fission reaction in a uranium-fuel nuclear reactor and can be recovered for future use.

Pneumatic device: A device moved or worked by air pressure.

Pole-mile: A unit of measuring the simple length of an electric transmission/distribution line/feeder carrying electric conductors, without regard to the number of conductors carried.

Polystyrene: A polymer of styrene that is a rigid, transparent thermoplastic with good physical and electrical insulating properties, used in molded products, foams, and sheet materials.

Polyvinyl chloride (PVC): A polymer of vinyl chloride. Tasteless. odorless, insoluble in most organic solvents. A member of the family vinyl resin, used in soft flexible films for food packaging and in molded rigid products, such as pipes, fibers, upholstery, and bristles.

Pondage: The amount of water stored behind a hydroelectric dam of relatively small storage capacity; the dam is usually used for daily or weekly control of the flow of the river.

Pool: In general, a reservoir. In certain situations, a pool may consist of more than one reservoir.

Pool site: One or more spent fuel storage pools that has a single cask loading area. Each dry cask storage area is considered a separate site.

Population-weighted degree-days: Heating or cooling degree-days weighted by the population of the area in which the degree-days are recorded. To compute national population-weighted degree-days, the Nation is divided into nine Census regions comprised of from three to eight states that are assigned weights based on the ratio of the population of the region to the total population of the Nation. Degree-day readings for each region are multiplied by the corresponding population weight for each region, and these products are then summed to arrive at the national population weighted degree-day figure.

Pore space: The open spaces or voides of a rock taken collectively. It is a measure of the amount of liquid or gas that may be absorbed or yielded by a particular formation.

Portable electric heater: A heater that uses electricity and that can be picked up and moved.

Portable fan: Box fans, oscillating fans, table or floor fans, or other fans that can be moved.

Portable kerosene heater: A heater that uses kerosene and that can be picked up and moved.

Post-mining emissions: Emissions of methane from coal occurring after the coal has been mined, during transport or pulverization.

Potential consumption: The total amount of consumption that would have occurred had the intensity of consumption remained the same over a period of time.

Potential peak reduction: The potential annual peak load reduction (measured in kilowatts) that can be deployed from Direct Load Control, Interruptible Load, Other Load Management, and Other DSM Program activities. (Please note that Energy Efficiency and Load Building are not included in Potential Peak Reduction.) It represents the load that can be reduced either by the direct control of the utility system operator or by the consumer in response to a utility request to curtail load. It reflects the installed load reduction capability, as opposed to the Actual Peak Reduction achieved by participants, during the time of annual system peak load.

Pounds (district heat): A weight quantity of steam, also used to denote a quantity of energy in the form of steam. The amount of usable energy obtained from a pound of steam depends on its temperature and pressure at the point of consumption and on the drop in pressure after consumption.

Power: The rate at which energy is transferred. Electrical energy is usually measured in watts. Also used for a measurement of capacity.

Power (electrical): An electric measurement unit of power called a voltampere is equal to the product of 1 volt and 1 ampere. This is equivalent to 1 watt for a direct current system, and a unit of apparent power is separated into real and reactive power. Real power is the work-producing part of apparent power that measures the rate of supply of energy and is denoted as kilowatts (kW). Reactive power is the portion of apparent power that does no work and is referred to as kilovars;

this type of power must be supplied to most types of magnetic equipment, such as motors, and is supplied by generator or by electrostatic equipment. Voltamperes are usually divided by 1,000 and called kilovoltamperes (kVA). Energy is denoted by the product of real power and the length of time utilized; this product is expressed as kilowathours.

Power ascension: The period of time between a plant's initial fuel loading date and its date of first commercial operation (including the low-power testing period). Plants in the first operating cycle (the time from initial fuel loading to the first refueling), which lasts approximately 2 years, operate at an average capacity factor of about 40 percent.

Power exchange: An entity providing a competitive spot market for electric power through day- and/or hour-ahead auction of generation and demand bids.

Power exchange generation: Generation scheduled by the power exchange.

Power exchange load: Load that has been scheduled by the power exchange and is received through the use of transmission or distribution facilities owned by participating transmission owners.

Power factor: The ratio of real power (kilowatt) to apparent power kilovolt-ampere for any given load and time.

Power loss: The difference between electricity input and output as a result of an energy transfer between two points.

Power marketers: Business entities engaged in buying and selling electricity. Power marketers do not usually own generating or transmission facilities. Power marketers, as opposed to brokers, take ownership of the electricity and are involved in interstate trade. These entities file with the Federal Energy Regulatory Commission (FERC) for status as a power marketer.

Power pool: An association of two or more interconnected electric systems having an agreement to coordinate operations and planning for improved reliability and efficiencies.

Power production plant: All the land and land rights, structures and improvements, boiler or reactor vessel equipment, engines and engine-driven generator, turbogenerator units, accessory electric equipment, and miscellaneous power plant equipment are grouped together for each individual facility.

Power transfer limit: The maximum power that can be transferred from one electric utility system to another without overloading any facility in either system.

Powerhouse: A structure at a hydroelectric plant site that contains the turbine and generator.

PP&E, additions to: The current year's expenditures on property, plant, and equipment (PP&E). The amount is predicated upon each reporting company's accounting practice. That is, accounting practices with regard to capitalization of certain items may differ across companies, and therefore this figure in FRS (Financial Reporting System) will be a function of each reporting company's policy.

Prediscovery costs: All costs incurred in an extractive industry operation prior to the actual discovery of minerals in commercially recoverable quantities; normally includes prospecting, acquisition, and exploration costs and may include some development costs.

Pregnant solution: A solution containing dissolved extractable mineral that was leached from the ore; uranium leach solution pumped up from the underground ore zone though a production hole.

Preliminary permit (hydroelectric power): A single site permit granted by the FERC (Federal Energy Regulatory Commission), which gives the recipient priority over anyone else to apply for a hydroelectric license. The preliminary permit enables the recipient to prepare a license application and conduct various studies for economic feasibility and environmental impacts. The period for a preliminary permit may extend to 3 years.

Premium gasoline: Gasoline having an antiknock index (R+M/2) greater than 90. Includes both leaded premium gasoline as well as unleaded premium gasoline.

Preparation plant: A mining facility at which coal is crushed, screened, and mechanically cleaned.

Preproduction costs: Costs of prospecting for, acquiring, exploring, and developing mineral reserves incurred prior to the point when production of commercially recoverable quantities of minerals commences.

Pressurized-water reactor (PWR): A nuclear reactor in which heat is transferred from the core to a heat exchanger via water kept under high pressure, so that high temperatures can be maintained in the primary system without boiling the water. Steam is generated in a secondary circuit.

Preventive maintenance program for heating and/or cooling equipment: A HVAC conservation feature consisting of a program of routine inspection and service for the heating and/or cooling equipment. The inspection is performed on a regular basis, even if there are no apparent problems.

Price: The amount of money or consideration-in-kind for which a service is bought, sold, or offered for sale.

Primary coal: All coal milled and, when necessary, washed and sorted.

Primary energy: All energy consumed by end users, excluding electricity but including the energy consumed at electric utilities to generate electricity. (In estimating energy expenditures, there are no fuel-associated expenditures for hydroelectric power, geothermal energy, solar energy, or wind energy, and the quantifiable expenditures for process fuel and intermediate products are excluded.)

Primary energy consumption: Primary energy consumption is the amount of site consumption, plus losses that occur in the generation, transmission, and distribution of energy.

Primary energy consumption expenditures: Expenditures for energy consumed in each of the four major end-use sectors, excluding energy in the form of electricity, plus expenditures by the electric utilities sector for energy used to generate electricity. There are no fuel-associated expenditures for associated expenditures for hydroelectric power, geothermal energy, photovoltaic and solar energy, or wind energy. Also excluded are the quantifiable consumption expenditures that are an integral part of process fuel consumption.

Primary fuels: Fuels that can be used continuously. They can sustain the boiler sufficiently for the production of electricity.

Primary recovery: The crude oil or natural gas recovered by any method that may be employed to produce them where the fluid enters the well bore by the action of natural reservoir pressure (energy or gravity).

Primary transportation: Conveyance of large shipments of petroleum raw materials and refined products usually by pipeline, barge, or ocean-going vessel. All crude oil transportation is primary, including the small amounts moved by truck. All refined product transportation by pipeline, barge, or ocean-going vessel is primary transportation.

Prime mover: The engine, turbine, water wheel, or similar machine that drives an electric generator; or, for reporting purposes, a device that converts energy to electricity directly (*e.g.*, photovoltaic solar and fuel cells).

Prime supplier: A firm that produces, imports, or transports selected petroleum products across State boundaries and local marketing areas, and sells the product to local distributors, local retailers, or end users.

Private fueling facility: A fueling facility which normally services only fleets and is not open to the general public.

Privately owned electric utility: A class of ownership found in the electric power industry where the utility is regulated and authorized to achieve an allowed rate of return.

Probable energy reserves: Estimated quantities of energy sources that, on the basis of geologic evidence that supports projections from **proved reserves** It can reasonably be expected to exist and be recoverable under existing economic and operating conditions. Site information is insufficient to establish with confidence the location, quality, and grades of the energy source.

Probable (indicated) reserves, coal: Reserves or resources for which tonnage and grade are computed partly from specific measurements, samples, or production data and partly from projection for a reasonable distance on the basis of geological evidence. The sites available are too widely or otherwise inappropriately spaced to permit the mineral bodies to be outlined completely or the grade established throughout.

Process cooling and refrigeration: The direct process end use in which energy is used to lower the temperature of substances involved in the manufacturing process. Examples include freezing processed meats for later sale in the food industry and lowering the temperature of chemical feedstocks below ambient temperature for use in reactions in the chemical industries. Not included are uses such as air-conditioning for personal comfort and cafeteria refrigeration.

Process fuel: All energy consumed in the acquisition, processing, and transportation of energy. Quantifiable process fuel includes three categories: natural gas lease and plant operations, natural gas pipeline operations, and oil refinery operations.

Process heating or cooling demand-side management (DSM) program: A DSM program designed to promote increased electric energy efficiency applications in industrial process heating or cooling.

Process heating or cooling waste heat recovery: An energy conservation system whereby some space heating or water heating is done by actively capturing byproduct heat that would otherwise be ejected into the environment. In nonresidential buildings, sources of waste heat include refrigeration/air-conditioner compressors, manufacturing or other processes, data processing centers, lighting fixtures, ventilation exhaust air, and the occupants themselves. Not to be considered is the passive use of radiant heat from lighting, workers, motors, ovens, etc., when there are no special systems for collecting and redistributing heat.

Processed gas: Natural gas that has gone through a processing plant.

Processing: Uranium-recovery operations whether at a mill, an in situ leach, byproduct plant, or other ype of recovery operation.

Processing gain: The volumetric amount by which total output is greater than input for a given period of time. This difference is due to the processing of crude oil into products which, in total, have a lower specific gravity than the crude oil processed.

Processing loss: The volumetric amount by which total refinery output is less than input for a given period of time. This difference is due to the processing of crude oil into products which, in total, have a higher specific gravity than the crude oil processed.

Processing of uranium: The recovery of uranium produced by non-conventional mining methods, *i.e.*, in situ leach mining, as a byproduct of copper or phosphate mining, or heap leaching.

Processing plant: A surface installation designed to separate and recover natural gas liquids from a stream of produced natural gas through the processes of condensation, absorption, adsorption, refrigeration, or other methods and to control the quality of natural gas marketed and/or returned to oil or gas reservoirs for pressure maintenance, repressuring, or cycling.

Producer: A company engaged in the production and sale of natural gas from gas or oil wells with delivery generally at a point at or near the wellhead, the field, or the tailgate of a gas processing plant. For the purpose of company classification, a company primarily engaged in the exploration for, development of, and/or production of oil and/or natural gas.

Producer and distributor coal stocks: Producer and distributor coal stocks consist of coal held in stock by producers/distributors at the end of a reporting period.

Producer contracted reserves: The volume of recoverable salable gas reserves committed to or controlled by the reporting pipeline company as the buyer in gas purchase contracts with the independent producer as seller, including warranty contracts, and which are used for acts and services for which the company has received certificate authorization from the Federal Energy Regulatory Commission.

Producing property: A term often used in reference to a property, well, or mine that produces wasting natural resources. The term means a property that produces in paying quantities (that is, one for which proceeds from production exceed operating expenses).

Product supplied, crude oil: Crude oil burned on leases and by pipelines as fuel.

Production capacity: The amount of product that can be produced from processing facilities.

Production costs: Costs incurred to operate and maintain wells and related equipment and facilities, including depreciation and applicable operating costs of support equipment and facilities and other costs of operating and maintaining those wells and related equipment and facilities. They become part of the cost of oil and gas produced. The following are examples of production costs (sometimes called lifting costs):

Costs of labor to operate the wells and related equipment and facilities; repair and maintenance costs; the costs of materials, supplies, and fuels consumed and services utilized in operating the wells and related equipment and facilities; the costs of property taxes and insurance applicable to proved properties and wells and related equipment and facilities; the costs of severance taxes.

Depreciation, depletion, and amortization (DD&A) of capitalized acquisition, exploration, and development costs are not production costs, but also become part of the cost of oil and gas produced along with production (lifting) costs identified above. Production costs include the following subcategories of costs:

Well workers and maintenance; operating fluid injections and improved recovery programs; operating gas processing plants; ad valorem taxes; production or severance taxes; other, including overhead.

Production, crude oil: The volumes of crude oil that are extracted from oil reservoirs. These volumes are determined through measurement of the volumes delivered from lease storage tanks or at the point of custody transfer, with adjustment for (1) net differences between opening and closing lease inventories and (2) basic sediment and water. Crude oil used on the lease is considered production.

Production expenses: Costs incurred in the production of electric power that conform to the accounting requirements of the Operation and Maintenance Expense Accounts of the FERC Uniform System of Accounts.

Production, lease condensate: The volume of lease condensate produced. Lease condensate volumes include only those volumes recovered from lease or field separation facilities.

Production, natural gas: The volume of natural gas withdrawn from reservoirs less (1) the volume returned to such reservoirs in cycling, repressuring of oil reservoirs, and conservation operations; less (2) shrinkage resulting from the removal of lease condensate; and less (3) nonhydrocarbon gases where they occur in sufficient quantity to render the gas unmarketable. Volumes of gas withdrawn from gas storage reservoirs and native gas, which has been transferred to the storage category, are not considered production. Flared and vented gas is also considered production. (This differs from "Marketed Production" which excludes flared and vented gas.)

Production, natural gas, dry: The volume of natural gas withdrawn from reservoirs during the report year less (1) the volume returned to such reservoirs in cycling, repressuring of oil reservoirs, and conservation operations; less (2) shrinkage resulting from the removal of lease condensate and plant liquids; and less (3) nonhydrocarbon gases where they occur in sufficient quantity to render the gas unmarketable. Volumes of gas withdrawn from gas storage reservoirs and native gas, which has been transferred to the storage category, are not considered production. This is not the same as marketed production, because the latter also excludes vented and flared gas, but contains plant liquids.

Production, natural gas liquids: Production of natural gas liquids is classified as follows:

Contract Production. Natural gas liquids accruing to a company because of its ownership of liquids extraction facilities that it uses to extract liquids from gas belonging to others, thereby earning a portion of the resultant liquids.

Leasehold Production. Natural gas liquids produced, extracted, and credited to a company's interest.

Contract Reserves. Natural gas liquid reserves corresponding to the contract production defined above.

Leasehold Reserves. Natural gas liquid reserves corresponding to leasehold production defined above.

Production, natural gas, wet after lease separation: The volume of natural gas withdrawn from reservoirs less (1) the volume returned to such reservoirs in cycling, repressuring of oil reservoirs, and conservation operations; less (2) shrinkage resulting from the removal of lease condensate; and less (3) nonhydrocarbon gases where they occur in sufficient quantity to render the gas unmarketable.

Production, oil and gas: The lifting of oil and gas to the surface and gathering, treating, field processing (as in the case of processing gas to extract liquid hydrocarbons), and field storage. The production function shall normally be regarded as terminating at the outlet valve on the lease or field production storage tank. If unusual physical or operational circumstances exist, it may be more appropriate to regard the production function as terminating at the first point at which oil, gas, or gas liquids are delivered to a main pipeline, a common carrier, a refinery, or a marine terminal.

Production payments: A contractual arrangement providing a mineral interest that gives the owner a right to receive a fraction of production, or of proceeds from the sale of production, until a specified quantity of minerals (or a definite sum of money, including interest) has been received.

Production plant liquids: The volume of liquids removed from natural gas in natural gas processing plants or cycling plants during the year.

Productive capacity: The maximum amount of coal that a mining operation can produce or process during a period with the existing mining equipment and/or preparation plant in place, assuming that the labor and materials sufficient to utilize the plant and equipment are available, and that the market exists for the maximum production.

Products supplied: Approximately represents consumption of petroleum products because it measures the disappearance of these products from primary sources, *i.e.*, refineries, natural gas-processing plants, blending plants, pipelines, and bulk terminals. In general, product supplied of each product in any given period is computed as follows: field production, plus refinery production, plus imports, plus unaccounted-for crude oil (plus net receipts when calculated on a PAD District basis) minus stock change, minus crude oil losses, minus refinery inputs, and minus exports.

Profit: The income remaining after all business expenses are paid.

Program cost: Utility costs that reflect the total cash expenditures for the year, reported in nominal dollars, that flowed out to support DSM (demand-side management) programs. They are reported in the year they are incurred, regardless of when the actual effects occur.

Propane (C_3H_8): A normally gaseous straight-chain hydrocarbon. It is a colorless paraffinic gas that boils at a temperature of -43.67 degrees Fahrenheit. It is extracted from natural gas or refinery gas streams. It includes all products designated in ASTM Specification D1835 and Gas Processors Association Specifications for commercial propane and HD-5 propane.

Propane air: A mixture of propane and air resulting in a gaseous fuel suitable for pipeline distribution.

Propane, consumer grade: A normally gaseous paraffinic compound (C3H8), which includes all products covered by Natural Gas Policy Act Specifications for commercial and HD-5 propane and ASTM Specification D 1835. Excludes: feedstock propanes, which are propanes not classified as consumer grade propanes, including the propane portion of any natural gas liquid mixes, *i.e.*, butane-propane mix.

Proportional interest in investee reserves: The proportional interest at the end of the year in the reserves of investees that are accounted for by the equity method.

Proposed rates: New electric rate schedule proposed by an applicant to become effective at a future date.

Propylene (C_3H_6): An olefinic hydrocarbon recovered from refinery processes or petrochemical processes.

Prospecting: The search for an area of probable mineralization; the search normally includes topographical, geological, and geophysical studies of relatively large areas undertaken in an attempt to locate specific areas warranting detailed exploration. Prospecting usually occurs prior to the acquisition of mineral rights.

Prospecting costs: Direct and indirect costs incurred to identify areas of interest that may warrant detailed exploration. Such costs include those incurred for topographical, geological, and geophysical studies; rights of access to properties in order to conduct such studies, salaries, equipment, instruments, and supplies for geologists, including geophysical crews, and others conducting such studies; and overhead that can be identified with those activities.

Proved energy reserves: Estimated quantities of energy sources that analysis of geologic and engineering data demonstrates with reasonable certainty are recoverable under existing economic and operating conditions. The location, quantity, and grade of the energy source are usually considered to be well established in such reserves. **Proved (measured) reserves, coal:** Reserves or resources for which tonnage is computed from dimensions revealed in outcrops, trenches, workings, and drill holes and for which the grade is computed from the results of detailed sampling. The sites for inspection, sampling, and measurement are spaced so closely and the geologic character is so well defined that size, shape, and mineral content are well established. The computed tonnage and grade are judged to be accurate within limits that are stated, and no such limit is judged to be different from the computed tonnage or grade by more than 20 percent.

Public authorities: Electricity supplied to municipalities, divisions, or agencies of state and Federal governments, usually under special contracts or agreements that are applicable only to public authorities.

Public authority service to public authorities: Public authority service includes electricity supplied and services rendered to municipalities or divisions or agencies of State or Federal governments under special contracts, agreements, or service classifications applicable only to public authorities.

Public street and highway lighting: Electricity supplied and services rendered for the purpose of lighting streets, highways, parks, and other public places; or for traffic or other signal system service, for municipalities or other divisions or agencies of State or Federal governments.

Public utility: Enterprise providing essential public services, such as electric, gas, telephone, water, and sewer under legally established monopoly conditions.

Public utility district: Municipal corporations organized to provide electric service to both incorporated cities and towns and unincorporated rural areas.

Public Utility Holding Company Act of 1935 (PUHCA): This act prohibits acquisition of any wholesale or retail electric business through a holding company unless that business forms part of an integrated public utility system when combined with the utility's other electric business. The legislation also restricts ownership of an electric business by non-utility corporations.

Public Utility Regulatory Policies Act (PURPA) of 1978: One part of the National Energy Act, PURPA contains measures designed to encourage the conservation of energy, more efficient use of resources, and equitable rates. Principal among these were suggested retail rate reforms and new incentives for production of electricity by cogenerators and users of renewable resources. The Commission has primary authority for implementing several key PURPA programs.

Publicly owned electric utility: A class of ownership found in the electric power industry. This group includes those utilities operated by municipalities and State and Federal power agencies.

Pulp chips: Timber or residues processed into small pieces of wood of more or less uniform dimensions with minimal amounts of bark.

Pulp wood: Roundwood, whole-tree chips, or wood residues.

Pulping liquor (black liquor): The alkaline spent liquor removed from the digesters in the process of chemically pulping wood. After evaporation, the liquor is burned as a fuel in a recovery furnace that permits the recovery of certain basic chemicals.

Pumped-storage hydroelectric plant: A plant that usually generates electric energy during peak load periods by using water previously pumped into an elevated storage reservoir during off-peak periods when excess generating capacity is available to do so. When additional generating capacity is needed, the water can be released from the reservoir through a conduit to turbine generators located in a power plant at a lower level.

Purchase-contract imports of uranium: The amount of foreign-origin uranium material that enters the United States during a survey year as reported on the "Uranium Industry Annual Survey (UIAS), Form EIA-858, as purchases of uranium ore, U3O8, natural UF6, or enriched UF6. The amount of foreign-origin uranium materials that enter the country during a survey year under other types of contracts, *i.e.*, loans and exchanges, is excluded.

Purchased: Receipts into transportation, storage, and/or distribution facilities within a state under gas purchase contracts or agreements whether or not billing or payment occurred during the report year.

Purchased power: Power purchased or available for purchase from a source outside the system.

Purchased power adjustment: A clause in a rate schedule that provides for adjustments to the bill when energy from another electric system is acquired and its cost varies from a specified unit base amount.

Pure pumped-storage hydroelectric plant: A plant that produces power only from water that has previously been pumped to an upper reservoir.

PURPA: The Public Utility Regulatory Policies Act of 1978, passed by the U.S. Congress. This statute requires States to implement utility conservation programs and create special markets for co-generators and small producers who meet certain standards, including the requirement that States set the prices and quantities of power the utilities must buy from such facilities.

PVCs that convert sunlight directly into energy: A method for producing energy by converting sunlight using photovoltaic cells (PVCs) that are solid-state single converter devices. Although currently not in wide usage, commercial customers have a growing interest in usage and, therefore, DOE has a growing interest in the impact of PVCs on energy consumption. Economically, PVCs are competitive with other sources of electricity.

Pyrolysis: The thermal decomposition of biomass at high temperatures (greater than 400° F, or 200° C) in the absence of air. The end product of pyrolysis is a mixture of solids (char), liquids (oxygenated oils), and gases (methane, carbon monoxide, and carbon dioxide) with proportions determined by operating temperature, pressure, oxygen content, and other conditions.

Quad: Quadrillion (10 to the 15th power).

Quadrillion: The quantity 1,000,000,000,000,000 (10 to the 15th power).

Qualifying facility (QF): A cogeneration or small power production facility that meets certain ownership, operating, and efficiency criteria established by the Federal Energy Regulatory Commission (FERC) pursuant to the Public Utility Regulatory Policies Act (PURPA).

Quality or grade (of coal): An informal classification of coal relating to its suitability for use for a particular purpose. Refers to individual measurements such as heat value, fixed carbon, moisture, ash, sulfur, major, minor, and trace elements, coking properties, petrologic properties, and particular organic constituents. The individual quality elements may be aggregated in various ways to classify coal for such special purposes as metallurgical, gas, petrochemical, and blending usages.

Quantity wires charge: A fee for moving electricity over the transmission and/or distribution system that is based on the quantity of electricity that is transmitted.

value: A measure of a material's resistance to heat flow in units of Fahrenheit degrees x hours x square feet per Btu. The higher the R-value of a material, the greater its insulating capability. The R-value of some insulating materials is 3.7 per inch for fiberglass and cellulose,2.5 per inch for vermiculite, and more than 4 per inch for foam. All building materials have some R-value. For example, a 4-inch brick has an R-value of 0.8, and half-inch plywood has an R-value of 0.6. The below table converts the most common "R" values to inches. For other "R" values, divide the "R" value by 3 to get the number of inches.

"R"-Value	*Inches*
3	1
11	3.5
19	6
52	18

Rack sales: Wholesale truckload sales or smaller of gasoline where title transfers at a terminal.

Radiant barrier: A thin, reflective foil sheet that exhibits low radiant energy transmission and under certain conditions can block radiant heat transfer; installed in attics to reduce heat flow through a roof assembly into the living space.

Radiant ceiling panels: Ceiling panels that contain electric resistance heating elements embedded within them to provide radiant heat to a room.

Radiant energy: Energy that transmits away from its source in all directions.

Radiation: The transfer of heat through matter or space by means of electromagnetic waves.

Radiative forcing: A change in average net radiation at the top of the troposphere (known as the tropopause) because of a change in either incoming solar or exiting infrared radiation. A positive radiative forcing tends on average to warm the earth's surface; a negative radiative forcing on average tends to cool the earth's surface. Greenhouse gases, when emitted into the atmosphere, trap infrared energy radiated from the earth's surface and therefore tend to produce positive radiative forcing.

Radiatively active gases: Gases that absorb incoming solar radiation or outgoing infrared radiation, affecting the vertical temperature profile of the atmosphere.

Radiator: A heating unit usually exposed to view within the room or space to be heated; it transfers heat by radiation to objects within visible range and by conduction to the surrounding air, which in turn is circulated by natural convection; usually fed by steam or hot water.

Radioactive waste: Materials left over from making nuclear energy. Radioactive waste can destroy living organisms if it is not stored safely.

Radioactivity: The spontaneous emission of radiation from the nucleus of an atom. Radionuclides lose particles and energy through this process.

Radioisotope: A radioactive isotope.

Radon: A naturally occurring radioactive gas found in the United States in nearly all types of soil, rock, and water. It can migrate into most buildings. Studies have linked high concentrations of radon to lung cancer.

Rail (method of transportation to consumers): Shipments of coal moved to consumers by rail (private or public/commercial). Includes coal hauled to or away from a railroad siding by truck.

Railroad and railway electric service: Electricity supplied to railroads and interurban and street railways, for general railroad use, including the propulsion of cars or locomotives, where such electricity is supplied under separate and distinct rate schedules.

Railroad locomotive: Self-propelled vehicle that runs on rails and is used for moving railroad cars.

Railroad use: Sales to railroads for any use, including that used for heating buildings operated by railroads.

Range top: The range burners or stove top and the oven are considered two separate appliances. Counted also with range tops are stand-alone "cook tops."

Rankine cycle: The thermodynamic cycle that is an ideal standard for comparing performance of heat-engines, steam power plants, steam turbines, and heat pump systems that use a condensable vapor as the working fluid. Efficiency is measured as work done divided by sensible heat supplied.

Rankine cycle engine: The Rankine cycle system uses a liquid that evaporates when heated and expands to produce work, such as turning a turbine, which when connected to a generator, produces electricity. The exhaust vapor expelled from the turbine condenses and the liquid is pumped back to the boiler to repeat the cycle. The working fluid most commonly used is water, though other liquids can also be used. Rankine cycle design is used by most commercial electric power plants. The traditional steam locomotive is also a common form of the Rankine cycle engine. The Rankine engine itself can be either a piston engine or a turbine.

Rate base: The value of property upon which a utility is permitted to earn a specified rate of return as established by a regulatory authority. The rate base generally represents the value of property used by the utility in providing service and may be calculated by any one or a combination of the following accounting methods: fair value, prudent investment, reproduction cost, or original cost. Depending on which method is used, the rate base includes cash, working capital, materials and supplies, deductions for accumulated provisions for depreciation, contributions in aid of construction, customer advances for construction, accumulated deferred income taxes, and accumulated deferred investment tax credits.

Rate case: A proceeding, usually before a regulatory commission, involving the rates to be charged for a public utility service.

Rate features: Special rate schedules or tariffs offered to customers by electric and/or natural gas utilities.

Rate of return: The ratio of net operating income earned by a utility is calculated as a percentage of its rate base.

Rate of return on rate base: The ratio of net operating income earned by a utility, calculated as a percentage of its rate base.

Rate schedule (electric): A statement of the financial terms and conditions governing a class or classes of utility services provided to a customer. Approval of the schedule is given by the appropriate rate-making authority.

Ratemaking authority: A utility commission's legal authority to fix, modify, approve, or disapprove rates as determined by the powers given the commission by a State or Federal legislature.

Rates: The authorized charges per unit or level of consumption for a specified time period for any of the classes of utility services provided to a customer.

Rating: A manufacturer's guaranteed performance of a machine, transmission line, or other electrical apparatus, based on design features and test data. The rating will specify such limits as load, voltage, temperature, and frequency. The rating is generally printed on a nameplate attached to equipment and is commonly referred to as the nameplate rating or nameplate capacity.

Ratio estimate: The ratio of two population aggregates (totals). For example, "average miles traveled per vehicle" is the ratio of total miles driven by all vehicles, over the total number of vehicles, within any subgroup. There are two types of ratio estimates: those computed using aggregates for vehicles and those computed using aggregates for households.

Ratoon crop: A crop cultivated from the shoots of a perennial plant.

Rayleigh frequency distribution: A mathematical representation of the frequency or ratio that specific wind speeds occur within a specified time interval.

Reactance: A phenomenon associated with AC power characterized by the existence of a time difference between voltage and current variations.

Reactive power: The electrical power that oscillates between the magnetic field of an inductor and the electrical field of a capacitor. Reactive power is never converted to nonelectrical power. It is calculated as the square root of the difference between the square of the kilovolt-amperes and the square of the kilowatts and is expressed as reactive volt-amperes.

Real dollars: These are dollars that have been adjusted for inflation.

Real price: A price that has been adjusted to remove the effect of changes in the purchasing power of the dollar. Real prices, which are expressed in constant dollars, usually reflect buying power relative to a base year.

Reasonably assured resources (RAR): The uranium that occurs in known mineral deposits of such size, grade, and configuration that it could be recovered within the given production cost ranges, with currently proven mining and processing technology. Estimates of tonnage and grade are based on specific sample data and measurements of the deposits and on knowledge of deposit characteristics. RAR correspond to DOE's Reserves category.

Rebate program: A utility company-sponsored conservation program whereby the utility company returns a portion of the purchase price cost when a more energy-efficient refrigerator, water heater, air conditioner, or other appliance is purchased.

Reburn: An advanced co-firing technique using natural gas to reduce pollution from electric power plants.

Receipts:

1. Deliveries of fuel to an electric plant
2. Purchases of fuel
3. All revenues received by an exporter for the reported quantity exported

Receivables from municipality: All charges by the utility department against the municipality or its other departments that are subject to current settlement.

Received: Gas (and other fuels) physically transferred into the responding company's transportation, storage, and/or distribution facilities.

Reclamation: Process of restoring surface environment to acceptable pre-existing conditions. Includes surface contouring, equipment removal, well plugging, revegetation, etc.

Reclamation expenses: In the context of the coal operation statement of income, refers to all payments made by the company attributable to reclamation, including taxes.

Recoverability: In reference to accessible coal resources, the condition of being physically, technologically, and economically minable. Recovery rates and recovery factors may be determined or estimated for coal resources without certain knowledge of their economic minability; therefore, the availability of recovery rates or factors does not predict recoverability.

Recoverable coal: Coal that is, or can be, extracted from a coal bed during mining.

Recoverable proved reserves: The proved reserves of natural gas as of December 31 of any given year are the estimated quantities of natural gas which geological and engineering data demonstrates with reasonable certainty to be recoverable in the future from known natural oil and gas reservoirs under existing economic and operating conditions.

Recoverable reserves: The amount of coal that can be recovered (mined) from the coal deposits at active producing mines as of the end of the year.

Recovery factor (coal): The percentage of total tons of coal estimated to be recoverable from a given area in relation to the total tonnage estimated to be in the demonstrated reserve base. The estimated recovery factors for the demonstrated reserve base generally are 50 percent for underground mining methods and 80 percent for surface mining methods. More precise recovery factors can be computed by determining the total coal in place and the total recoverable in any specific locale.

Recovery percentage (coal): The percentage of coal that can be recovered from the coal deposits at existing mines.

Rectifier: A device for converting alternating current to direct current.

Recycled feeds: Feeds that are continuously fed back for additional processing.

Recycling: The process of converting materials that are no longer useful as designed or intended into a new product.

Redox potential: A measurement of the state of oxidation of a system.

Redrill footage: Occasionally, a hole is lost or junked and a second hole may be drilled from the surface in close proximity to the first. Footage drilled for the second hole is defined as "redrill footage." Under these circumstances, the first hole is reported as a dry hole (explanatory or developmental) and the total footage is reported as dry hole footage. The second hole is reported as an oil well, gas well, or dry hole according to the result. The redrill footage is included in the appropriate classification of total footage, but is not reported as a separate classification.

Reduced use-off hours: A conservation feature consisting of manually or automatically reducing the amount of heating or cooling produced during the hours a building is not in full use.

Reference month: The calendar month and year to which the reported cost, price, and volume information relates.

Reference year: The calendar year to which the reported sales volume information relates.

Refined petroleum products: Refined petroleum products include but are not limited to gasolines, kerosene, distillates (including No. 2 fuel oil), liquefied petroleum gas, asphalt, lubricating oils, diesel fuels, and residual fuels.

Refiner: A firm or the part of a firm that refines products or blends and substantially changes products, or refines liquid hydrocarbons from oil and gas field gases, or recovers liquefied petroleum gases incident to petroleum refining and sells those products to resellers, retailers, reseller/retailers or ultimate consumers. "Refiner" includes any owner of products that contracts to have those products refined and then sells the refined products to resellers, retailers, or ultimate consumers.

Refiner acquisition cost of crude oil: The cost of crude oil, including transportation and other fees paid by the refiner. The composite cost is the weighted average of domestic and imported crude oil costs.

Refinery: An installation that manufactures finished petroleum products from crude oil, unfinished oils, natural gas liquids, other hydrocarbons, and oxygenates.

Refinery capacity utilization: Ratio of the total amount of crude oil, unfinished oils, and natural gas plant liquids run through crude oil distillation units to the operable capacity of these units.

Refinery fuel: Crude oil and petroleum products consumed at the refinery for all purposes.

Refinery gas: Noncondensate gas collected in petroleum refineries.

Refinery input, crude oil: Total crude oil (domestic plus foreign) input to crude oil distillation units and other refinery processing units (cokers, etc.).

Refinery input, total: The raw materials and intermediate materials processed at refineries to produce finished petroleum products. They include crude oil, products of natural gas processing plants, unfinished oils, other hydrocarbons and oxygenates, motor gasoline and aviation gasoline blending components and finished petroleum products.

Refinery losses and gains: Processing gain and loss that takes place during the refining process itself. Excludes losses that do not take place during the refining process, *e.g.*, spills, fire losses, and contamination during blending, transportation, or storage.

Refinery output: The total amount of petroleum products produced at a refinery. Includes petroleum consumed by the refinery.

Refinery production: Petroleum products produced at a refinery or blending plant. Published production of these products equals refinery production minus refinery input. Negative production will occur when the amount of a product produced during the month is less than the amount that is reprocessed (input) or reclassified to become another product during the same month. Refinery production of unfinished oils and motor and aviation gasoline blending components appear on a net basis under refinery input.

Refinery utilization rate: Represents the use of the atmospheric crude oil distillation units. The rate is calculated by dividing the gross input to these units by the operable refining capacity of the units.

Refinery yield: Refinery yield (expressed as a percentage) represents the percent of finished product produced from input of crude oil and net input of unfinished oils. It is calculated by dividing the sum of crude oil and net unfinished input into the individual net production of finished products. Before calculating the yield for finished motor gasoline, the input of natural gas liquids, other hydrocarbons and oxygenates, and net input of motor gasoline blending components must be subtracted from the net production of finished aviation gasoline.

Reflective film: Transparent covering for glass that helps keep out heat from the sun.

Reflectivity: The ratio of the energy carried by a wave after reflection from a surface to its energy before reflection.

Reforestation: Replanting of forests on lands that have recently been harvested or otherwise cleared of trees.

Reformulated gasoline: Finished motor gasoline formulated for use in motor vehicles, the composition and properties of which meet the requirements of the reformulated gasoline regulations promulgated by the U.S. Environmental Protection Agency under Section 211(k) of the Clean Air Act.

Refrigeration unit: Lowers the temperature through a mechanical process. In a typical refrigeration unit, electricity powers a motor that runs a pump to compress the refrigerant to maintain proper pressure. (A "refrigerant" is a substance that changes between liquid and gaseous states under desirable temperature and pressure conditions.) Heat from the compressed liquid is removed and discharged from the unit and the refrigerant then evaporates when pressure is reduced. The refrigerant picks up heat as it evaporates and it returns to the compressor to repeat the cycle. A few refrigeration units use gas (either natural gas or LPG) in an absorption process that does not use a compressor. The gas is burned to heat a chemical solution in which the refrigerant has been absorbed. Heating drives off the refrigerant which is later condensed. The condensed refrigerant evaporates by a release of pressure, and it picks up heat as it evaporates. The evaporated refrigerant is then absorbed back into the chemical solution, the heat is removed from the solution and discharged as waste heat, and the process repeats itself. By definition, refrigerators, freezers, and air-conditioning equipment all contain refrigeration units.

Refunding: Retirement of one security issue with proceeds received from selling another. Refunding provides for retiring maturing debt by taking advantage of favorable money market conditions.

Refuse bank: A repository for waste material generated by the coal cleaning process.

Refuse-derived fuel (RDF): A fuel produced by shredding municipal solid waste (MSW). Noncombustible materials such as glass and metals are generally removed prior to making RDF. The residual material is sold as-is or compressed into pellets, bricks, or logs. RDF processing facilities are typically located near a source of MSW, while the RDF combustion facility can be located elsewhere.

Refuse mine: A surface mine where coal is recovered from previously mined coal. It may also be known as a silt bank, culm bank, refuse bank, slurry dam, or dredge operation.

Regional Transmission Group: A utility industry concept that the Federal Energy Regulatory Commission (FERC) embraced for the certification of voluntary groups that would be responsible for transmission planning and use on a regional basis.

Regular grade gasoline: A grade of unleaded gasoline with a lower octane rating (approximately 82) than other grades. Octane boosters are added to gasoline to control engine pre-ignition or "knocking" by slowing combustion rates.

Regulated streamflow: The rate of flow past a given point during a specified period that is controlled by reservoir water release operation.

Regulation: The governmental function of controlling or directing economic entities through the process of rulemaking and adjudication.

Regulation, procedures, and practices: A utility commission carries out its regulatory functions through rulemaking and adjudication. Under rulemaking, the utility commission may propose a general rule of regulation change. By law, it must issue a notice of the proposed rule and a request for comments is also made; the Federal Energy Regulatory Commission publishes this in the Federal Register. The final decision must be published. A utility commission may also work on a case-by-case basis from submissions from regulated companies or others. Objections to a proposal may come from the commission or intervenors, in which case the proposal must be presented to a hearing presided over by an administrative law judge. The judge's decision may be adopted, modified, or reversed by the utility commissioners, in which case those involved can petition for a rehearing and may appeal a decision through the courts system to the U.S. Supreme Court.

Reheating coils: A part of some air-conditioning systems. Electric coils in air ducts used primarily to raise the temperature of circulated air after it was over-cooled to remove moisture. Some buildings have reheating coils as their sole heating source.

Reinjected: The forcing of gas under pressure into an oil reservoir in an attempt to increase recovery.

Reinserted fuel: Irradiated fuel that is discharged in one cycle and inserted in the same reactor during a subsequent refueling. In a few cases, fuel discharged from one reactor has been used to fuel a different reactor.

Reinsertion: The process of returning nuclear fuel that has been irradiated and then removed from a reactor back into a reactor for further irradiation. Reinserted assemblies are assemblies that have been irradiated in a cycle, were not in the core in the prior cycle (cycle N), and which are in the core in the current cycle (cycle N+1).

Reliability (electric system): A measure of the ability of the system to continue operation while some lines or generators are out of service. Reliability deals with the performance of the system under stress.

Relocation of tailings: Relocation of tailings is sometimes necessary if the pile poses a threat to inhabitants or the environment, for example, through being situated too close to populated areas, on top of aquifers or other sources of water, or in unstable areas such as flood plains or faults near earthquake zones.

Renewable energy resources: Energy resources that are naturally replenishing but flow-limited. They are virtually inexhaustible in duration but limited in the amount of energy that is available per unit of time. Renewable energy resources include: biomass, hydro, geothermal, solar, wind, ocean thermal, wave action, and tidal action.

Replacement energy source for primary heating: For the CBECS, the heating energy source to which the building could switch within one week without major modifications to the main heating equipment, without substantially reducing the area heated, and without substantially reducing the temperature maintained in the heated area.

Replacement vehicle: A vehicle which is acquired in order to take the place of a vehicle which is being retired from service. These acquisitions do not increase the size of the company fleet.

Report State: The State, including adjacent offshore continental shelf areas in the Federal domain, in which a company operated natural gas gathering, transportation, storage, and/or distribution facilities or a synthetic natural gas plant covered by the individual report.

Report week: A calendar week beginning at 12:01 a.m. on Sunday and ending at midnight on Saturday.

Report year (calendar): The 12-month period, January 1 through December 31.

Report year (fiscal): A 12-month period for which an organization plans the use of its funds. The fiscal year is designated by the calendar year in which it ends.

Reporting: The average number of Btu per cubic foot of gas at 60 degrees Fahrenheit and 14.73 psia delivered directly to consumers. Where billing is on a thermal basis, the heat content values used for billing purposes are to be used to determine the annual average heat content.

Repowering: Refurbishment of a plant by replacement of the combustion technology with a new combustion technology, usually resulting in better performance and greater capacity.

Repressuring: The injection of gas into oil or gas formations to effect greater ultimate recovery.

Reprocessing: Synonymous with chemical separations.

Requirements power: The firm service needs required by designated load plus losses from the points of supply.

Reregulation: The design and implementation of regulatory practices to be applied to the remaining regulated entities after restructuring of the vertically-integrated electric utility. The remaining regulated entities would be those that continue to exhibit characteristics of a natural monopoly, where imperfections in the market prevent the realization of more competitive results, and where, in light of other policy considerations, competitive results are unsatisfactory in one or more respects. Regulation could employ the same or different regulatory practices as those used before restructuring.

Resale (wholesale) sales: Resale or wholesale sales are electricity sold (except under exchange agreements) to other electric utilities or to public authorities for resale distribution. (This includes sales to requirements and nonrequirements consumers.)

Research and development (R&D): Basic and applied research in the sciences and engineering and the design and development of prototypes and processes, excluding quality control, routine product testing, market research, sales promotion, sales service, research in the social sciences or psychology, and other non-technological activities or technical services.

Reseller: A firm (other than a refiner) that is engaged in a trade or business that buys refined petroleum products and then sells them to a purchaser who is not the ultimate consumer of those refined products.

Reserve: That portion of the demonstrated reserve base that is estimated to be recoverable at the time of determination. The reserve is derived by applying a recovery factor to that component of the identified coal resource designated as the demonstrated reserve base.

Reserve additions: The estimated original, recoverable, salable, and new proved reserves credited to new fields, new reservoirs, new gas purchase contracts, amendments to old gas purchase contracts, or purchase of gas reserves in-place that occurred during the year and had not been previously reported. Reserve additions refer to domestic in-the-ground natural gas reserve additions and do not refer to interstate pipeline purchase agreements; contracts with foreign suppliers; coal gas, SNG, or LNG purchase arrangements.

Reserve cost categories of $15, $30, $50, and $100 per pound U_3O_8: Classification of uranium reserves estimated by using break-even cutoff grades that are calculated based on forward-operating costs of less than $15, $30, $50, and $100 per pound U_3O_8.

Reserve generating capacity: Amount of generating capacity available to meet peak or abnormally high demands for power and to generate power during scheduled or unscheduled outages.

Reserve margin (operating): The amount of unused available capability of an electric power system (at peak load for a utility system) as a percentage of total capability.

Reserve revisions: Changes to prior year-end proved reserves estimates, either positive or negative, resulting from new information other than an increase in proved acreage (extension). Revisions include increases of proved reserves associated with the installation of improved recovery techniques or equipment. They also include correction of prior year arithmetical or clerical errors and adjustments to prior year-end production volumes to the extent that these alter reserves estimates.

Reserves, coal: Quantities of unextracted coal that comprise the demonstrated base for future production, including both proved and probable reserves.

Reserves, net: Includes all proved reserves associated with the company's net working interests.

Reserves changes: Positive and negative revisions, extensions, new reservoir discoveries in old fields, and new field discoveries that occurred during the report year.

Reservoir: A porous and permeable underground formation containing an individual and separate natural accumulation of producible hydrocarbons (crude oil and/or natural gas) which is confined by impermeable rock or water barriers and is characterized by a single natural pressure system.

Reservoir capacity: The present total developed capacity (base and working) of the storage reservoir, excluding contemplated future development.

Reservoir repressuring: The injection of a pressurized fluid (such as air, gas, or water) into oil and gas reservoir formations to effect greater ultimate recovery.

Residential building: A structure used primarily as a dwelling for one or more households.

Residential/commercial (consumer category): Housing units, wholesale or retail businesses (except coal wholesale dealers); health institutions (hospitals, social and educational institutions (schools and universities); and Federal, state, and local governments (military installations, prisons, office buildings, etc.). Excludes shipments to Federal power projects, such as TVA, and rural electrification cooperatives, power districts, and state power projects.

Residential consumers: Consumers using gas for heating, air conditioning, cooking, water heating, and other residential uses in single and multi-family dwellings and apartments and mobile homes.

Residential energy consumption survey (RECS): A national multistage probability sample survey conducted by the Energy End Use Division of the Energy Information Administration. The RECS provides baseline information on how households in the United States use energy. The Residential Transportation Energy Consumption Survey (RTECS) sample is a subset of the RECS. Household demographic characteristics reported in the RTECS publication are collected during the RECS personal interview.

Residential heating oil price: The price charged for home delivery of No. 2 heating oil, exclusive of any discounts such as those for prompt cash payment. Prices do not include taxes paid by the consumer.

Residential propane price: The "bulk keep full" price for home delivery of consumer-grade propane intended for use in space heating, cooking, or hot water heaters in residences.

Residential sector: An energy-consuming sector that consists of living quarters for private households. Common uses of energy associated with this sector include space heating, water heating, air conditioning, lighting, refrigeration, cooking, and running a variety of other appliances. The residential sector excludes institutional living quarters.

Residential type central air conditioner: There are four basic parts to a residential central air-conditioning system: (1) a condensing unit, (2) a cooling coil, (3) ductwork, and (4) a control mechanism such as a thermostat. There are two basic configurations of residential central systems: (1) a "split system" where the condensing unit is located outside and the other components are inside, and (2) a packaged-terminal air-encased in one unit and is usually found in a "utility closet."

Residential vehicles: Motorized vehicles used by U.S. households for personal transportation. Excluded are motorcycles, mopeds, large trucks, and buses. Included are automobiles, station wagons, passenger vans, cargo vans, motor homes, pickup trucks, and jeeps or similar vehicles. In order to be included, vehicles must be (1) owned by members of the household, or (2) company cars not owned by household members but regularly available to household members for their personal use and ordinarily kept at home, or (3) rented or leased for 1 month or more.

Residual fuel oil: A general classification for the heavier oils, known as No. 5 and No. 6 fuel oils, that remain after the distillate fuel oils and lighter hydrocarbons are distilled away in refinery operations. It conforms to ASTM Specifications D 396 and D 975 and Federal Specification VV-F-815C. No. 5, a residual fuel oil of medium viscosity, is also known as Navy Special and is defined in Military Specification MIL-F-859E, including Amendment 2 (NATO Symbol F-770). It is used in steam-powered vessels in government service and inshore powerplants. No. 6 fuel oil includes Bunker C fuel oil and is used for the production of electric power, space heating, vessel bunkering, and various industrial purposes.

Residue gas: Natural gas from which natural gas processing plant liquid products and, in some cases, nonhydrocarbon components have been extracted.

Residuum: Residue from crude oil after distilling off all but the heaviest components, with a boiling range greater than 1,000 degrees Fahrenheit.

Respondent: A company or individual who completes and returns a report or survey form.

Restoration: The returning of all affected groundwater to its premining quality for its premining use by employing the best practical technology.

Restoration time: The time when the major portion of the interrupted load has been restored and the emergency is considered to be ended. However, some of the loads interrupted may not have been restored due to local problems.

Restricted-universe census: This is the complete enumeration of data from a specifically defined subset of entities including, for example, those that exceed a given level of sales or generator nameplate capacity.

Restructuring: The process of replacing a monopoly system of electric utilities with competing sellers, allowing individual retail customers to choose their electricity supplier but still receive delivery over the power lines of the local utility. It includes the reconfiguration of the vertically-integrated electric utility.

Retail motor gasoline prices: Motor gasoline prices calculated each month by the Bureau of Labor Statistics (BLS) in conjunction with the construction of the Consumer Price Index.

Retail wheeling: The process of moving electric power from a point of generation across third-party-owned transmission and distribution systems to a retail customer.

Retailer: A firm (other than a refiner, reseller, or reseller/retailer) that carries on the trade or business of purchasing refined petroleum products and reselling them to ultimate consumers without substantially changing their form.

Retained earnings: The balance, either debit or credit, of appropriated or unappropriated retained earnings of the utility department arising from earnings.

Retire from service: A vehicle is retired from service if that vehicle is placed out of service and there are no future plans to return that vehicle to service.

Retired hydropower plant sites: The site of a plant that formerly produced electrical or mechanical power but is now out of service. Includes plants that have been abandoned, damaged by flood or fire, inundated by new reservoirs, or dismantled.

Return on common equity: The net income less preferred stock dividends, divided by the average common stock equity.

Return on common stock equity: An equity's earnings available for common stockholders calculated as a percentage of its common equity capital

Revenue–(electricity): The total amount of money received by an entity from sales of its products and/or services; gains from the sales or exchanges of assets, interest, and dividends earned on investments; and other increases in the owner's equity, except those arising from capital adjustments.

Revenue requirement: The total revenue that the utility is authorized an opportunity to recover, which includes operating expenses and a reasonable return on rate base.

Reversible turbine: A hydraulic turbine, normally installed in a pumped-storage plant, which can be used alternatively as a pump or as an engine, turbine, water wheel, or other apparatus that drives an electrical generator.

Revisions and additions (gross change in reserves): The difference (plus or minus) between the year-end reserves plus production for a given year and the year-end reserves for the previous year.

Ribbon silicon: Crystalline silicon that is used in photovoltaic cells. Ribbon silicon is fabricated by a variety of solidification (crystallization) methods that withdraw thin silicon sheets from pools of relatively pure molten silicon.

Right-of-way: The land and legal right to use and service the land along which a transmission line is located. Transmission line right-of-way is usually acquired in widths that vary with the kilovolt (kV) size of the line.

Rip rap: Cobblestone or coarsely broken rock used for protection against erosion of embankment or gully.

River (method of transportation to consumers–coal): Shipments of coal moved to consumers via river by barge. Shipments to Great Lakes coal loading docks or Tidewater pier or coastal points are not included.

Road oil: Any heavy petroleum oil, including residual asphaltic oil used as a dust pallative and surface treatment on roads and highways. It is generally produced in six grades, from 0, the most liquid, to 5, the most viscous.

Rodlet or GAD basket: An open garbage and debris (GAD) basket that may have contain pieces of fuel rods, disassembled fuel rods, and other fuel and nonfuel components.

Roll front: A type of uranium deposition localized as a roll or interface separating an oxidized interior from a reduced exterior. The reduced side of this interface is significantly enriched in uranium.

Roof (coal): The rock immediately above a coal seam. The roof is commonly a shale, often carbonaceous and softer than rocks higher up in the roof strata.

Roof insulation: Insulating materials placed underneath the roof or on the roof (building).

Roof or ceiling insulation: A building shell conservation feature consisting of insulation placed in the roof (below the waterproofing layer) or in the ceiling of the top floor in the building.

Roof or ceiling insulation, insulation in exterior walls: Any material that when placed between the interior surface of the building and the exterior surface of the building, reduces the rate of heat loss to the environment or heat gain from the environment. Roof or ceiling insulation refers to insulation placed in the roof or ceiling of the top occupied floor in the building. Wall insulation refers to insulation placed between the exterior and interior walls of the building.

Roof pond: A solar energy collection device consisting of containers of water located on a roof that absorb solar energy during the day so that the heat can be used at night or that cools a building by evaporation at night.

Room air conditioner: Air-conditioning units that typically fit into the window or wall and are designed to cool only one room.

Room heater burning gas, oil, and kerosene: Any of the following heating equipment: circulating heaters, convectors, radiant gas heaters, space heaters, or other nonportable room heaters that may or may not be connected to a flue, vent, or chimney.

Room-and-pillar mining: The most common method of underground mining in which the mine roof is supported mainly by coal pillars left at regular intervals. Rooms are places where the coal is mined; pillars are areas of coal left between the rooms. Room-and-pillar mining is done either by conventional or continuous mining.

Rotary rig: A machine used for drilling wells that employs a rotating tube attached to a bit for boring holes through rock.

Round test mesh: A sieving screen with round holes, the dimensions of which are of specific sizes to allow certain sizes of coal to pass through while retaining other sizes.

Roundwood: Wood cut specifically for use as a fuel.

Royalty: A contractual arrangement providing a mineral interest that gives the owner a right to a fractional share of production or proceeds therefrom, that does not contain rights and obligations of operating a mineral property, and that is normally free and clear of exploration, developmental and operating costs, except production taxes.

Royalty cost: A share of the profit or product reserved by the grantor of a mining lease, such as a royalty paid to a lessee.

Royalty interest: An interest in a mineral property provided through a royalty contract.

Royalty interest (including overriding royalty): These interests entitle their owner(s) to a share of the mineral production from a property or to a share of the proceeds therefrom. They do not contain the rights and obligations of operating the property and normally do not bear any of the costs of exploration, development, and operation of the property.

Rulemaking (regulations): The authority delegated to administrative agencies by Congress or State legislative bodies to make rules that have the force of law. Frequently, statutory laws that express broad terms of a policy are implemented more specifically by administrative rules, regulations, and practices.

Run off: That portion of the precipitation that flows over the land surface and ultimately reaches streams to complete the water cycle. Melting snow is an important source of this water as well as all amounts of surface water that move to streams or rivers through any given area of a drainage basin.

Run-of-mine coal: Coal as it comes from the mine prior to screening or any other treatment.

Run-of-river hydroelectric plant: A low-head plant using the flow of a stream as it occurs and having little or no reservoir capacity for storage.

Running and quick-start capability: The net capability of generating units that carry load or have quick-start capability. In general, quick-start capability refers to generating units that can be available for load within a 30-minute period.

Rural Electrification Administration (REA): A lending agency of the U. S. Department of Agriculture, the REA makes self-liquidating loans to qualified borrowers to finance electric and telephone service to rural areas. The REA finances the construction and operation of generating plants, electric transmission and distribution lines, or systems for the furnishing of initial and continued adequate electric services to persons in rural areas not receiving central station service.

Salable coal: The shippable product of a coal mine or preparation plant. Depending on customer specifications, salable coal may be run-of-mine, crushed-and-screened (sized) coal, or the clean coal yield from a preparation plant.

Salable natural gas: Natural gas marketed under controlled quality conditions.

Sales for resale: A type of wholesale sales covering energy supplied to other electric utilities, cooperatives, municipalities, and Federal and state electric agencies for resale to ultimate consumers.

Sales to end users: Sales made directly to the consumer of the product. Includes bulk consumers, such as agriculture, industry, and utilities, as well as residential and commercial consumers.

Sales type: Sales categories of sales to end-users and sales for resale.

Sales volume (coal): The reported output from Federal and/or Indian lands, the basis of royalties. It is approximately equivalent to production, which includes coal sold, and coal added to stockpiles.

Salt dome: A domical arch (anticline) of sedimentary rock beneath the earth's surface in which the layers bend downward in opposite directions from the crest and that has a mass of rock salt as its core.

Salt gradient solar ponds: These consist of three main layers. The top layer is near ambient and has low salt content. The bottom layer is hot, typically 160° F to 212° F (71° C to 100° C), and is very salty. The important gradient zone separates these zones. The gradient zone acts as a transparent insulator, permitting the sunlight to be trapped in the hot bottom layer (from which useful heat is withdrawn). This is because the salt gradient, which increases the brine density with depth, counteracts the buoyancy effect of the warmer water below (which would otherwise rise to the surface and lose its heat to the air). An organic Rankine cycle engine is used to convert the thermal energy to electricity.

Sample (coal): A representative fraction of a coal bed collected by approved methods, guarded against contamination or adulteration, and analyzed to determine the nature; chemical, mineralogic, and (or) petrographic composition; percentage or parts-per-million content of specified constituents; heat value; and possibly the reactivity of the coal or its constituents.

Schedule: A statement of the pricing format of electricity and the terms and conditions governing its applications.

Scheduled outage: The shutdown of a generating unit, transmission line, or other facility for inspection or maintenance, in accordance with an advance schedule.

Scheduling coordinators: Entities certified by the Federal Energy Regulatory Commission (FERC) that act on behalf of generators, supply aggregators (wholesale marketers), retailers, and customers to schedule the distribution of electricity.

Scoop loading: An underground loading method by which coal is removed from the working face by a tractor unit equipped with a hydraulically operated bucket attached to the front; also called a front-end loader.

Screenings: The undersized coal from a screening process, usually one-half inch or smaller.

Seam: A bed of coal lying between a roof and floor. Equivalent term to bed, commonly used by industry.

Seasonal energy efficiency ratio (SEER): Ratio of the cooling output divided by the power consumption. It is the Btu of cooling output during its normal annual usage divided by the total electric energy input in watt hours during the same period. This is a measure of the cooling performance for rating central air conditioners and central heat pumps. The appliance standards required a minimum SEER of 10 for split-system central air conditioners and for split-system central heat pumps in 1992.

Seasonal pricing: A special electric rate feature under which the price per unit of energy depends on the season of the year.

Seasonal rates: Different seasons of the year are structured into an electric rate schedule whereby an electric utility provides service to consumers at different rates. The electric rate schedule usually takes into account demand based on weather and other factors.

Seasonal units: Housing units intended for occupancy at only certain seasons of the year. Seasonal units include units intended only for recreational use, such as beach cottages and hunting cabins. It is not likely that this type of unit will be the usual residence for a household, because it may not be fit for living quarters for more than half of the year.

Seasoned wood: Wood, used for fuel, that has been air dried so that it contains 15 to 20 percent moisture content (wet basis).

Secondary heating equipment: Space-heating equipment used less often than the main space-heating equipment.

Secondary heating fuel: Fuels used in secondary space-heating equipment.

Securitization: A proposal for issuing bonds that would be used to buy down existing power contracts or other obligations. The bonds would be repaid by designating a portion of future customer bill payments. Customer bills would be lowered, since the cost of bond payments would be less than the power contract costs that would be avoided.

Securitize: To aggregate contracts into one pool, which then offers shares for sale in the investment market. This strategy diversifies project risks from what they would be if each project were financed individually, thereby reducing the cost of financing.

Selective absorber: A solar absorber surface that has high absorbtance at wavelengths corresponding to that of the solar spectrum and low emittance in the infrared range.

Seller type: Categories of major refiners and other refiners and gas plant operators.

Semiconductor: Any material that has a limited capacity for conducting an electric current. Certain semiconductors, including silicon, gallium arsenide, copper indium diselenide, and cadmium telluride, are uniquely suited to the photovoltaic conversion process.

Separate metering: Measurement of electricity or natural gas consumption in a building using a separate meter for each of several tenants or establishments in the building.

Separative work unit (SWU): The standard measure of enrichment services. The effort expended in separating a mass F of feed of assay x_f into a mass P of product assay x_p and waste of mass W and assay x_w is expressed in terms of the number of separative work units needed, given by the expression $SWU = WV(x_w) + PV(x_p) - FV(x_f)$, where V(x) is the "value function," defined as $V(x) = (1-2x)\ 1n((1-x)/x)$.

Septic tank: A tank in which the solid matter of continuously flowing sewage is disintegrated by bacteria.

Series connection: A way of joining photovoltaic cells by connecting positive leads to negative leads; such a configuration increases the voltage.

Series resistance: Parasitic resistance to current flow in a cell due to mechanisms such as resistance from the bulk of the semiconductor material, metallic contacts, and interconnections.

Service area: The territory in which a utility system or distributor is authorized to provide service to consumers.

Service well: A well drilled, completed, or converted for the purpose of supporting production in an existing field. Wells of this class also are drilled or converted for the following specific purposes: gas injection (natural gas, propane, butane or fuel-gas); water injection; steam injection; air injection; salt water disposal; water supply for injection; observation; and injection for in-situ combustion.

Shaft mine: A mine that reaches the coal bed by means of a vertical shaft.

Shakes/shingles: Flat pieces of weatherproof material laid with others in a series of overlapping rows as covering for roofs and sometimes the sides of buildings. Shakes are similar to wood shingles, but instead of having a cut and smoothly planed surface, shakes have textured grooves and a rough or "split" appearance to give a rustic feeling.

Shallow pitting: Testing a potential mineral deposit by systematically sinking small shafts into the earth and analyzing the material recovered.

Shell storage capacity: The design capacity of a petroleum storage tank that is always greater than or equal to working storage capacity.

Short circuit: An electric current taking a shorter or different path than intended.

Short circuit current: The current flowing freely through an external circuit that has no load or resistance; the maximum current possible.

Short purchases: A single shipment of fuel or volumes of fuel purchased for delivery within 1 year. Spot purchases are often made by a user to fulfill a certain portion of energy requirements, to meet unanticipated energy needs, or to take advantage of low-fuel prices.

Short term sales: Any short-term purchase covering a time period of 2 years or less. Purchases from intrastate pipelines pursuant to Section 311(b) of the NGPA of 1978 are classified as short-term sales, regardless of the stated contract term.

Short ton: A unit of weight equal to 2,000 pounds.

Short-term debt or borrowings: Debt securities or borrowings having a maturity of less than one year.

Short-term purchase: A purchase contract under which all deliveries of materials are scheduled to be completed by the end of the first calender year following the contract-signing year. Deliveries can be made during the contract year, but deliveries are not scheduled to occur beyond the first calendar year thereafter.

Shortwall mining: A form of underground mining that involves the use of a continuous mining machine and movable roof supports to shear coal panels 150 to 200 feet wide and more than half a mile long. Although similar to longwall mining, shortwall mining is generally more flexible because of the smaller working area. Productivity is lower than with longwall mining because the coal is hauled to the mine face by shuttle cars as opposed to conveyors.

Shrinkage: The volume of natural gas that is transformed into liquid products during processing, primarily at natural gas liquids processing plants.

Shut in: Closed temporarily; wells and mines capable of production may be shut in for repair, cleaning, inaccessibility to a market, etc.

Shut-in royalty: A royalty paid by a lessee as compensation for a lessor's loss of income because the lessee has deferred production from a property that is known to be capable of producing minerals. Shut in may be caused by a lack of a ready market, by a lack of transportation facilities, or by other reasons. A shut-in royalty may or may not be recoverable out of future production.

Shutdown date: Month and year of shutdown for fuel discharge and refueling. The date should be the point at which the reactor became subcritical.

Sidetrack drilling: This is a remedial operation that results in the creation of a new section of well bore for the purpose of (1) detouring around junk, (2) redrilling lost holes, or (3) straightening key seats and crooked holes. Directional "side-track" wells do not include footage in the common bore that is reported as footage for the original well.

Siding: An exterior wall covering material made of wood, plastic (including vinyl), or metal. Siding is generally produced in the shape of boards and is applied to the outside of a building in overlapping rows.

Silicon: A semiconductor material made from silica, purified for photovoltaic applications.

Silt: Waste from Pennsylvania anthracite preparation plants, consisting of coarse rock fragments containing as much as 30 percent small-sized coal; sometimes defined as including very fine coal particles called silt. Its heat value ranges from 8 to 17 million Btu per short ton.

Silt, culm, refuse bank, or slurry dam mining: A mining operation producing coal from these sources of coal.

Single-circuit line: A transmission line with one electric circuit. For three-phase supply, a single circuit requires at least three conductors, one per phase.

Single crystal silicon: An extremely pure form of crystalline silicon produced by dipping a single crystal seed into a pool of molten silicon under high vacuum conditions and slowly withdrawing a solidifying single crystal boule (rod) of silicon. The boule is sawed into thin silicon wafers and fabricated into single-crystal photovoltaic cells.

Single crystal silicon (Czochralsky): Silicon cells with a well-ordered crystalline structure consisting of one crystal (usually obtained by means of the Czochralsky growth technique and involving ingot slicing), composing a module. Ribbon silicon is excluded.

Single purpose project: A hydroelectric project constructed only to generate electricity.

Site characterization: An onsite investigation at a known or suspected contaminated waste or release site to determine the extent and type(s) of contamination.

Site energy: The Btu value of energy at the point it enters the home, sometimes referred to as "delivered" energy. The site value of energy is used for all fuels, including electricity.

Site energy consumption: The Btu value of energy at the point it enters the home, building, or establishment, sometimes referred to as "delivered" energy.

Site-specific information DSM program assistance: A DSM (demand-side management) assistance program that provides quidance on energy efficiency and load management options tailored to a particular customer'sfacility; it often involves an on-site inspection of the customer facility to identify cost-effective DSM actions that could be taken. They include audits, engineering design calculations on information provided about the building, and technical assistance to architects and engineers who design new facilities.

Sinter: A chemical sedimentary rock deposited by precipitation from mineral waters, especially siliceous sinter and calcareous sinter.

Slope mine: A mine that reaches the coal bed by means of an inclined opening.

Slot: A physical position in a rack in a storage pool that is intended to be occupied by an intact assembly or equivalent (that is, a canister or an assembly skeleton).

Sludge: A dense, slushy, liquid-to-semifluid product that accumulates as an end result of an industrial or technological process designed to purify a substance. Industrial sludges are produced from the processing of energy-related raw materials, chemical products, water, mined ores, sewerage, and other natural and man-made products. Sludges can also form from natural processes, such as the run off produced by rain fall, and accumulate on the bottom of bogs, streams, lakes, and tidelands.

Slurry: A viscous liquid with a high solids content.

Slurry dam: A repository for the silt or culm from a preparation plant.

Small pickup truck: A pickup truck weighing under 4,500 lbs GVW.

Small power producer (SPP): Under the Public Utility Regulatory Policies Act (PURPA), a small power production facility (or small power producer) generates electricity using waste, renewable (biomass, conventional hydroelectric, wind and solar, and geothermal) energy as a primary energy source. Fossil fuels can be used, but renewable resource must provide at least 75 percent of the total energy input.

Sodium lights: A type of high intensity discharge light that has the most lumens per watt of any light source.

Sodium silicate: A grey-white powder soluble in alkali and water, insoluble in alcohol and acid. Used to fireproof textiles, in petroleum refining and corrugated paperboard manufacture, and as an egg preservative. Also referred to as liquid gas, silicate of soda, sodium metasilicate, soluble glass, and water glass.

Sodium tripolyphosphate: A white powder used for water softening and as a food additive and texturizer.

Solar constant: The average amount of solar radiation that reaches the earth's upper atmosphere on a surface perpendicular to the sun's rays; equal to 1353 Watts per square meter or 492 Btu per square foot.

Solar cooling: The use of solar thermal energy or solar electricity to power a cooling appliance. There are five basic types of solar cooling technologies: absorption cooling, which can use solar thermal energy to vaporize the refrigerant; desiccant cooling, which can use solar thermal energy to regenerate (dry) the desiccant; vapor compression cooling, which can use solar thermal energy to operate a Rankine-cycle heat engine; and evaporative coolers ("swamp" coolers), and heat-pumps and air conditioners that can by powered by solar photovoltaic systems.

Solar declination: The apparent angle of the sun north or south of the earth's equatorial plane. The earth's rotation on its axis causes a daily change in the declination.

Solar energy: The radiant energy of the sun, which can be converted into other forms of energy, such as heat or electricity.

Solar pond: A body of water that contains brackish (highly saline) water that forms layers of differing salinity (stratifies) that absorb and trap solar energy. Solar ponds can be used to provide heat for industrial or agricultural processes, building heating and cooling, and to generate electricity.

Solar power tower: A solar energy conversion system that uses a large field of independently adjustable mirrors (heliostats) to focus solar rays on a near single point atop a fixed tower (receiver). The concentrated energy may be used to directly heat the working fluid of a Rankine cycle engine or to heat an intermediary thermal storage medium (such as a molten salt).

Solar radiation: A general term for the visible and near visible (ultraviolet and near-infrared) electromagnetic radiation that is emitted by the sun. It has a spectral, or wavelength, distribution that corresponds to different energy levels; short wavelength radiation has a higher energy than long-wavelength radiation.

Solar spectrum: The total distribution of electromagnetic radiation emanating from the sun. The different regions of the solar spectrum are described by their wavelength range. The visible region extends from about 390 to 780 nanometers (a nanometer is one billionth of one meter). About 99 percent of solar radiation is contained in a wavelength region from 300 nm (ultraviolet) to 3,000 nm (near-infrared). The combined radiation in the wavelength region from 280 nm to 4,000 nm is called the broadband, or total, solar radiation.

Solar thermal collector: A device designed to receive solar radiation and convert it to thermal energy. Normally, a solar thermal collector includes a frame, glazing, and an absorber, together with appropriate insulation. The heat collected by the solar collector may be used immediately or stored for later use. Solar collectors are used for space heating; domestic hot water heating; and heating swimming pools, hot tubs, or spas.

Solar thermal collector, high temperature: A collector that generally operates at temperatures above 180 degrees Fahrenheit.

Solar thermal collector, low-temperature: A collector that generally operates at temperatures below 110 degrees Fahrenheit. Typically, it has no glazing or insulation and is made of plastic or rubber, although some are made of metal.

Solar thermal collector, medium-temperature: A collector that generally operates at temperatures of 140 degrees F to 180 degrees Fahrenheit, but can also operate at temperatures as low as 110 degrees Fahrenheit. Typically, it has one or two glazings, a metal frame, a metal absorption panel with integral flow channels or attached tubing (liquid collector) or with integral ducting (air collector) and insulation on the sides and back of the panel.

Solar thermal collector, special: An evacuated tube collector or a concentrating (focusing) collector. Special collectors operate in the temperature range from just above ambient temperature (low concentration for pool heating) to several hundred degrees Fahrenheit (high concentration for air conditioning and specialized industrial processes).

Solar thermal panels: A system that actively concentrates thermal energy from the sun by means of solar collector panels. The panels typically consist of fat, sun-oriented boxes with transparent covers, containing water tubes of air baffles under a blackened heat absorbent panel. The energy is usually used for space heating, for water heating, and for heating swimming pools.

Solar thermal parabolic dishes: A solar thermal technology that uses a modular mirror system that approximates a parabola and incorporates two-axis tracking to focus the sunlight onto receivers located at the focal point of each dish. The mirror system typically is made from a number of mirror facets, either glass or polymer mirror, or can consist of a single stretched membrane using a polymer mirror. The concentrated sunlight may be used directly by a Stirling, Rankine, or Brayton cycle heat engine at the focal point of the receiver or to heat a working fluid that is piped to a central engine. The primary applications include remote electrification, water pumping, and grid-connected generation.

Source material: The term "source material" means (1) uranium, thorium, or any other material that is determined by the Atomic Energy Commission pursuant to the provisions of section 61 of the Atomic Energy Act of 1954, as amended, to be source material; or (2) ores containing one or more of the foregoing materials, in such concentration as the Commission may by regulation determine from time to time.

Space heating: The use of energy to generate heat for warmth in housing units using space-heating equipment. The equipment could be the main space-heating equipment or secondary space-heating equipment. It does not include the use of energy to operate appliances (such as lights, televisions, and refrigerators) that give off heat as a byproduct.

Special collector: An evacuated tube collector or a concentrating (focusing) collector. Special collectors operate in the temperature range from just above ambient temperature (low concentration for pool heating) to several hundred degrees Fahrenheit (high concentration for air conditioning and specialized industrial processes).

Special contract rate schedule: An electric rate schedule for an electric service agreement between a utility and another party in addition to, or independent of, any standard rate schedule.

Special naphthas: All finished products within the naphtha boiling range that are used as paint thinners, cleaners, or solvents. These products are refined to a specified flash point. Special naphthas include all commercial hexane and cleaning solvents conforming to ASTM Specification D1836 and D484, respectively. Naphthas to be blended or marketed as motor gasoline or aviation gasoline or that are to be used as petrochemical and synthetic natural gas (SNG) feedstocks are excluded.

Special nuclear material: The term "special nuclear material" means (1) plutonium, uranium enriched in the isotope 233 or in the isotope 235, and any other material that the Atomic Energy Commission, pursuant to the provisions of section 51 of the Atomic Energy Act of 1954, as amended, determines to be special nuclear material, but does not include source material; or (2) any material artificially enriched by any of the foregoing, but does not include source material.

Special purpose rate schedule: An electric rate schedule limited in its application to some particular purpose or process within one, or more than one, type of industry or business.

Specular reflectors: Specular reflectors have mirrorlike characteristics (the word "specular" is derived from the Greek word meaning mirror). The most common materials used for ballasts, the devices that turn on and operate Fluorescent tubes, are aluminum and silver. Silver has the highest reflectivity; aluminum has the lowest cost. The materials and shape of the reflector are designed to reduce absorption of light within the fixture while delivering light in the desired angular pattern. Adding (or retrofitting) specular reflectors to an existing light fixture is frequently implemented as a conservation measure.

Speculative resources (coal): Undiscovered coal in beds that may occur either in known types of deposits in a favorable geologic setting where no discoveries have been made, or in deposits that remain to be recognized. Exploration that confirms their existence and better defines their quantity and quality would permit their reclassification as identified resources.

Speculative resources (uranium): Uranium in addition to Estimated Additional Resources (EAR) that is thought to exist, mostly on the basis of indirect evidence and geological extrapolations, in deposits discoverable with existing exploration techniques. The locations of deposits in this category can generally be specified only as being somewhere within given regions or geological trends. The existence and size of such deposits are speculative. The estimates in this category are less reliable than estimates of EAR. SR corresponds to DOE's Possible Potential Resources plus Speculative Potential Resources categories.

Spent fuel: Irradiated fuel that is permanently discharged from a reactor. Except for possible reprocessing, this fuel must eventually be removed from its temporary storage location at the reactor site and placed in a permanent repository. Spent fuel is typically measured either in metric tons of heavy metal (*i.e.*, only the heavy metal content of the spent fuel is considered) or in metric tons of initial heavy metal (essentially, the initial mass of the fuel before irradiation). The difference between these two quantities is the weight of the fission products.

Spent fuel disassembly hardware: The skeleton of a fuel assembly after the fuel rods have been removed. Generally, SFD hardware for PWR assemblies includes guide tubes; instrument tubes, top and bottom nozzles; grid spacers; hold-down springs; and attachment components, such as nuts and locking caps. For BWR fuel assemblies, SFD hardware includes the top and bottom tie plates, compression springs for individual fuel rods, grid spacers, and water rods.

Spent liquor: The liquid residue left after an industrial process; can be a component of waste materials used as fuel.

Spillway: A passage for surplus water to flow over or around a dam.

Spinning reserve: That reserve generating capacity running at a zero load and synchronized to the electric system.

Split system: When applied to electric air-conditioning equipment, it means a two-part system–an indoor unit and an outdoor unit. The indoor unit is an evaporator coil mounted in the indoor circulating air system, and the outdoor unit is an air-cooled condensing unit containing an electric motor-driven compressor, a condenser fan, and a fan motor.

Split tails: Use of one tails assay for transaction of enrichment services and a different tails assay for operation of the enrichment plant. This mode of operations typically increases the use of uranium, which is relatively inexpensive, while decreasing the use of separative work, which is expensive.

Spontaneous combustion, or self-heating, of coal: A naturally occurring process caused by the oxidation of coal. It is most common in low-rank coals and is a potential problem in storing and transporting coal for extended periods. Factors involved in spontaneous combustion include the size of the coal (the smaller sizes are more susceptible), the moisture content, and the sulfur content. Heat buildup in stored coal can degrade the quality of coal, cause it to smolder, and lead to a fire.

Spot market (natural gas): A market in which natural gas is bought and sold for immediate or very near-term delivery, usually for a period of 30 days or less. The transaction does not imply a continuing arrangement between the buyer and the seller. A spot market is more likely to develop at a location with numerous pipeline interconnections, thus allowing for a large number of buyers and sellers. The Henry Hub in southern Louisiana is the best known spot market for natural gas.

Spot market (uranium): Buying and selling of uranium for immediate or very near-term delivery. It typically involves transactions for delivery of up to 500,000 pounds U_3O_8 within a year of contract execution.

Spot price: The price for a one-time open market transaction for immediate delivery of a specific quantity of product at a specific location where the commodity is purchased "on the spot" at current market rates.

Spot purchases: A single shipment of fuel or volumes of fuel purchased for delivery within 1 year. Spot purchases are often made by a user to fulfill a certain portion of energy requirements, to meet unanticipated energy needs, or to take advantage of low-fuel prices.

Stabilization lagoon: A shallow artificial pond used for the treatment of wastewater. Treatment includes removal of solid material through sedimentation, the decomposition of organic material by bacteria, and the removal of nutrients by algae.

Stability: The property of a system or element by virture of which its output will ultimately attain a steady state. The amount of power that can be transferred from one machine to another following a disturbance. The stability of a power system is its ability to develop restoring forces equal to or greater than the disturbing forces so as to maintain a state of equilibrium.

Stack: A tall, vertical structure containing one or more flues used to discharge products of combustion to the atmosphere.

Stand-alone generator: A power source / generator that operates independently of or is not connected to an electric transmission and distribution network; used to meet a load(s) physically close to the generator.

Standard contract: The agreement between the Department of Energy (DOE) and the owners or generators of spent nuclear fuel and high-level radioactive waste, under which DOE will make available nuclear waste disposal services to those owners and generators.

Standard fluorescent: A light bulb made of a glass tube coated on the inside with fluorescent material, which produces light by passing electricity through mercury vapor causing the fluorescent coating to glow or fluoresce.

Standard Industrial Classification (SIC): Replaced with North American Industry Classification System.

Standby charge: A charge for the potential use of a utility service, usually done by an agreement with another electric utility service. These services include system backup support and other running and quick-start capabilities.

Standby electricity generation: Involves use of generators during times of high demand on utilities to avoid extra "peak-demand" charges.

Standby facility: A facility that supports a utility system and is generally running under no-load. It is available to replace or supplement a facility normally in service.

Standby heat loss: A term used to describe heat energy lost from a water heater tank.

Standby service: Support service that is available as needed to supplement a customer, a utility system, or another utility if a schedule or an agreement authorizes the transaction. The service is not regularly used.

Startup test phase of nuclear power plant: A nuclear power plant that has been licensed by the Nuclear Regulatory Commission to operate but is still in the initial testing phase, during which the production of electricity may not be continuous. In general, when the electric utility is satisfied with the plant's performance, it formally accepts the plant from the manufacturer and places it in commercial operation status. A request is then submitted to the appropriate utility rate commission to include the power plant in the rate base calculation.

Startup/flame stabilization fuel: Any fuel used to initiate or sustain combustion or used to stabilize the height of flames once combustion is underway.

State: One of the 50 States, including adjacent outer continental shelf areas, or the District of Columbia.

State permit/license/mine number: Code assigned to a mining operation by the state in which the operation is located.

State severance taxes: Any severance, production, or similar tax, fee, or other levy imposed on the production of crude oil, natural gas, or coal by any State, local government acting under authority of State law, or by an Indian tribe recognized as eligible for services by the Secretary of the Interior.

Station (electric): A plant containing prime movers, electric generators, and auxiliary equipment for converting mechanical, chemical, and/or nuclear energy into electric energy.

Station use: Energy that is used to operate an electric generating plant. It includes energy consumed for plant lighting, power, and auxiliary facilities, regardless of whether the energy is produced at the plant or comes from another source.

Steam: Water in vapor form; used as the working fluid in steam turbines and heating systems.

Steam (purchased): Steam, purchased for use by a refinery, that was not generated from within the refinery complex.

Steam boiler: A type of furnace in which fuel is burned and the heat is used to produce steam.

Steam coal: All nonmetallurgical coal.

Steam electric power plant (conventional): A plant in which the prime mover is a steam turbine. The steam used to drive the turbine is produced in a boiler where fossil fuels are burned.

Steam expenses: The cost of labor, materials, fuel, and other expenses incurred in production of steam for electric generation.

Steam from other sources: Steam purchased, transferred from another department of the utility, or acquired from others under a joint-facility operating agreement.

Steam or hot water radiators or baseboards: A distribution system where steam or hot water circulates through cast-iron radiators or baseboards. Some other types of equipment in the building may be used to produce the steam or hot water or it may enter the building already heated as part of a district hot water system. Hot water does not include domestic hot water used for cooking and cleaning.

Steam or hot-water system: Either of two types of a central space-heating system that supplies steam or hot water to radiators, convectors, or pipes. The more common type supplies either steam or hot water to conventional radiators, baseboard radiators, convectors, heating pipes embedded in the walls or ceilings, or heating coils or equipment that are part of a combined heating/ventilating or heating/air-conditioning system.The other type supplies radiant heat through pipes that carry hot water and are held in a concrete slab floor.

Steam transferred-credit: The expenses of producing steam are charged to others or to other utility departments under a joint operating arrangement.

Steam turbine: A device that converts high-pressure steam, produced in a boiler, into mechanical energy that can then be used to produce electricity by forcing blades in a cylinder to rotate and turn a generator shaft.

Still gas (refinery gas): Any form or mixture of gases produced in refineries by distillation, cracking, reforming, and other processes. The principal constituents are methane, ethane, ethylene, normal butane, butylene, propane, propylene, etc. Still gas is used as a refinery fuel and a petrochemical feedstock. The conversion factor is 6 million Btus per fuel oil equivalent barrel.

Stock change: The difference between stocks at the beginning of the reporting period and stocks at the end of the reporting period.

Stocks: Inventories of fuel stored for future use.

Storage additions: Volumes of gas injected or otherwise added to underground natural gas reservoirs or liquefied natural gas storage.

Storage agreement: Any contractual arrangement between the responding company and a storage operator under which gas was stored for, or gas storage service was provided to, the responding company by the storage operator, irrespective of any responding company ownership interest in either the storage facilities or stored gas.

Storage capacity: The amount of energy an energy storage device or system can store.

Storage hydroelectric plant: A hydroelectric plant with reservoir storage capacity for power use.

Storage site: Spent nuclear fuel storage pool or dry cask storage facility, usually located at the reactor site, as licensed by (or proposed to be licensed by) the Nuclear Regulatory Commission (NRC).

Storage withdrawals: Total volume of gas withdrawn from underground storage or from liquefied natural gas storage over a specified amount of time.

Storm door: A second door installed outside or inside a prime door creating an insulating air space. Included are sliding glass doors made of double glass or of insulating glass such as thermopane and sliding glass doors with glass or Plexiglas placed on either the outside or inside of the door to create an insulating air space. Not included are doors or sliding glass doors covered by plastic sheets or doors with storm window covering on just the glass portion of the door.

Storm or multiple glazing: A building shell conservation feature consisting of storm windows, storm doors, or double- or triple-paned glass that are placed on the exterior of the building to reduce the rate of heat loss.

Storm window: A window or glazing material placed outside or inside a window creating an insulating air space. Plastic material over windows is counted a storm window if the same plastic material can be used year after year or if the plastic is left in place year-round and is in good condition (no holes or tears). If the plastic material must be put up new each year, it is not counted as a storm window. It is counted as "plastic coverings." Glass or Plexiglas placed over windows on either the interior or exterior side is counted as storm windows.

Stranded benefits: Benefits associated with regulated retail electric service which may be at risk under open market retail competition. Examples include conservation programs, fuel diversity, reliability of supply, and tax revenues based on utility revenues.

Stranded costs: Costs incurred by a utility which may not be recoverable under market-based retail competition. Examples include undepreciated generating facilities, deferred costs, and long-term contract costs.

Strategic Petroleum Reserve (SPR): Petroleum stocks maintained by the Federal Government for use during periods of major supply interruption.

Stratigraphic test well: A geologically directed drilling effort to obtain information pertaining to a specific geological condition that might lead toward the discovery of an accumulation of hydrocarbons. Such wells are customarily drilled without the intention of being completed for hydrocarbon production. This classification also includes tests identified as core tests and all types of expendable holes related to hydrocarbon exploration.

Stratosphere: The region of the upper atmosphere extending from the tropopause (8 to 15 kilometers altitude) to about 50 kilometers. Its thermal structure, which is determined by its radiation balance, is generally very stable with low humidity.

Stream-flow: The rate at which water passes a given point in a stream, usually expressed in cubic feet per second.

Strip mine: An open cut in which the overburden is removed from a coal bed prior to the removal of coal.

Strip mining (surface): A method used on flat terrain to recover coal by mining long strips successively; the material excavated from the strip being mined is deposited in the strip previously mined.

Strip or stripping ratio: The amount of overburden that must be removed to gain access to a unit amount of coal. A stripping ratio may be expressed as (1) thickness of overburden to thickness of coal, (2) volume of overburden to volume coal, (3) weight of overburden to weight of coal, or (4) cubic yards of overburden to tons of coal. A stripping ratio commonly is used to express the maximum thickness, volume, or weight of overburden that can be profitably removed to obtain a unit amount of coal.

Stripper well: An oil or gas well that produces at relatively low rates. For oil, stripper production is usually defined as production rates of between 5 and 15 barrels of oil per day. Stripper gas production would generally be anything less than 60 thousand cubic feet per day.

Styrene: A colorless, toxic liquid with a strong aromatic aroma. Insoluble in water, soluble in alcohol and ether; polymerizes rapidly; can become explosive. Used to make polymers and copolymers, polystyrene plastics, and rubber.

Subbituminous coal: A coal whose properties range from those of lignite to those of bituminous coal and used primarily as fuel for steam-electric power generation. It may be dull, dark brown to black, soft and crumbly, at the lower end of the range, to bright, jet black, hard, and relatively strong, at the upper end. Subbituminous coal contains 20 to 30 percent inherent moisture by weight. The heat content of subbituminous coal ranges from 17 to 24 million Btu per ton on a moist, mineral-matter-free basis. The heat content of subbituminous coal consumed in the United States averages 17 to 18 million Btu per ton, on the as-received basis (*i.e.*, containing both inherent moisture and mineral matter).

Subcompact/compact passenger car: A passenger car containing less than 109 cubic feet of interior passenger and luggage volume.

Subdivision: A prescribed portion of a given State or other geographical region.

Submetered data: End-use consumption data obtained for individual appliances when a recording device has been attached to the appliance to measure the amount of energy consumed by the appliance.

Subsidiary: An entity directly or indirectly controlled by a parent company which owns 50% or more of its voting stock.

Substation: Facility equipment that switches, changes, or regulates electric voltage.

Subtransmission:: A set of transmission lines of voltages between transmission voltages and distribution voltages. Generally, lines in the voltage range of 69 kV to 138 kV.

Sulfur: A yellowish nonmetallic element, sometimes known as "brimstone." It is present at various levels of concentration in many fossil fuels whose combustion releases sulfur compounds that are considered harmful to the environment. Some of the most commonly used fossil fuels are categorized according to their sulfur content, with lower sulfur fuels usually selling at a higher price. *Note*: No. 2 Distillate fuel is currently reported as having either a 0.05 percent or lower sulfur level for on-highway vehicle use or a greater than 0.05 percent sulfur level for off-highway use, home heating oil, and commercial and industrial uses. Residual fuel, regardless of use, is classified as having either no more than 1 percent sulfur or greater than 1 percent sulfur. Coal is also classified as being low-sulfur at concentrations of 1 percent or less or high-sulfur at concentrations greater than 1 percent.

Sulfur dioxide (SO_2): A toxic, irritating, colorless gas soluble in water, alcohol, and ether. Used as a chemical intermediate, in paper pulping and ore refining, and as a solvent.

Sulfur hexafluoride (SF_6): A colorless gas soluble in alcohol and ether, and slightly less soluble in water. It is used as a dielectric in electronics. It possesses the highest 100-year Global Warming Potential of any gas (23,900).

Sulfur oxides (SO_x): Compounds containing sulfur and oxygen, such as sulfur dioxide (SO_2) and sulfur trioxide (SO_3).

Summer and winter peaking: Having the annual peak demand reached both during the summer months (May through October) and during the winter months (November through April).

Sunk cost: Part of the capital costs actually incurred up to the date of reserves estimation minus depreciation and amortization expenses. Items such as exploration costs, land acquisition costs, and costs of financing can be included.

Superconductivity: The abrupt and large increase in electrical conductivity exhibited by some metals as the temperature approaches absolute zero.

Supplemental gas: Any gaseous substance introduced into or commingled with natural gas that increased the volume available for disposition. Such substances include, but are not limited to, propane-air, refinery gas, coke-oven gas, still gas, manufactured gas, biomass gas, or air or inerts added for Btu stabilization.

Supplemental gaseous fuels supplies: Synthetic natural gas, propane-air, coke oven gas, refinery gas, biomass gas, air injected for Btu stabilization, and manufactured gas commingled and distributed with natural gas.

Supply, petroleum: A set of categories used to account for how crude oil and petroleum products are transferred, distributed, or placed into the supply stream. The categories include field production, refinery production, and imports. Net receipts are also included on a Petroleum Administration for Defense (PAD) District basis to account for shipments of crude oil and petroleum products across districts.

Supply source: May be a single completion, a single well, a single field with one or more reservoirs, several fields under a single gas-purchase contract, miscellaneous fields, a processing plant, or a field area; provided, however, that the geographic area encompassed by a single supply source may not be larger than the state in which the reserves are reported.

Support equipment and facilities: These include, but are not limited to, seismic equipment, drilling equipment, construction and grading equipment, vehicles, repair shops, warehouses, supply points, camps, and division, district, or field offices.

Supporting structure: The main supporting unit (usually a pole or tower) for transmission line conductors, insulators, and other auxiliary line equipment.

Surface drilling expenses (uranium): These include drilling, drilling roads, site preparation, geological and other technical support, sampling, and drill-hole logging costs.

Surface mine: A coal-producing mine that is usually within a few hundred feet of the surface. Earth above or around the coal (overburden) is removed to expose the coalbed, which is then mined with surface excavation equipment, such as draglines, power shovels, bulldozers, loaders, and augers. It may also be known as an area, contour, open-pit, strip, or auger mine.

Surface mining equipment:

1. An **Auger machine** is a large, horizontal drill, generally 3 feet or more in diameter and up to about 100 feet long. It can remove coal at a rate of more than 25 tons per minute.
2. A **bucket-wheel** excavator is a continuous digging machine equipped with a broom on which is mounted a rotating wheel with buckets along its edge. The buckets scoop up material, then empty onto a conveyor leading to a spoil bank. It is best suited for removing overburden that does not require blasting. This excavator is not widely used in the United States.
3. A **bulldozer** is a tractor with a movable steel blade mounted on the front. It can be used to remove overburden that needs little or no blasting.
4. A **carryall scraper** (or **pan scraper**) is a self-loading machine, usually self-propelled, with a scraper-like retractable bottom. It is used to excavate and haul overburden.
5. A **continuous surface miner**, used in some lignite mines, is equipped with crawlers, a rotating cutting head, and a conveyor. It travels over the bed, excavating a swath up to 13 feet wide and 2 feet deep.
6. A **dragline excavator** removes overburden to expose the coal by means of a scoop bucket that is suspended from a long boom. The dragline digs by pulling the bucket toward the machine by means of a wire rope.
7. A **walking dragline** is equipped with large outrigger platforms, or walking beams, instead of crawler tracks. It "walks" by the alternate movement of the walking beams.
8. A **drilling rig** is used to determine the amount and type of overburden overlying a coal deposit and the extent of the deposit, tc delineate major geologic features, and to drill holes for explosives to fragment the overburden for easier removal.
9. A **front-end loader** is a tractor with a digging bucket mounted and operated on the front. It is often used to remove overburden in contour mining and to load coal.
10. An **hydraulic shovel** excavates and loads by means of a bucket attached to a rigid arm that is hinged to a broom.
11. A **power shovel** removes overburden and loads coal by means of a digging bucket mounted at the end of an arm suspended from a broom. The shovel digs by pushing the bucket forward and upward. It does not dig below the level at which it stands.
12. A **thin-seam miner** resembles an auger machine but has a drum-type cutting head that cuts a rectangular cross section.

Surface mining methods:

1. **Auger mining** recovers coal through the use of a large-diameter drill driven into a coalbed in the side of a surface mine pit. It usually follows contour surface mining, particularly when the overburden is too costly to excavate.
2. **Area mining** is practiced on relatively flat or gently rolling terrain and recovers coal by mining long strips successively; the material excavated from the strip being mined is deposited in the strip pit previously mined.

3. **Contour mining** is practiced when the coal is mined on hillsides. The mining follows the contour of the hillside until the overburden becomes uneconomical to remove. This method creates a shelf, or bench, on the hillside. Several variations of contour mining have been developed to control environmental problems. These methods include slope reduction (overburden is spread so that the angle of the slope on the hillside is reduced), head-of-hollow fill (overburden is placed in narrow V-shaped valleys to control erosion), and block-cut (overburden from current mining is backfilled into a previously mined cut).
4. **Explosives casting** is a technique designed to blast up to 65 percent of the overburden into the mine pit for easier removal. It differs from conventional overburden blasting, which only fractures the overburden before it is removed by excavating equipment.
5. **Mountaintop mining**, sometimes considered a variation of contour mining, refers to the mining of a coalbed that underlies the top of a mountain. The overburden, which is the mountaintop, is completely removed so that all of the coal can be recovered. The overburden material is later replaced in the mined-out area. This method leaves large plateaus of level land.
6. **Open-pit** coal mining is essentially a combination of contour and area mining methods and is used to mine thick, steeply inclined coalbeds. The overburden is removed by power shovels and trucks.

Surface rights: Fee ownership in surface areas of land. Also used to describe a lessee's right to use as much of the surface of the land as may be reasonably necessary for the conduct of operations under the lease.

Surplus energy: Energy generated that is beyond the immediate needs of the producing system. This energy may be supplied by spinning reserve and sold on an interruptible basis.

Suspended rates: New rates that have been accepted for review by a utility commission. When these rates are suspended, they do not go into effect for a designated period of time. Charges under the new rate may be refunded after the resolution of the rate proceeding.

Swamp coolers (evaporative coolers): Air-conditioning equipment that removes heat by evaporating water. Evaporative cooling techniques are most commonly found in warm, dry climates such as in the Southwest, although they are found throughout the country. They usually work by spraying cool water into the air ducts, cooling the air as the spray evaporates.

Switching station: Facility equipment used to tie together two or more electric circuits through switches. The switches are selectively arranged to permit a circuit to be disconnected or to change the electric connection between the circuits.

Synthetic coal: Coal that has been processed by a coal synfuel plant; and coal-based fuels such as briquettes, pellets, or extrusions, which are formed by binding materials and processes that recycle material.

Synthetic natural gas (SNG): (Also referred to as substitute natural gas) A manufactured product, chemically similar in most respects to natural gas, resulting from the conversion or reforming of petroleum hydrocarbons that may easily be substituted for or interchanged with pipeline-quality natural gas.

System (gas): An interconnected network of pipes, valves, meters, storage facilities, and auxiliary equipment used in the transportation, storage, and/or distribution of natural gas or commingled natural and supplemental gas.

System (electric): Physically connected generation, transmission, and distribution facilities operated as an integrated unit under one central management or operating supervision.

System interconnection: A physical connection between two electric systems that permits the transfer of electric energy in either direction.

Tailgate: The outlet of a natural gas processing plant where dry residue gas is delivered or re-delivered for sale or transportation.

Tailings: The remaining portion of a metal-bearing ore consisting of finely ground rock and process liquid after some or all of the metal, such as uranium, has been extracted.

Tall oil: The oily mixture of rosin acids, fatty acids, and other materials obtained by acid treatment of the alkaline liquors from the digesting (pulping) of pine wood.

Tangible development costs: Costs incurred during the development stage for access, mineral-handling, and support facilities having a physical nature. In mining, such costs would include tracks, lighting equipment, ventilation equipment, other equipment installed in the mine to facilitate the extraction of minerals, and supporting facilities for housing and care of work forces. In the oil and gas industry, tangible development costs would include well equipment (such as casing, tubing, pumping equipment, and well heads), as well as field storage tanks and gathering systems.

Tank farm: An installation used by trunk and gathering pipeline companies, crude oil producers, and terminal operators (except refineries) to store crude oil.

Tanker and barge: Vessels that transport crude oil or petroleum products.

Tar sands: Naturally occurring bitumen-impregnated sands that yield mixtures of liquid hydrocarbon and that require further processing other than mechanical blending before becoming finished petroleum products.

Tariff: A published volume of rate schedules and general terms and conditions under which a product or service will be supplied.

Tax-cost: A deduction (allowance) under U.S. Federal income taxation normally calculated under a formula whereby the adjusted basis of the mineral property is multiplied by a fraction, the numerator of which is the number of units of minerals sold during the tax year and the denominator of which is the estimated number of units of unextracted minerals remaining at the end of the tax year plus the number of units of minerals sold during the tax year.

Temperature coefficient (of a solar photovoltaic cell): The amount that the voltage, current, and/or power output of a solar cell changes due to a change in the cell temperature.

Temporarily discharged fuel: Fuel that was irradiated in the previous fuel cycle (cycle N) and not in the following fuel cycle (cycle N+1) and that will be irradiated in a subsequent fuel cycle.

Tennessee Valley Authority (TVA): A federal agency established in 1933 to develop the Tennessee river valley region of the southeastern U.S.

Terawatthour: One trillion watthours.

Term agreement: Any written or unwritten agreement between two parties in which one party agrees to supply a commodity on a continuing basis to a second party for a price or for other considerations.

Tertiary amyl methyl ether–$(CH_3)_2(C_2H_5)COCH_3$: An oxygenate blend stock formed by the catalytic etherification of isoamylene with methanol.

Tertiary butyl alcohol–$(CH_3)_3COH$: An alcohol primarily used as a chemical feedstock or a solvent or feedstock, for isobutylene production for MTBE (methyl tertiary butyl ether) and produced as a co-product of propylene oxide production or by direct hydration of isobutylene.

Test well contribution: A payment made to the owner of an adjacent or nearby tract who has drilled an exploratory well on that tract in exchange for information obtained from the drilling effort.

Therm: One hundred thousand (100,000) Btu.

Thermal: A term used to identify a type of electric generating station, capacity, capability, or output in which the source of energy for the prime mover is heat.

Thermal cracking: A refining process in which heat and pressure are used to break down, rearrange, or combine hydrocarbon molecules. Thermal-cracking includes gas oil, visbreaking, fluid coking, delayed coking, and other thermal cracking processes (*e.g.*, flexicoking).

Thermal efficiency: A measure of the efficiency of converting a fuel to energy and useful work; useful work and energy output divided by higher heating value of input fuel times 100 (for percent).

Thermal energy storage: The storage of heat energy during utility off-peak times at night, for use during the next day without incurring daytime peak electric rates.

Thermal limit: The maximum amount of power a transmission line can carry without suffering heat-related deterioration of line equipment, particularly conductors.

Thermal resistance (R-Value): This designates the resistance of a material to heat conduction. The greater the R-value the larger the number.

Thermal storage: Storage of heat or heat sinks (coldness) for later heating or cooling. Examples are the storage of solar energy for night heating; the storage of summer heat for winter use; the storage of winter ice for space cooling in the summer; and the storage of electrically-generated heat or coolness when electricity is less expensive, to be released in order to avoid using electricity when the rates are higher. There are four basic types of thermal storage systems: ice storage; water storage; storage in rock, soil or other types of solid thermal mass; and storage in other materials, such as glycol (antifreeze).

Thermocouple: A device consisting of two dissimilar conductors with their ends connected together. When the two junctions are at different temperatures, a small voltage is generated.

Thermodynamics: A study of the transformation of energy from one form to another, and its practical application.

Thermophotovoltaic cell: A device where sunlight concentrated onto a absorber heats it to a high temperature, and the thermal radiation emitted by the absorber is used as the energy source for a photovoltaic cell that is designed to maximize conversion efficiency at the wavelength of the thermal radiation.

Thermosiphon system: A solar collector system for water heating in which circulation of the collection fluid through the storage loop is provided solely by the temperature and density difference between the hot and cold fluids.

Thermostat: A device that adjusts the amount of heating and cooling produced and/or distributed by automatically responding to the temperature in the environment.

Third-party transactions: Third-party transactions are arms-length transactions between nonaffiliated firms. Producing country-to-company transactions are not considered to be third-party transactions.

Third-party DSM program sponsor: An energy service company (ESCO) which promotes a program sponsored by a manufacturer or distributor of energy products such as lighting or refrigeration whose goal is to encourage consumers to improve energy efficiency, reduce energy costs, change the time of usage, or promote the use of a different energy source.

Thorium: An element that is a byproduct of the decay of uranium.

Three-phase power: Power generated and transmitted from generator to load on three conductors.

Tidewater piers and coastal ports (method of transportation to consumers): Shipments of coal moved to tidewater piers and coastal ports for further shipments to consumers via coastal water or ocean.

Tie line: A transmission line connecting two or more power systems.

Time clocks or timed switches: Time clocks are automatic controls, which turn lights off and on at predetermined times.

Time-of-day lock-out or limit: A special electric rate feature under which electricity usage is prohibited or restricted to a reduced level at fixed times of the day in return for a reduction in the price per kilowatthour.

Time-of-day pricing: A special electric rate feature under which the price per kilowatthour depends on the time of day.

Time-of-day rate: The rate charged by an electric utility for service to various classes of customers. The rate reflects the different costs of providing the service at different times of the day.

Timing differences: Differences between the periods in which transactions affect taxable income and the periods in which they enter into the determination of pretax accounting income. Timing differences originate in one period and reverse or "turn around" in one or more subsequent periods. Some timing differences reduce income taxes that would otherwise be payable currently; others increase income taxes that would otherwise be payable currently.

Tinted or reflective glass or shading films: Types of glass or a shading film applied to glass that, when installed on the exterior of a building, reduces the rates of solar penetration into the building. Includes Low E Glass.

Tipping fee: Price charged to deliver municipal solid waste to a landfill, waste-to-energy facility, or recycling facility.

Tipple: A central facility used in loading coal for transportation by rail or truck.

Tolling arrangement: Contract arrangement under which a raw material or intermediate product stream from one company is delivered to the production facility of another company in exchange for the equivalent volume of finished products and payment of a processing fee.

Toluene ($C_6H_5CH_3$): Colorless liquid of the aromatic group of petroleum hydrocarbons, made by the catalytic reforming of petroleum naphthas containing methyl cyclohexane. A high-octane gasoline-blending agent, solvent, and chemical intermediate, and a base for TNT (explosive).

Ton mile: The product of the distance that freight is hauled, measured in miles, and the weight of the cargo being hauled, measured in tons. Thus, moving one ton for one mile generates one ton mile.

Total discoveries: The sum of extensions, new reservoir discoveries in old fields, and new field discoveries, that occurred during the report year.

Total gas in storage: The sum of base gas and working gas.

Total liquid hydrocarbon reserves: The sum of crude oil and natural gas liquids reserves volumes.

Total operated basis: The total reserves or production associated with the wells operated by an individual operator. This is also commonly known as the "gross operated" or "8/8ths" basis.

Transfer capability: The overall capacity of interregional or international power lines, together with the associated electrical system facilities, to transfer power and energy from one electrical system to another.

Transfer price: The monetary value assigned to products, services, or rights conveyed or exchanged between related parties, including those occurring between units of a consolidated entity.

Transformer: An electrical device for changing the voltage of alternating current.

Transmission and distribution loss: Electric energy lost due to the transmission and distribution of electricity. Much of the loss is thermal in nature.

Transmission (electric) (verb): The movement or transfer of electric energy over an interconnected group of lines and associated equipment between points of supply and points at which it is transformed for delivery to consumers or is delivered to other electric systems. Transmission is considered to end when the energy is transformed for distribution to the consumer.

Transmission circuit: A conductor used to transport electricity from generating stations to load.

Transmission line: A set of conductors, insulators, supporting structures, and associated equipment used to move large quantities of power at high voltage, usually over long distances between a generating or receiving point and major substations or delivery points.

Transmission network: A system of transmission or distribution lines so cross-connected and operated as to permit multiple power supply to any principal point.

Transmission system (electric): An interconnected group of electric transmission lines and associated equipment for moving or transferring electric energy in bulk between points of supply and points at which it is transformed for delivery over the distribution system lines to consumers or is delivered to other electric systems.

Transmission type (engine): The transmission is the part of a vehicle that transmits motive force from the engine to the wheels, usually by means of gears for different speeds using either a hydraulic "torque-converter" (automatic) or clutch assembly (manual). On front-wheel drive cars, the transmission is often called a "transaxle." Fuel efficiency is usually higher with manual rather than automatic transmissions, although modern, computer-controlled automatic transmissions can be efficient.

Transmitting utility: A regulated entity which owns and may construct and maintain wires used to transmit wholesale power. It may or may not handle the power dispatch and coordination functions. It is regulated to provide non-discriminatory connections, comparable service, and cost recovery. According to the Energy Policy Act of 1992, it includes any electric utility, qualifying co-generation facility, qualifying small power production facility, or Federal power marketing agency which owns or operates electric power transmission facilities which are used for the sale of electric energy at wholesale.

Transport: Movement of natural, synthetic, and/or supplemental gas between points beyond the immediate vicinity of the field or plant from which produced except (1) for movements through well or field lines to a central point for delivery to a pipeline or processing plant within the same state or (2) movements from a citygate point of receipt to consumers through distribution mains.

Transportation agreement: Any contractual agreement for the transportation of natural and/or supplemental gas between points for a fee.

Transportation sector: An energy-consuming sector that consists of all vehicles whose primary purpose is transporting people and/or goods from one physical location to another. Included are automobiles; trucks; buses; motorcycles; trains, subways, and other rail vehicles; aircraft; and ships, barges, and other waterborne vehicles. Vehicles whose primary purpose is not transportation (*e.g.*, construction cranes and bulldozers, farming vehicles, and warehouse tractors and forklifts) are classified in the sector of their primary use.

Transported gas: Natural gas physically delivered to a building by a local utility, but not purchased from that utility. A separate transaction is made to purchase the volume of gas, and the utility is paid for the use of its pipeline to deliver the gas. Also called "Direct-Purchase Gas," "Spot Market Gas," "Spot Gas," "Gas for the Account of Others", and "Self-Help Gas."

Transporter: The party or parties, other than buyer or seller, owning the facilities by which gas or LNG is physically transferred between buyer and seller.

Transshipment: A method of ocean transportation whereby ships off-load their oil cargo to a deepwater terminal, floating storage facility, temporary storage, or to one or more smaller tankers from which or in which the oil is then transported to a market destination.

Treating plant: A plant designed primarily to remove undesirable impurities from natural gas to render the gas marketable.

Trillion Btu: Equivalent to 1,000,000,000,000 or 10 to the 12th power Btu.

Troposphere: The inner layer of the atmosphere below about 15 kilometers, within which there is normally a steady decrease of temperature with increasing altitude. Nearly all clouds form and weather conditions manifest themselves within this region. Its thermal structure is caused primarily by the heating of the earth's surface by solar radiation, followed by heat transfer through turbulent mixing and convection.

Trough: High-temperature (180+) concentrator with one axis-tracking.

Trunk line: A main pipeline.

Turbine: A machine for generating rotary mechanical power from the energy of a stream of fluid (such as water, steam, or hot gas). Turbines convert the kinetic energy of fluids to mechanical energy through the principles of impulse and reaction, or a mixture of the two.

Type of drive (vehicle): Refers to which wheels the engine power is delivered to, the so-called "drive wheels." Rear-wheel drive has drive wheels on the rear of the vehicle. Front-wheel drive, a newer technology, has drive wheels on the front of the vehicle. Four-wheel drive uses all four wheels as drive wheels and is found mostly on Jeep-like vehicles and trucks, though it is becoming increasingly more common on station wagons and vans.

Ultimate customer: A customer that purchases electricity for its own use and not for resale.

Ultraviolet: Electromagnetic radiation in the wavelength range of 4 to 400 nanometers.

Unaccounted for (crude oil): Represents the arithmetic difference between the calculated supply and the calculated disposition of crude oil. The calculated supply is the sum of crude oil production plus imports minus changes in crude oil stocks. The calculated disposition of crude oil is the sum of crude oil input to refineries, crude oil exports, crude oil burned as fuel, and crude oil losses.

Unaccounted for (natural gas): Represents differences between the sum of the components of natural gas supply and the sum of components of natural gas disposition. These differences may be due to quantities lost or to the effects of data reporting problems. Reporting problems include differences due to the net result of conversions of flow data metered at varying temperatures and pressure bases and converted to a standard temperature and pressure base; the effect of variations in company accounting and billing practices; differences between billing cycle and calendar-period time frames; and imbalances resulting from the merger of data reporting systems that vary in scope, format, definitions, and type of respondents.

Uncompleted wells, equipment, and facilities costs: The costs incurred to (1) drill and equip wells that are not yet completed, and (2) acquire or construct equipment and facilities that are not yet completed and installed.

Unconsolidated entity: A firm directly or indirectly controlled by a parent but not consolidated with the parent for purposes of financial statements prepared in accordance with generally accepted accounting principles. An unconsolidated entity includes any firm consolidated with the unconsolidated entity for purposes of financial statements prepared in accordance with generally accepted accounting principles historically and consistently applied. An individual shall be deemed to control a firm that is directly or indirectly controlled by him or by his father, mother, spouse, children, or grandchildren.

Unbundling: Separating vertically integrated monopoly functions into their component parts for the purpose of separate service offerings.

Underground gas storage: The use of sub-surface facilities for storing gas that has been transferred from its original location. The facilities are usually hollowed-out salt domes, geological reservoirs (depleted oil or gas fields) or water-bearing sands topped by an impermeable cap rock (aquifer).

Underground gas storage reservior capacity: Interstate company reservoir capacities are those certificated by the Federal Energy Regulatory Commission. Independent producer and intrastate company reservior capacities are reported as developed capacity.

Underground mine: A mine where coal is produced by tunneling into the earth to the coalbed, which is then mined with underground mining equipment such as cutting machines and continuous, longwall, and shortwall mining machines. Underground mines are classified according to the type of opening used to reach the coal, *i.e.*, drift (level tunnel), slope (inclined tunnel), or shaft (vertical tunnel).

Underground mining equipment:

A **coal-cutting machine** is used in conventional mining to undercut, topcut, or shear the coal face so that coal can be fractured easily when blasted. It cuts 9 to 13 feet into the bed.

Continuous auger machine is used in mining coalbeds less than 3 feet thick. The auger has a cutting depth of about 5 feet and is 20 to 28 inches in diameter. Continuous auger mining usually uses a conveyor belt to haul the coal to the surface.

Continuous mining machine, used during continuous mining, cuts or rips coal from the face and loads it into shuttle cars or conveyors in one operation. It eliminates the use of blasting devices and performs many functions of other equipment such as drills, cutting machines, and loaders. A continuous mining machine typically has a turning "drum" with sharp bits that cut and dig out the coal for 16 to 22 feet before mining stops so that the mined area can be supported with roof bolts. This machine can mine coal at the rate of 8 to 15 tons per minute.

There are of two types of **conveyor systems**:

1. A **mainline conveyor**, which is usually a permanent installation that carries coal to the surface.
2. A **section conveyor**, which connects the working face to the mainline conveyor.

Face drill is used in conventional mining to drill shotholes in the coalbed for explosive charges.

Loading machine is used in conventional mining to scoop broken coal from the working area and load it into a shuttle car, which hauls the coal to mine cars or conveyors for delivery to the surface.

Longwall mining machine shears coal from a long straight coal face (up to about 700 feet) by working back and forth across the face under a movable, hydraulic-jack roof-support system. The broken coal is transported by converyor. Longwall machines can mine coal at the rate of 1,000 tons per shift.

Mine locomotive, operating on tracks, is used to haul mine cars containing coal and other material, and to move personnel in specially designed "mantrip" cars. Large locomotives can haul more than 20 tons at a speed of about 10 miles per hour. Most mine locomotives run on electricity provided by a trolley wire; some are battery-powered.

Ram car or shuttle ram is a rubber-tired haulage vehicle that is unloaded through the use of a movable steel plate located at the back of the haulage bed.

Roof-bolting machine, or roof bolter, is used to drill holes and place bolts to support the mine roof. Roof bolting units can be installed on a continuous mining machine.

Scoop is a rubber-tired haulage vehicle used in thin coalbeds.

Shortwall mining machine generally is a continuous-mining machine used with a powered, self-advancing roof support system. It shears coal from a short coal face (up to about 150 feet long). The broken coal is hauled by shuttle cars to a conveyor belt.

Shuttle car is a rubber-tired haulage vehicle that is unloaded by a built-in conveyor.

Underground mining methods:

1. A **drift mine** is driven horizontally into coal that is exposed or accessible in a hillside. In a hydraulic mine, high-pressure water jets break the coal from a steeply inclined, thick coalbed that would be difficult to mine with the usual underground methods. The coal is then transported to the surface by a system of flumes or by pipeline. Although currently not in commercial use in the United States, hydraulic mining is used in western Canada.
2. A **punch mine** is a type of small drift mine used to recover coal from strip-mine highwalls or from small, otherwise uneconomical, coal deposits. A shaft mine is driven vertically to the coal deposit. A slope mine is driven at an angle to reach the coal deposit.
3. In a **room-and-pillar mining** system, the most common method, the mine roof, is supported mainly by coal pillars left at regular intervals. Rooms are places where the coal is mined;

pillars are areas of coal left between the rooms. Room-and-pillar mining is done either by 1) conventional mining, which involves a series of operations that require cutting the working face of the coalbed so that it breaks easily when blasted with explosives or high-pressure air, and then loading the broken coal or 2) continuous mining, in which a continuous mining machine extracts and removes coal from the working face in one operation. When a section of a mine has been fully developed, additional coal is extracted by mining the supportive pillars until the roof caves in; the procedure is called room-and-pillar retreat mining.

4. In a **longwall mining** system, long sections of coal, up to about 700 feet, are removed and no pillars are left to support the mined-out areas. The working area is protected by a movable, powered roof support system. The caved area (gob) compacts and, after initial subsidence, supports the overlying strata. Longwall mining is used where the coalbed is thick and generally flat, where surface subsidence is acceptable.
5. A **shortwall mining** system generally refers to the room-and-pillar mining in which the working face is wider than usual but smaller (less than 150 feet) than that in longwall mining.

   Roof support and mine ventilation are paramount in all underground mining operations. Roof bolting is the principal method of supporting the mine roof. In roof bolting, long bolts, 2 to 10 feet long with an expansion shell or with resin grouting are placed in the mine roof. The bolts reinforce the roof by pulling together rock strata to make a strong beam or by fastening weak strata to strong strata. Mine ventilation, accomplished with fans, is essential to supply fresh air and to remove gases and dust from the mine. To reduce the possibility of coal dust explosions, rock dust is sprayed in an underground coal mine. Rock dust is a very fine noncombustible material (pulverized limestone).

Underground storage: The storage of natural gas in underground reservoirs at a different location from which it was produced.

Underground storage injections: Gas from extraneous sources put into underground storage reservoirs.

Underground storage withdrawals: Gas removed from underground storage reservoirs.

Undifferentiated/unspecified reserves and production: Reserves and production that are not separable by FERC production areas or by states. Undifferentiated and unspecified reserves consist only of company-owned gas in underground storage.

Undiscovered recoverable reserves (crude oil and natural gas): Those economic resources of crude oil and natural gas, yet undiscovered, that are estimated to exist in favorable geologic settings.

Undiscovered resources (coal): Unspecified bodies of coal surmised to exist on the basis of broad geologic knowledge and theory. Undiscovered resources include beds of bituminous coal and anthracite 14 inches or more thick and beds of subbituminous coal and lignite 30 inches or more thick that are presumed to occur in unmapped and unexplored areas to depths of 6,000 feet. The speculative and hypothetical resource categories comprise undiscovered resources.

Unfilled requirements: Requirements not covered by usage of inventory or supply contracts in existence as of January 1 of the survey year.

Unfinished oils: All oils requiring further processing, except those requiring only mechanical blending. Unfinished oils are produced by partial refining of crude oil and include naphthas and lighter oils, kerosene and light gas oils, heavy gas oils, and **residuum**.

Unfractionated streams: Mixtures of unsegregated natural gas liquid components, excluding those in plant condensate. This product is extracted from natural gas.

Unglazed solar collector: A solar thermal collector that has an absorber that does not have a glazed covering. Solar swimming pool heater systems usually use unglazed collectors because they circulate relatively large volumes of water through the collector and capture nearly 80 percent of the solar energy available.

Uniform system of accounts: Prescribed financial rules and regulations established by the Federal Energy Regulatory Commission for utilities subject to its jurisdiction under the authority granted by the Federal Power Act.

Unit price: Total revenue derived from the sale of product during the reference month divided by the total volume sold; also known as the weighted average price. Total revenue should exclude all taxes but include transportation costs that were paid as part of the purchase price.

Unit value, consumption: Total price per specified unit, including all taxes, at the point of consumption.

Unit value, wellhead: The wellhead sales price, including charges for natural gas plant liquids subsequently removed from the gas; gathering and compression charges; and state production, severance, and/or similar charges.

United States: The 50 States and the District of Columbia.

Unleaded gasoline: Contains not more than 0.05 gram of lead per gallon and not more than 0.005 gram of phosphorus per gallon. Premium regular and intermediate grades are included, depending on the octane rating.

Unprocessed gas: Natural gas that has not gone through a processing plant.

Unscheduled outage service: Power received by a system from another system to replace power from a generating unit forced out of service.

Uranium (U): A heavy, naturally radioactive, metallic element (atomic number 92). Its two principally occurring isotopes are uranium-235 and uranium-238. Uranium-235 is indispensable to the nuclear industry because it is the only isotope existing in nature, to any appreciable extent, that is fissionable by thermal neutrons. Uranium-238 is also important because it absorbs neutrons to produce a radioactive isotope that subsequently decays to the isotope plutonium-239, which also is fissionable by thermal neutrons.

Uranium concentrate: A yellow or brown powder obtained by the milling of uranium ore, processing of in situ leach mining solutions, or as a byproduct of phosphoric acid production.

Uranium deposit: A discrete concentration of uranium mineralization that is of possible economic interest.

Uranium endowment: The uranium that is estimated to occur in rock with a grade of at least 0.01 percent U_3O_8. The estimate of the uranium endowment is made before consideration of economic availability of any associated uranium resources.

Uranium hexaflouride (UF_6): A white solid obtained by chemical treatment of U_3O_8 and which forms a vapor at temperatures above 56 degrees Centigrade. UF_6 is the form of uranium required for the enrichment process.

Uranium importation: The actual physical movement of uranium from a location outside the United States to a location inside the United States.

Uranium mill: A plant where uranium is separated from ore taken from mines.

Uranium mill tailings: The sand-like materials left over from the separation of uranium from its ore. More than 99 percent of the ore becomes tailings.

Uranium Mill Tailings Radiation Control Act (UMTRA) of 1978: The act that directed the Department of Energy to provide for stabilization and control of the uranium mill tailings from inactive sites in a safe and environmentally sound manner to minimize radiation health hazards to the public. It authorized the Department to undertake remedial actions at 24 designated inactive uranium-processing sites and at an estimated 5,048 vicinity properties.

Uranium ore: Rock containing uranium mineralization in concentrations that can be mined economically, typically one to four pounds of U_3O_8 per ton or 0.05 percent to 0.2 percent U_3O_8.

Uranium oxide: Uranium concentrate or yellowcake. Abbreviated as U_3O_8.

Uranium property: A specific piece of land with uranium reserves that is held for the ultimate purpose of economically recovering the uranium. The land can be developed for production or undeveloped.

Uranium reserves: Estimated quantities of uranium in known mineral deposits of such size, grade, and configuration that the uranium could be recovered at or below a specified production cost with currently proven mining and processing technology and under current law and regulations. Reserves are based on direct radiometric and chemical measurements of drill holes and other types of sampling of the deposits. Mineral grades and thickness, spatial relationships, depths below the surface, mining and reclamation methods, distances to milling facilities, and amenability of ores to processing are considered in the evaluation. The amount of uranium in ore that could be exploited within the chosen forward-cost levels are estimated in accordance with conventional engineering practices.

Uranium resource categories (international): Three categories of uranium resources defined by the international community to reflect differing levels of confidence in the existence of the resources. Reasonably assured resources (RAR), estimated additional resources (EAR), and speculative resources (SR) are described below:

1. **Reasonably assured resources (RAR):** Uranium that occurs in known mineral deposits of such size, grade, and configuration that it could be recovered within the given production cost ranges, with currently proven mining and processing technology. Estimates of tonnage and grade are based on specific sample data and measurements of the deposits and on knowledge of deposit characteristics.
2. **Estimated additional resources (EAR):** Uranium in addition to **RAR** that is expected to occur, mostly on the basis of geological evidence, in extensions of well-explored deposits, in little-explored deposits, and in undiscovered deposits believed to exist along well-defined geological trends with known deposits. This uranium can subsequently be recovered within the given cost ranges. Estimates of tonnage and grade are based on available sampling data and on knowledge of the deposit characteristics, as determined in the best-known parts of the deposit or in similar deposits.
3. **Speculative resources (SR):** Uranium in addition to **EAR** that is thought to exist, mostly on the basis of indirect evidence and geological extrapolations, in deposits discoverable with existing exploration techniques. The location of deposits in this category can generally be specified only as being somewhere within given regions or geological trends. The estimates in this category are less reliable than estimates of **RAR** and **EAR**.

U.S. refiner acquisition cost of imported crude oil: The average price paid by U.S. refiners for imported, that is, non-U.S., crude oil booked into their refineries in accordance with accounting procedures generally accepted and consistently and historically applied by the refiners concerned. The refiner acquisition cost of imported crude oil includes transportation and other fees paid by the refiner.

Useage agreement: Contracts held by enrichment customers that allow feed material to be stored at the enrichment plant site in advance of need.

Used and useful: A concept used by regulators to determine whether an asset should be included in the utility's rate base. This concept requires that an asset currently provide or be capable of providing a needed service to customers.

Useful thermal output: The thermal energy made available in a combined-heat-and-power system for use in any industrial or commercial process, heating or cooling application, or delivered to other end users, *i.e.*, total thermal energy made available for processes and applications other than electrical generation.

Utility demand-side management costs: The costs incurred by the utility to achieve the capacity and energy savings from the Demand-Side Management (DSM) Program. Costs incurred by consumers or third parties are to be excluded. The costs are to be reported in nominal dollars in the year in which they are incurred, regardless of when the savings occur. The utility costs are all the annual expenses (labor, administrative, equipment, incentives, marketing, monitoring and evaluation, and other) incurred by the utility for operation of the DSM Program, regardless of whether the costs are expensed or capitalized. Lump-sum capital costs (typically accrued over several years prior to start up) are not to be reported. Program costs associated with strategic load growth activities are also to be excluded.

Utility distribution companies: The entities that will continue to provide regulated services for the distribution of electricity to customers and serve customers who do not choose direct access. Regardless of where a consumer chooses to purchase power, the customer's current utility, also known as the utility distribution company, will deliver the power to the consumer.

Utility generation: Generation by electric systems engaged in selling electric energy to the public.

Utility-sponsored conservation program: Any program sponsored by an electric and/or natural gas utility to review equipment and construction features in buildings and advise on ways to increase the energy efficiency of buildings. Also included are utility-sponsored programs to encourage the use of more energy-efficient equipment. Included are programs to improve the energy efficiency in the lighting system or building equipment or the thermal efficiency of the building shell.

Vacuum distillation: Distillation under reduced pressure (less the atmospheric), which lowers the boiling temperature of the liquid being distilled. This technique with its relatively low temperatures prevents cracking or decomposition of the charge stock.

Value (of shipments): The value received for the complete systems at the company's net billing price, freight-on-board factory, including charges for cooperative advertising and warranties. This does not include excise taxes, freight or transportation charges, or installation charges.

Value added by manufacture: A measure of manufacturing activity that is derived by subtracting the cost of materials (which covers materials, supplies, containers, fuel, purchased electricity, and contract work) from the value of shipments. This difference is then adjusted by the net change in finished goods and work-in-progress between the beginning- and end-of-year inventories.

Vapor displacement: The release of vapors that had previously occupied space above liquid fuels stored in tanks. These releases occur when tanks are emptied and filled.

Vapor-dominated geothermal system: A conceptual model of a hydrothermal system where steam pervades the rock and is the pressure-controlling fluid phase.

Vapor retarder: A material that retards the movement of water vapor through a building element (walls, ceilings) and prevents insulation and structural wood from becoming damp and metals from corroding. Often applied to insulation batts or separately in the form of treated papers, plastic sheets, and metallic foils.

Variable air volume (VAV) system on the heating and cooling system: A means of varying the amount of conditioned air to a space. A variable air volume system maintains the air flow at a constant temperature, but supplies varying quantities of conditioned air in different parts of the building according to the heating and cooling needs.

Variable-speed wind turbines: Turbines in which the rotor speed increases and decreases with changing wind speed, producing electricity with a variable frequency.

Vehicle fuel consumption: Vehicle fuel consumption is computed as the vehicle miles traveled divided by the fuel efficiency reported in miles per gallon (MPG). Vehicle fuel consumption is derived from the actual vehicle mileage collected and the assigned MPGs obtained from EPA certification files adjusted for on-road driving. The quantity of fuel used by vehicles.

Vehicle fuel expenditures: The cost, including taxes, of the gasoline, gasohol, or diesel fuel added to the vehicle's tank. Expenditures do not include the cost of oil or other items that may have been purchased at the same time as the vehicle fuel.

Vehicle identification number (VIN): A set of codes, usually alphanumeric characters, assigned to a vehicle at the factory and inscribed on the vehicle. When decoded, the VIN provides vehicle characteristics. The VIN is used to help match vehicles to the EPA certification file for calculating MPGs.

Vehicle importer: An original vehicle manufacturer (of foreign or domestic ownership) that imports vehicles as finished products into the United States.

Vehicle miles traveled (VMT): The number of miles traveled nationally by vehicles for a period of 1 year. VMT is either calculated using two odometer readings or, for vehicles with less than two odometer readings, imputed using a regression estimate.

Vented: Gas released into the air on the production site or at processing plants.

Vented/Flared: Gas that is disposed of by releasing (venting) or burning (flaring).

Ventilation system: A method for reducing methane concentrations in coal mines to non-explosive levels by blowing air across the mine face and using large exhaust fans to remove methane while mining operations proceed.

Vertical-axis wind turbine (VAWT): A type of wind turbine in which the axis of rotation is perpendicular to the wind stream and the ground.

Vertical integration: The combination within a firm or business enterprise of one or more stages of production or distribution. In the electric industry, it refers to the historical arrangement whereby a utility owns its own generating plants, transmission system, and distribution lines to provide all aspects of electric service.

Vessel: A ship used to transport crude oil, petroleum products, or natural gas products. Vessel categories are as follows: Ultra Large Crude Carrier (ULCC), Very Large Crude Carrier (VLCC), Other Tanker, and Specialty Ship (LPG/LNG).

Vessel bunkering: Includes sales for the fueling of commercial or private boats, such as pleasure craft, fishing boats, tugboats, and ocean-going vessels, including vessels operated by oil companies.

VIN (vehicle identification number): A set of about 17 codes, combining letters and numbers, assigned to a vehicle at the factory and inscribed on a small metal label attached to the dashboard and visible through the windshield. The VIN is a unique identifier for the vehicle and therefore is often found on insurance cards, vehicle registrations, vehicle titles, safety or emission certificates, insurance policies, and bills of sale. The coded information in the VIN describes characteristics of the vehicle such as engine size and weight.

Virgin coal: Coal that has not been accessed by mining.

Visbreaking: A thermal cracking process in which heavy atmospheric or vacuum-still bottoms are cracked at moderate temperatures to increase production of distillate products and reduce viscosity of the distillation residues.

Volatile matter: Those products, exclusive of moisture, given off by a material as gas or vapor. Volatile matter is determined by heating the coal to 950 degrees Centigrade under carefully controlled conditions and measuring the weight loss, excluding weight of moisture driven off at 105 degrees Centigrade.

Volatile organic compounds (VOCs): Organic compounds that participate in atmospheric photochemical reactions.

Volatile solids: A solid material that is readily decomposable at relatively low temperatures.

Volt (V): The volt is the International System of Units (SI) measure of electric potential or electromotive force. A potential of one volt appears across a resistance of one ohm when a current of one ampere flows through that resistance. Reduced to SI base units, 1 V = 1 kg times m^2 times s^{-3} times A^{-1} (kilogram meter squared per second cubed per ampere).

Voltage: The difference in electrical potential between any two conductors or between a conductor and ground. It is a measure of the electric energy per electron that electrons can acquire and/or give up as they move between the two conductors.

Voltage reduction: Any intentional reduction of system voltage by 3 percent or greater for reasons of maintaining the continuity of service of the bulk electric power supply system.

Wafer: A thin sheet of semiconductor made by cutting it from a single crystal or ingot.

Walk-in refrigeration units: Refrigeration/freezer units within a building that are large enough to walk into. They may be portable or permanent, such as a meat storage locker in a butcher store. Walk-in units may or may not have a door, plastic strips, or other flexible covers.

Wall insulation: Insulating materials within or on the walls between heated areas of the building and unheated areas or the outside. The walls may separate air-conditioned areas from areas not air-conditioned.

Warranty contracts: Gas purchase agreements for the sale of natural gas by a producer to a pipeline company wherein the producer warrants it will have available sufficient gas supplies to meet its commitments over the life of the contract. Generally, the producer does not dedicate gas reserves underlying any specific acreage, lease, or fields to the agreement. Substitution of various sources of gas supply may be permitted according to the terms of the contract. Warranty contracts, by their terms, may vary from the above.

Waste coal: Usable coal material that is a byproduct of previous processing operations or is recaptured from what would otherwise be refuse. Examples include anthracite culm, bituminous gob, fine coal, lignite waste, coal recovered from a refuse bank or slurry dam, and coal recovered by dredging.

Waste energy: Municipal solid waste, landfill gas, methane, digester gas, liquid acetonitrile waste, tall oil, waste alcohol, medical waste, paper pellets, sludge waste, solid byproducts, tires, agricultural byproducts, closed loop biomass, fish oil, and straw used as fuel.

Waste heat boiler: A boiler that receives all or a substantial portion of its energy input from the combustible exhaust gases from a separate fuel-burning process.

Waste heat recovery: Any conservation system whereby some space heating or water heating is done by actively capturing byproduct heat that would otherwise be ejected into the environment. In commercial buildings, sources of water- heat recovery include refrigeration/air-conditioner compressors, manufacturing or other processes, data processing centers, lighting fixtures, ventilation exhaust air, and the occupants themselves. Not to be considered is the passive use of radiant heat from lighting, workers, motors, ovens, etc., when there are no special systems for collecting and redistributing heat.

Waste materials: Otherwise discarded combustible materials that, when burned, produce energy for such purposes as space heating and electric power generation. The size of the waste may be reduced by shredders, grinders, or hammermills. Noncombustible materials, if any, may be removed. The waste may be dried and then burned, either alone or in combination with fossil fuels.

Waste oils and tar: Petroleum-based materials that are worthless for any purpose other than fuel use.

Wastewater, domestic and commercial: Wastewater (sewage) produced by domestic and commercial establishments.

Wastewater, industrial: Wastewater produced by industrial processes.

Water bed heater: An appliance that uses an electric resistance coil to maintain the temperature of the water in a water bed at a comfortable level.

Water conditions: The status of the water supply and associated water in pondage and reservoirs at hydroelectric plants.

Water heated in furnace: Some furnaces provide hot water as well as heat the home. The water is heated by a coil that is part of the furnace. There is no separate hot water tank.

Water heater: An automatically controlled, thermally insulated vessel designed for heating water and storing heated water at temperatures less than 180 degrees Fahrenheit.

Water heating DSM programs: These are demand-side management (DSM) programs designed to promote increased efficiency in water heating, including water heater insulation wraps.

Water heating equipment: Automatically controlled, thermal insulated equipment designed for heating and storing heated water at temperatures less than 180 degrees Fahrenheit for other than space heating purposes.

Water pollution abatement equipment: Equipment used to reduce or eliminate waterborne pollutants, including chlorine, phosphates, acids, bases, hydrocarbons, sewage, and other pollutants.

Examples of water pollution abatement structures and equipment include those used to treat thermal pollution; cooling, boiler, and cooling tower blowdown water; coal pile runoff; and fly ash waste water. Water pollution abatement excludes expenditures for treatment of water prior to use at the plant.

Water pumping: Photovoltaic modules/cells used for pumping water for agricultural, land reclamation, commercial, and other similar applications where water pumping is the main use.

Water reservoir: A large inland body of water collected and stored above ground in a natural or artificial formation.

Water source heat pump: A type of (geothermal) heat pump that uses well (ground) or surface water as a heat source. Water has a more stable seasonal temperature than air thus making for a more efficient heat source.

Water turbine: A turbine that uses water pressure to rotate its blades; the primary types are the Pelton wheel, for high heads (pressure); the Francis turbine, for low to medium heads; and the Kaplan for a wide range of heads. Primarily used to power an electric generator.

Water vapor: Water in a vaporous form, especially when below boiling temperature and diffused (*e.g.*, in the atmosphere).

Water well: A well drilled to (1) obtain a water supply to support drilling or plant operations, or (2) obtain a water supply to be used in connection with an improved recovery program.

Water wheel: A wheel that is designed to use the weight and/or force of moving water to turn it, primarily to operate machinery or grind grain.

Waterway: A river, channel, canal, or other navigable body of water used for travel or transport.

Watt (W): The unit of electrical power equal to one ampere under a pressure of one volt. A Watt is equal to 1/746 horsepower.

Watthour (Wh): The electrical energy unit of measure equal to one watt of power supplied to, or taken from, an electric circuit steadily for one hour.

Wattmeter: A device for measuring power consumption.

Wax: A solid or semi-solid material derived from petroleum distillates or residues by such treatments as chilling, precipitating with a solvent, or de-oiling. It is a light-colored, more-or-less translucent crystalline mass, slightly greasy to the touch, consisting of a mixture of solid hydrocarbons in which the paraffin series predominates. Includes all marketable wax, whether crude scale or fully refined. The three grades included are microcrystalline, crystalline-fully refined, and crystalline-other. The conversion factor is 280 pounds per 42 U.S. gallons per barrel.

Weather stripping or caulking: Any of several kinds of crack-filling material around any windows or doors to the outside used to reduce the passage of air and moisture around moveable parts of a door or window. Weather stripping is available in strips or rolls of metal, vinyl, or foam rubber and can be applied on the inside or outside of a building.

Weir: A dam in a waterway over which water flows and that serves to raise the water level or to direct or regulate flow.

Well: A hole drilled in the earth for the purpose of (1) finding or producing **crude oil or natural gas**; or (2) producing services related to the production of crude or natural gas.

Well water for cooling: A means of cooling that uses water from a well drilled specifically for that purpose. The subterranean temperature of the water stays at a relatively constant temperature. Where water is abundant, it provides a means of getting 55-degree Fahrenheit water with no mechanical cooling. Used usually for heat rejection in a water source heat pump.

Wellhead: The point at which the crude (and/or natural gas) exits the ground. Following historical precedent, the volume and price for crude oil production are labeled as "wellhead," even though the cost and volume are now generally measured at the lease boundary. In the context of domestic crude price data, the term "wellhead" is the generic term used to reference the production site or lease property.

Wellhead price: The value at the mouth of the well. In general, the wellhead price is considered to be the sales price obtainable from a third party in an arm's length transaction. Posted prices, requested prices, or prices as defined by lease agreements, contracts, or tax regulations should be used where applicable.

Wet bottom boiler: Slag tanks are installed usually at the furnace throat to contain and remove molten ash.

Wet natural gas: A mixture of hydrocarbon compounds and small quantities of various nonhydrocarbons existing in the gaseous phase or in solution with crude oil in porous rock formations at reservoir conditions. The principal hydrocarbons normally contained in the mixture are methane, ethane, propane, butane, and pentane. Typical nonhydrocarbon gases that may be present in reservoir natural gas are water vapor, carbon dioxide, hydrogen sulfide, nitrogen and trace amounts of helium. Under reservoir conditions, natural gas and its associated liquefiable portions occur either in a single gaseous phase in the reservoir or in solution with crude oil and are not distinguishable at the time as separate substances.

Wheeling charge: An amount charged by one electrical system to transmit the energy of and for, another system or systems.

Wheeling service: The movement of electricity from one system to another over transmission facilities of interconnecting systems. Wheeling service contracts can be established between two or more systems.

White spirit: A highly refined distillate with a boiling point range of about 150 degrees to 200 degrees Centigrade. It is used as a paint solvent and for dry-cleaning purposes.

Whole-house colling fan: A mechanical/electrical device used to pull air out of an interior space; usually located in the highest location of a building, in the ceiling, and venting to the attic or directly to the outside.

Wholesale competition: A system whereby a distributor of power would have the option to buy its power from a variety of power producers, and the power producers would be able to compete to sell their power to a variety of distribution companies.

Wholesale electric power market: The purchase and sale of electricity from generators to resellers (retailers), along with the ancillary services needed to maintain reliability and power quality at the transmission level.

Wholesale power market: The purchase and sale of electricity from generators to resellers (who sell to retail customers), along with the ancillary services needed to maintain reliability and power quality at the transmission level.

Wholesale price: The rack sales price charged for No. 2 heating oil; that is, the price charged customers who purchase No. 2 heating oil free-on-board at a supplier's terminal and provide their own transportation for the product.

Wholesale sales: Energy supplied to other electric utilities, cooperatives, municipals, and Federal and state electric agencies for resale to ultimate consumers.

Wholesale transmission services: The transmission of electric energy sold, or to be sold, in the wholesale electric power market.

Wholesale wheeling: An arrangement in which electricity is transmitted from a generator to a utility through the transmission facilities of an intervening system.

Wind energy: Kinetic energy present in wind motion that can be converted to mechanical energy for driving pumps, mills, and electric power generators.

Wind energy conversion system (WECS) or device: An apparatus for converting the energy available in the wind to mechanical energy that can be used to power machinery (grain mills, water pumps) and to operate an electrical generator.

Wind power plant: A group of wind turbines interconnected to a common utility system through a system of transformers, distribution lines, and (usually) one substation. Operation, control, and maintenance functions are often centralized through a network of computerized monitoring systems, supplemented by visual inspection. This is a term commonly used in the United States. In Europe, it is called a generating station.

Wind turbine: Wind energy conversion device that produces electricity; typically three blades rotating about a horizontal axis and positioned up-wind of the supporting tower.

Wires charge: A broad term referring to fees levied on power suppliers or their customers for the use of the transmission or distribution wires.

Wood and waste (as used at electric utilities): Wood energy, garbage, bagasse (sugarcane residue), sewerage gas, and other industrial, agricultural, and urban refuse used to generate electricity for distribution.

Wood conversion to Btu: Converting cords of wood into a Btu equivalent is an imprecise procedure. The number of cords each household reports having burned is inexact, even with the more precise drawings provided, because the estimate requires the respondent to add up the use of wood over a 12-month period during which wood may have been added to the supply as well as removed. Besides errors of memory inherent in this task, the estimates are subject to problems in definition and perception of what a cord is. The nominal cord as delivered to a suburban residential buyer may differ from the dimensions of the standard cord. This difference is possible because wood is most often cut in lengths that are longer than what makes a third of a cord (16 inches) and shorter than what makes a half cord (24 inches).

In other cases, wood is bought or cut in unusual units (for example, pickup-truck load, or trunk load). Finally, volume estimates are difficult to make when the wood is left in a pile instead of being stacked. Other factors that make it difficult to estimate the Btu value of the wood burned is that the amount of empty space between the stacked logs may vary from 12 to 40 percent of the volume. Moisture content may vary from 20 percent in dried wood to 50 percent in green wood. (Moisture reduces the useful Btu output because energy is used in driving off the moisture). Finally, some tree species contain twice the Btu content of species with the lowest Btu value.

Generally, hard woods have greater Btu value than soft woods. Wood is converted to Btu at the rate of 20 million Btu per cord, which is a rough average that takes all these factors into account.

Wood energy: Wood and wood products used as fuel, including round wood (cord wood), limb wood, wood chips, bark, sawdust, forest residues, charcoal, pulp waste, and spent pulping liquor.

Wood pellets: Sawdust compressed into uniform diameter pellets to be burned in a heating stove.

Working (top storage) gas: The volume of gas in the reservoir that is in addition to the cushion or base gas. It may or may not be completely withdrawn during any particular withdrawal season. Conditions permitting, the total working capacity could be used more than once during any season.

Working interest: An interest in a mineral property that entitles the owner of that interest to all of share of the mineral production from the property, usually subject to a royalty.

A working interest permits the owner to explore, develop, and operate the property. The working-interest owner bears the costs of exploration, development, and operation of the property and, in return, is entitled to a share of the mineral production from the property or to a share of the proceeds therefrom. It may be assigned to another party in whole or in part, or it may be divided into other special property interests.

Gross working interest. The reporting company's working interest plus the proportionate share of any basic royalty interest or overriding royalty interest related to the working interest.

Net working interest. The reporting company's working interest is not including any basic royalty or overriding royalty interests.

Working storage capacity: The difference in volume between the maximum safe fill capacity and the quantity below which pump suction is ineffective (bottoms).

Xylene ($C_6H_4(CH_3)_2$): Colorless liquid of the aromatic group of hydrocarbons made by the catalytic reforming of certain naphthenic petroleum fractions. Used as high-octane motor and aviation gasoline blending agents, solvents, chemical intermediates. Isomers are metaxylene, orthoxylene, paraxylene.

Yellowcake: A natural uranium concentrate that takes its name from its color and texture. Yellowcake typically contains 70 to 90 percent U_3O_8 (uranium oxide) by weight. It is used as feedstock for uranium fuel enrichment and fuel pellet fabrication.

References

______. 1996. *Oil Gas Coal and Electricity: Quarterly Statistics: International Energy Agency*. O E C D.

______. 1971. *Petroleum Conservation in the United States.* Baltimore, Md.: The Johns Hopkins University Press.

______. 1984. "Are Utilities Obsolete? A Troubled System Faces a Radical Change." May 21.

Abbasi, S A. 1995. *Alternative Energy: Development and Management*. International Book Distributors.

Abrams Dennis M. 1980. "The Rockefeller Amendment: Its Origins, Its Effect and Its Future." *West Virginia Law Review 82*:1241-1255.

Agarwal, S K. 2003. *Nuclear Energy: Principles Practices and Prospects*. Ashish Publishing House, New Delhi.

Allison Graham T. 1971. *Essence of Decision.* Boston: Little, Brown and Company.

Athparia, R P *et el* eds. 1999. *Renewable Energy Resources and Its Management*. Reliance.

Bala, B K. 1998. *Solar Drying Systems*. Agrotech Publishing Academy.

Bala, B K. 2003. *Renewable Energy*. Agrotech Publishing Academy.

Ball Ben C., Jr. 1979. "New Challenges to Management in the Synfuels Revolution." *Technology Review* (August/September).

Bardach Eugene, and Robert A. Kagan. 1982. *Going by the Book: The Problem of Regulatory Unreasonableness.* Philadelphia: Temple University Press.

Bardach Eugene. 1977. *The Implementation Game.* Cambridge, Mass.: The MIT Press.

Bhatia, S C. 1983. *Mineral Oils Petroleum and Lubricants.* Shree.

Browning.. *Global Energy and Water Crisis*. Cambridge Univ.

Brueden R. 1976. "Federalism and the Development of Outer Continental Shelf Mineral Resources." *Stanford Law Review* 28.

Buchanan James M., and Gordon Tullock. 1962. *The Calculus of Consent: Logical Foundations of Constitutional Democracy*. Ann Arbor: The University of Michigan Press.

Burton, Tony *et al.* 2001. *Wind Energy Handbook*. John Wiley.

Business Week. 1984. "Pacific Power and Light: Back to Basics After Swallowing its Losses from Whoops.'" March 5.

Chambliss Barbe. 1982. "Developing Hydropower in Your Community." In *Alternative Energy Sources: Barriers and Incentives,* Van R. Johnston, ed. American Society for Public Administration, Colorado Chapter, Autumn.

Chaturvedi, Pradeep ed.. *Financing of Energy Sector in Developing Countries*. Concept.

Chaturvedi, Pradeep. 1997. *Energy Management: Policy Planning and Utilization*. Concept.

Chaturvedi, Pradeep. 1998. *Rural Energy for Sustainable Development: Technology and Environmental Issues*. Concept.

Chaturvedi, Pradeep. 2000. *Energy Management: Challanges for the Next Millennium*. Concept.

Chaturvedi, Pradeep. 2003. *Energy Environment and Sustainable Development*. Concept.

Chhatwal, G R. 1998. *Encyclopaedia of Environmental Energy Resources in 2 Vols*. Anmol Publishing Company, New Delhi.

Chronis Peter. 1982. "Energy Crisis Not Over," *Denver Post*. Article quoting Richard Lamm, Governor of Colorado, September 9.

Code of Federal Regulations. 1983. Title 40, Sections 1502.4, 15066.

Colorado, State of. 1983. CRS 30-20-503 (3), 1983 Cumulative Supplement.

Comment, 1978. "The Outer Continental Shelf: Bridging the Information Gap Through Regulation." *Arizona State Law Journal*.

Comment, 1979. "CLF v. Andrus and Oil Drilling in Georges Bank: The First Circuit Attempts to Balance Conflicting Interests." *Boston College Environmental Affairs Law Review* 8 <27459684>.

Das, Pradip Kumar & Hrishikesh Baruah. 1997. *Petroleum and Coal*. M D Publications.

Davis Kenneth C. 1969. *Discretionary Justice: A Preliminary Inquiry*. Baton Rouge: Louisiana State University Press.

Davis, K H & F S Berner. 2004. *Handbook of Industrial Chemistry Vol 1: Unit Operations Heavy Inorganic Chemicals Explosives and Propellants Catalysis*. C B S Pub.

Davis, K H & F S Berner. 2004. *Handbook of Industrial Chemistry Vol 2: Petroleum and Petrochemicals Essential Oil Perfumery Chemicals & Medicinal Plant*. C B S Pub.

Dewan, K K & M Mustafa. 2004. *Management of Modern City Transportation System: A Long Term Strategy for Energy and Environment Efficient Transport*. Deep & Deep.

Edgmon Terry, and Donald C. Menzel. 1981. "The Regulation of Coal Surface Mining in a Federal System." *Natural Resources Journal 21*:245-65.

Foster John L., and Judson H. Jones. 1978. "Rule Orientation and Bureaucratic Reform." *American Journal of Political Science* 22:348-63.

Franklin C. 1961. "The Law of the Sea: Some Recent Development." *U.S. Naval War College International Studies* 49 <27459725>- 62 <27459738>.

Garg, Anil K *et al.* eds. 2000. *Petroleum Geochemistry and Exploration in the Afro Asian Region*. B R Publishing.

Gaynor, Frank ed. 2003. *A to Z Illustrated Encyclopaedia of Atomic Energy*. Khel Sahitya Kendra.

Gopi, K N & Kalpana Markandey. 1994. *Energy Environment and Human Well Being*. Concept.

Graham Sandy. 1984. "Extra Power Boosters Rates in the West," *Rocky Mountain News*. February 19.

Gulick Luther. 1937. "Notes on the Theory of Administration." In Luther Gulick and L. Urwich, eds., *Papers on the Science of Administration*. New York: Institute of Public Administration.

Gupta, K C. 2002. *Energy and Environment in India: A Study of Energy Management*. Gyan Pub.

Gupta, V P & Sneh Gupta. 2002. *Energy Patterns in India*. Aavishkar Publishers Distributors.

Hollander, Jack M *et al.* eds. 1987. *Annual Review of Energy Vol 12*. Annual Reviews Inc.

Hollick Ann. 1981. *U.S. Foreign Policy and the Law of the Sea 18*. Princeton: Princeton University Press.

Hotelling Harold. 1931. "The Economics of Exhaustible Resources." *Journal of Political Economy* (April):137-75.

Jogi, Kripal Singh. 2000. *Encyclopaedia of Energy Management: Priorities for 21st Century in 2 Vols*. Sarup & Sons.

Johnston Van R. 1982. "The Developing Energy Situation in the United States." A paper delivered at the Japanese Executives' Seminar at the University of Denver, October. Program sponsored by Kansai Productivity Center of Osaka, Japan.

Johnston Van R., et al. 1982. *Alternative Energy Sources: Barriers and Incentives.* American Society for Public Administration, Colorado Chapter. Autumn.

Juda Lawrence. 1975. *Ocean Space Rights: Developing U.S. Policy 12*. New York: Praeger Publishers.

Justheim v. McKay, 229 F. 2d 1930 (D.C. Cir. 1956).

Kadian, Promila & Sushma Kaushik. 2003. *Rural Energy for Sustainable Development: Participatory Assessment of Energy Resources*. Deep & Deep.

Kakani, S L & C Hemrajani. 1990. *Recent Advances in Superconductivity*. T T P P.

Katz James E. 1984. "U.S. Energy Policy: Impact of the Reagan Administration," *Energy Policy*. 12, 3, June.

Kaufman Herbert. 1960. *The Forest Ranger.* Baltimore: Johns Hopkins University Press.

Kaushik, Sushma & Tej Verma. 1996. *Rural Energy Management*. Deep & Deep.

Kaushika, N D & Kshitij Kaushik. 2004. *Energy Ecology and Environment: A Technological Approach*. Capital Publishing Company.

Keshavendra. 1998. *Electricity Generation: Policy Technology Economics*. Venus.

Khan, Farooq. 1999. *Energy Management*. Anmol Publishing Company, New Delhi.

Krueger R. 1967. "Mineral Development on the Continental Shelf and Beyond." *California State Bar Journal 42*:515.

Krueger R. 1978. "The Management of Federal Petroleum Resources in the United States." *International and Comparative Law Quarterly 17*:69.

Krueger R., and M. Nordquist. 1979. "The Evolution of the 200-mile Exclusive Economic Zone: State Practice in the Pacific Basin." *Virginia Journal of International Law*.

Kumar, Harish. 2002. *Encyclopaedia of Energy Resources and their Management in 2 Vols*. IVY Pubilishing.

Kumar, Surinder. 2004. *Electricity Theft: Empowering People and Reforming Power Sector*. Manohar.

Kumari, Vinod. 1994. *Energy and Society*. Discovery.

Kunz J. 1956. "Continental Shelf and International Law: Confusion and Abuse." *American Journal of International Law 50*:829.

Lauterpacht H. 1950. "Sovereignty over Submarine Areas." *British Year Book of International Law* 53 (1961): 431.

Lehman D. 1960. "The Legal Status of the Continental Shelf." *Louisiana Law Review*.

Levine Alice. 1982. "Financing Community Energy Products." In *Alternative Energy Sources: Barriers and Incentives*, Van R. Johnston, ed. American Society for

MacLeish William H. 1981. "Resources: Oil, Fish and Georges Bank." *The Atlantic* (18 September).

Mahajan, V S *et al*. eds. 1999. *Energy and Energy Resource Management*. Deep & Deep.

Mahajan, V S. 1983. *Energy Development in India*. Deep & Deep.

Mahajan, V S. 1991. *Studies in Energy and Economic Development*. Deep & Deep.

Maheshwari, A & Geeta Parmar. 2002. *A Textbook of Energy Ecology Environment and Society*. Anmol Publishing Company, New Delhi.

Maheshwari, R C & Pradeep Chaturvedi. 1997. *Bioenergy for Rural Energisation*. Concept.

Marglin Stephen. 1963. "The Social Rate of Discount and the Optimal Rate of Investment." *Quarterly Journal of Economics* (February):95-111.

Mathur, A N & L N Verma. 1988. *Management and Utilisation of Biogas Plant Slurry*. Himanshu Publishing Company, Udaipur.

Mathur, A N & N S Rathore. 1990. *Renewable Energy and Environment*. Himanshu Publishing Company, Udaipur.

Mathur, A N & N S Rathore. 1992. *Biogas Production Management and Utilisation*. Himanshu Publishing Company, Udaipur.

Mathur, A N & N S Rathore. 1995. *Non Conventional Energy Sources: Theory and Practice*. Himanshu Publishing Company, Udaipur.

Mathur, A N *et al*. 1989. *Solar Drying*. Himanshu Publishing Company, Udaipur.

Mathur, A N *et al*. 1992. *Economics of Energy Plantation*. Himanshu Publishing Company, Udaipur.

Mathur, A N *et al*. 1995. *Urja Paristhitiki Vigyan evam Paryavaran (Energy Ecology and Environment)*. Himanshu Publishing Company, Udaipur.

Mathur, A N. 1987. *Bioenergy for Arid and Semi Arid Zones*. Himanshu Publishing Company, Udaipur.

Mazumdar, B. 1999. *A Textbook of Energy Technology: Bock Coventional and Renewable Source of Energy*. Ashish Publishing House, New Delhi.

McCrae Alexander, and Janice L. Dudas. 1977. *The Energy Source Book*. The Center for Compliance Information, Aspen Systems Corporation, Germantown, Maryland.

McDonald Stephen L. 1979. *The Leasing of Federal Lands for Fossil Fuels Production*. Baltimore, Md.: The Johns Hopkins University Press.

Melvin, Alec. 1988. *Natural Gas: Basic Science and Technology*. British Gas.

Menzel Donald C. 1981. "Implementation of the Federal Surface Mining Control and Reclamation Act of 1977." *Public Administration Review* 41:212-19.

Menzel Donald C. 1984. "Redirecting the Implementation of a Law: The Reagan Administration and Coal Surface Mining Regulation." *Public Administration Review* 43:411-20.

Meyer Richard, and Barbara Shirley. 1982. "The Use of Geothermal Energy to Achieve Economic Development and Diversification....," *Alternative Energy Sources: Barriers and Incentives*, Van R. Johnston, ed. American Society for Public Administration, Colorado Chapter, Autumn.

Mitchell Mark. 1982. "Solar Energy and Solar Access Protection." In *Alternative Energy Sources: Barriers and Incentives*. Van R. Johnston, ed. American Society for Public Administration, Colorado Chapter, Autumn.

Monga, Pradeep & T N Lakhanpal. 1988. *Rural Energy Alternatives in the Hilly Areas: Social Forestry and Biogas Systems*. T T P P.

Murphy J., and M. Belsky. 1980. "OCS Development: A New Law and a New Beginning." *Coastal Zone Management Journal* 7:300.

Murphy, W R & G McKay. 2001. *Energy Management*. B S Publications.

Naidu, B S K. 2003. *Hydropower: An Indian Perspective*. C B S Pub.

Nakicenovic.. *Global Energy Perspectives*. Cambridge Univ.

Niskamen William A. 1971. *Bureaucracy and Representative Government*. Chicago: Aldine Atherton.

Odland Bob, and Debra Perkins-Smith. 1982. "Developing Wind Power...." In *Alternative Energy Sources: Barriers and Incentives*, Van R. Johnston, ed. American Society for Public Administration, Colorado Chapter, Autumn.

Office of Technology. 1979. *Management of Fuel and Nonfuel Minerals in Federal Land*. Washington, D.C.: U.S. Government Printing Office.

Ostrom Vincent. 1973. *The Intellectual Crisis in American Public Administration*. University, Ala.: The University of Alabama Press.

Palmer, Joy. 1992. *Conservation 2000: Radiation and Nuclear Energy*. B T Batsford Ltd.

Pandey, G N *et al.* 1998. *Energy Conservation*. Anmol Publishing Company, New Delhi.

Paramar, Ashwin. 2001. *Energy Future: Coping with Critical Conditions*. Dominant P & D.

Pigou Alfred. 1946. *Economics of Warfare*. London: MacMillan Inc.

Pimental, David. 1980. *Handbook of Energy Utilization in Agriculture*. C R C Press, Boca Raton.

Plantenga, Margriet S W *et al.* eds. 1994. *Food Intake and Energy Expenditure*. C R C Press, Boca Raton.

Prabhakar, V K ed. 2001. *Energy Resources and Environment*. Anmol Publishing Company, New Delhi.

Prabhakar, V K ed. 2001. *Laws on Nuclear Issues*. Anmol Publishing Company, New Delhi.

Prasad, Kamta & R S Goel eds. 2001. *Environmental Management in Hydro Electric Projects*. Concept.

Prasad, Kamta ed. 2003. *Water Resources and Sustainable Development: Challenges of 21st Century*. Shipra.

Prindle David F. 1981. *Petroleum Politics and the Texas Railroad Commission*. Austin: The University of Texas Press.

Public Administration, Colorado Chapter, Autumn.

Raikhy, P S & P Singh. 1990. *Energy Consumption in India*. Deep & Deep.

Raj, Gurdeep. 1993. *Encyclopaedia of Energy in 5 Vols*. Anmol Publishing Company, New Delhi.

Raj, M G. 1999. *Problems in Electricity and Magnetism*. Anmol Publishing Company, New Delhi.

Rajan, Thillai A. 2002. *The Old the New and All in Between: A Process Study of the Pioneering Power Sector Reform in Orissa*. Classical.

Rao, Digumarti Bhaskara & D Harshita eds. 2001. *Energy Security*. Discovery.

Rao, Digumarti Bhaskara ed. 2001. *Nuclear Materials: Issues and Concerns in 2 Vols*. Discovery.

Rao, M V R Koteswara. 2004. *Energy Resources: Conventional and Non Conventional*. B S Publications.

Rathore, N S *et al.* 1994. *Integrated Rural Energy Planing*. Agrotech Publishing Academy.

Rehman, Hifzur. 2004. *Energy Use in Agricultural Productivity*. Concept.

Reynolds Stephen, and Jane Christopherson. 1984. "Public Policy and Price Per KWH." *California Management Review* 26, 2 (Winter).

Roy, Joyashree. 1992. *Demand for Energy in Indian Industries: A Quantitative Approach*. Daya.

Ruet, Joel ed. 2003. *Against the Current: Organisational Restructuring of State Electricity Boards*. Manohar.

Sah, S L. 1995. *Renewable and Novel Energy Sources*. M D Publications.

Sah, S L. 2003. *Encyclopaedia of Petroleum Science and Engineering in 5 Vols*. Kalpaz.

Sawhill John C. 1974. *Project Independence Report*. Federal Energy Administration, U.S. Government Printing Office, Washington, D.C.

Sawyer Stephen W. 1984. "State Energy Conditions and Policy Development." *Public Administration Review* 44, 3 (May/June).

Schmalensee Richard C. 1980. "Appropriate Government Policy Commercialization of New Energy Supply Technologies." *The Energy Journal* (April).

Sharma, L C. 1985. *Economic Development and Non Commercial Energy in India*. B S M P S.

Sharma, R S ed. 1995. *Environmental Energy Resources*. Commonwealth.

Shover Neal, Donald A. Clelland, and John Lynxwiler. 1982. *Developing a Regulatory Bureaucracy: The Office of Surface Mining Reclamation and Enforcement*. University of Tennessee: Department of Sociology.

Shukla, S K & P R Srivastava eds. 1992. *Environmental Energy Impact Analysis*. Commonwealth.

Singh, K K *et al.* 1992. *Advances in Biomass Gasification Technology*. Himanshu Publishing Company, Udaipur.

Singh, K K *et al.* 1993. *Action Research and Development in Biomass Gasification*. Himanshu Publishing Company, Udaipur.

Singh, M P & Soma Dey. 2004. *Natural Resources and Reneawable Energy*. Daya.

Singh, Naunihal. 2001. *Energy Crisis*. Authorspress.

Singh, Palwinder. 1992. *Changing Pattern and Energy Use*. Anmol Publishing Company, New Delhi.

Singh, Surendra & J P Mittal. 1992. *Energy in Production Agriculture*. Mittal Pubilcations, New Delhi.

Singh, V K & Abrar M Khan. 1986. *Aspects of Energy and Environment Studies*. T T P P.

Sinha, P C ed. 1998. *Energy Crisis*. Anmol Publishing Company, New Delhi.

Solo Robert A. 1959. *Synthetic Rubber: A Case Study in Technological Development under Government Direction*. Washington, D.C.: G.P.O.

Sorg James D. 1983. "A Typology of Implementation Behaviors of Street-Level Bureaucrats." *Policy Studies Review* 2:391-406.

Stone O. 1968. "Legal Aspects of Offshore Oil and Gas Operations." *Natural Resources Journal 8*:484-85.

Subrahmanya, K N. 1983. *Energy Problem and World Economic Development*. Deep & Deep.

Subudhi, R N. 1996. *Energy Environment and Economy*. Kanishka.

Thurow Lester. 1980. *The Zero Sum Society*. New York: Basic Books.

Tiwari, Satish ed. 2000. *Atomic and Non Conventional Energy*. Anmol Publishing Company, New Delhi.

Tiwari, Satish ed. 2000. *Petroleum and Natural Gas*. Anmol Publishing Company, New Delhi.

Tiwari, Satish ed. 2000. *Power: Potential and Development*. Anmol Publishing Company, New Delhi.

Tiwari, Satish ed. 2000. *Water and Energy Resources*. Anmol Publishing Company, New Delhi.

Toffler Alvin. 1981. "Reflections on the Third Wave." *National Forum*, The Phi Kappa Phi Journal (Summer).

Tomar, S S. 1995. *Energy Agriculture and Environment with Special Reference to Non Conventional Energy Sources in Development of Rural*. Mittal Pubilcations, New Delhi.

Trivedi, P R & B R Julka. 1997. *Energy Management*. Commonwealth.

Trivedi, P R *et al.* 1992. *Environmental Energy Resources*. Anmol Publishing Company, New Delhi.

U.S. Congress, House of Representatives, House Committee on Merchant Marine and Fisheries. 1981. *OCS Oversight—Part 2: Hearings on OCS Oversight and Related Issues, The National OCS Program and the Five-Year OCS Leasing Program Before the Subcommittee on the Panama Canal/Outer Continental Shelf of the House Committee on Merchant Marine and Fisheries*. 97th Congress, 1st Session. 215-16.

U.S. Congress, House of Representatives. 1948. *House of Representatives Report No. 1778*. 80th Congress. Reprinted in U.S. Code Congressional and Administrative News, 1426-27.

U.S. Congress, House of Representatives. 1953. *House of Representatives Report No. 413*. 83rd Congress, 1st Session, Page 2. Reprinted in U.S. Code Congressional and Administrative News, 2177-78.

U.S. Congress, House of Representatives. 1977. *House of Representatives Report No. 590*. 95th Congress, 1st Session 1977:74; reprinted in U.S. Code Congressional and Administrative News. 1978:1481.

U.S. Congress, House of Representatives. 1980. *House of Representatives Report No. 1214*. 96th Congress, 2nd Session, 113.

U.S. Department of the Interior. 1979. Office of Surface Mining, *Annual Report*.

United Press International. 1984. "Election 84: Presidential Vote/State by State," *Rocky Mountain News*, November 7.

United States Code a, Title 16, Sections 1451-1464, 1982.

United States Code b, Title 30, Sections 1979-287, 1982.

United States Code c, Title 42, Sections 4321-4370, 1982.

United States Code d, Title 43, Sections 1301-1842, 1982.

United States v. California. 332 U.S. 19 (1947).

United States, Office of the President. 1977. *The National Energy Plan*. U.S. Government Printing Office, Washington, D.C.

Van Donald S. Meter, and Carl Van Horn. 1975. "The Policy Implementation Process: A Conceptual Framework." *Administration and Society* 6:445-88.

Vandana, S. 2002. *Alternative Energy*. Ashish Publishing House, New Delhi.

Varadarajan, D B. 1994. *Economics of Wind Energy*. Ashish Publishing House, New Delhi.

Varma, Ajit & Basant Behera. 2003. *Green Energy: Biomass Processing and Technology*. Capital Publishing Company.

Vidyanath, V. 2000. *Environment Energy Health: Planning for Conservation*. Gyan Pub.

Virnve, S N. 2001. *Atomic Energy and India's Nuclear Capability*. Anmol Publishing Company, New Delhi.

Vyas, Ș K & N S Sandhar *et al*. 1998. *Biogas Technology: Problems and Prospects*. USG Pub.

Wall Street Journal. February 3, 1981. "Increased Drilling for Oil May Consume More Energy than It Gleans, Study Finds."

Williams Walter. 1980. *The Implementation Perspective*. Berkeley: University of California Press.

Zaborsky, Oskar R ed. 1981. *Handbook of Biosolar Resources in 3 Vols*. C R C Press, Boca Raton.

Index

A

Afghanistan, 5-8
- Afghanistan as an energy transit route, 7
- Angot oilfield, 7
- Dahla dam, 8
- Darunta dam, 8
- energy infrastructure at a glance, 7-8
  - electricity, 8
  - electricity, 8
  - natural gas, 8
  - oil, 7
- energy overview, 5-7
- general background, 5
- Kajaki dam, 8
- Local pipelines, 8
- Mahipur dam, 8
- Mazar-I-sharif power plant, 8
- Naghlu dam, 8
- Pipeline to mazar-I-sharif, 8
- Sheberghan area gas fields, 8

Algeria, 9-20
- crude oil production capacity, 18
- downstream, 13-14
- energy overview, 18
- environmental overview, 18
- ethnic groups, 17
- fuel share of energy consumption, 19
- general introduction, 9-10
  - country overview, 17
  - economic overview, 17=18
  - electricity, 16-17
- independence, 17
- liquefied natural gas (LNG) exports, 16
- major international environmental agreements, 19
- major state companies, 19
- merchandise exports, 17
- middle east Economic digest, 16
- natural gas, 14-15
  - reserves, 18
- oil, 11
  - and gas industry, 19-20
  - production, 11-12
- population, 17
- selected foreign energy company involvement, 20

Angola, 21-30
- country overview, 28
- crude oil production, 24-25
- currency, 28
- current account balance, 28
- defence, 28
- electric generation capacity, 29
- electricity, 27-28
- energy overview, 29
- environmental overview, 29-30
- exploration, 25-26
- exports, 24-25
- fuel share of carbon emissions, 29

general background, 21-22
gross domestic product (GDP), 21, 28
inflation rate, 28
International Monetary Fund (IMF), 22
major cities, 28
major environmental issues, 29
major oil fields, 30
major oil terminals, 30
natural gas, 27
oil, 22-23
and gas industries, 30
production, 23-24
organization, 30
proven oil reserves, 29
refining and downstream, 26
religion, 28
Antarctica, 31-34
Bahia Paraiso, 33
climate change, 33-34
energy use in Antarctica, 34
Exxon valdez, 33
in the 21st century, 34
oil exploration, 31-32
Arab Maghreb Union, 35-46
carbon dioxide emissions, 44
European Investment Bank (EIB), 46
International Petroleum Investment Company (IPIC), 40
Mauritania, 35-36
energy, 37
Morocco, 37-38
coal and electricity, 41
oil and natural gas, 39-40
Tunisia, 42-44
electricity, 45-46
natural gas, 45
oil, 42-43
World Trade Organization (WTO), 36
Argentina, 47-57
coal, 52
consumption, 56
country overview, 55
devaluation and utilities, 53-54
downstream, 50
electricity, 52
– industry, 57
– intensity, 56
– overview, 56
environment, 54-55
environmental overview, 56-57
exploration and production, 48-49
financial exchange rate, 55
fuel share of carbon emission, 56
Gas del Estado, 51
general background, 47-48
independence, 55
industry organization, 51, 52-53
major environmental issues, 57
– oil and gas producing basins, 57
natural gas, 50
reserve, 56
net oil exports, 56
nuclear power, 54
oil, 48
per capita energy consumption, 56
pipelines, 51-52
population, 55
regional interconnections, 54
sector organization, 48
status in climate change negotiations, 57
Transportadora de gas del norte (TGN), 51
Transportadora de gas del sur (TGS), 51
transportation, 49
Yacyreta, 54
Armenia, 198-202
Australia, 58-68
coal, 63-64
consumption, 66
industries, 68
country overview, 65-66
East Timor, 62-63
economic overview, 66
electric generation capacity, 66
electricity, 64
energy, 58-59
– intensity, 67
– overview, 66
– related carbon emissions, 66
environment, 65
environmental overview, 66-67
general background, 58
languages, 65
LNG, 63
major environmental issues, 67
major export ports, 68
– gas fields, 67
– imports products, 66
– oil and gas producing regions, 67
– trading parents, 66
market exchange rate, 66
natural gas, 61-62
oil, 59-60
and gas industries, 67-68
pipelines, 62
refining, 60-61
religions, 65
sector organization, 64-65
unemployment rate, 66
Azerbaijan, 69-79
Azeri, Chirag, deepwater Gunashli 72-73
bahrain, 80-85

coal, 76
Cocar, 71-72
country overview, 77
downstream/refining, 74-75
economic overview, 77-78
electricity, 76-77
energy overview, 79
environmental overview, 78-79
general background, 69-70
major environmental issues, 79
– international environmental agreements, 79
– oil refineries, 79
– power plants, 79
natural gas, 74
exports, 75-76
oil, 70
export pipelines, 79
exports, 73-74
organization, 79
Shah Deniz, 74-75
status in climate change negotiations, 79

B

Bahrain
carbon intensity, 84
country overview, 83
crude oil refining capacity, 84
currency, 83
current account balance, 84
economic overview, 83-84
electric generation capacity, 84
electricity production, 83-84
energy intensity, 84
– overview, 84
– related carbon emissions, 84
environmental overview-84, 85
ethnic groups, 82
external debt, 84
fixed exchange rate, 83
fuel share of energy consumption, 84
general background, 80-81
independence, 83
inflation rate (consumer prices), 84
languages, 83
location size, 83
main destinations of exports, 84
– origins of imports, 84
major cities, 83
– environmental issues, 85
– export products, 84
– foreign energy company involvement, 85
– import products, 84
– ports, 85
– refineries, 85
natural gas, 82
consumption, 84
production, 84
reserves, 84
net oil exports, 84
nominal gross domestic product, 83
oil, 81-82
and gas industries, 85
consumption, 84
production, 84
organization, 85
per capita carbon emissions, 84
– – energy consumption, 84
population, 83
producing oil fields, 85
proven oil reserves, 84
real (GDP) growth rate, 83
religion, 83
status in climate change negotiations, 84
total energy consumption, 84
Balkans region, 86-93
general background, 93-88
oil transit, 88-93
Albania-Macedonia-Bulgaria pipeline, 88-89
coal, 91
constant-Trieste SEEL pipeline, 89-90
Druzhba-Adria integration, 90
electricity, 92
hydroelectric, 93
natural gas, 91
nuclear, 92
oil, 90-91
regional energy issues, 88
Baltics region, 94-108
coal, 104-105
general background, 94-96
electricity, 105-108
Estonian ports, 100
Lativa, 94
Lativan ports and pipelines, 100
Lithuania, 94
Lithuanian ports and pipelines, 100-101
natural gas transit, 101-104
oil transit and export, 99-100
regional energy issues, 96
Russia's northern oil exports, 94
Bangladesh, 109-117
per capita energy consumption, 116
Bangladesh corporation, 117
Bangladesh petroleum exploration company, 117
coal, 113
country overview, 115
crude oil refining capacity, 116
currency, 115
current accounts balance, 115
defense, 115

economic overview, 115
electricity, 113-115
energy, 111
– intensity, 116
– overview, 115-116
environmental overview, 116-117
ethnic groups, 115
foreign energy company involvement, 117
fuel share of carbon emissions, 116
– – – energy consumption, 116
gas fields, 117
general background, 109-110
gross domestic products, 115
independence, 115
languages, 115
location size, 115
major cities, 115
– environmental issues, 116
– trading partners, 115
market exchange rate, 115
natural gas, 111-113
reserves, 116
oil, 111
and gas industries, 117
production, 115
organization, 117
population, 115
ports, 117
proven oil reserves, 115
refinery, 117
refining downstream, 111
religions, 115
Rupantarita Prakritik gas company, 117
status in climate change negotiations, 116
total energy consumption, 116
total foreign debt, 115

Bolivia, 118-126
country overview, 124
currency, 124
current account balance as per cent of GDP, 124
defense, 124
downstream, 120
economic overview, 124-125
electric generation capacity, 125
electricity, 123
energy overview, 125
energy-related carbon emissions, 125
environmental overview, 125126
ethnic groups, 124
general background, 118-120
generation, 123
gross domestic product, 124
industry organization, 120
inflation rate, 124
international Energy Agency, 126
liquefied natural gas (LNG) plans, 122-123
major cities languages, 124
– export products, 124
market exchange rate, 124
merchandise exports, 124
natural gas, 121-152
pipelines, 122
production, 125
net merchandise trade deficit, 124
oil, 120
and gas industries, 126
consumption, 125
production, 125
per capita energy consumption, 125
proven oil reserves, 125
real GDP growth rate, 124
religions, 124
rid rounds, 120
total energy consumption, 125
transmission and distribution, 123-124

Brazil, 127-142
accidents and the environment, 130
bid rounds, 132
Brazilian Labor Party, 127
coal, 136-137
Communist Party of Brazil, 127
country overview, 140
downstream, 132-133
economic overview, 140
electricity, 137
shortage, 137-138
energy crisis, 2001, 127
– industry overview, 141-142
– overview, 140-141
environment, 139
environmental overview, 141
exploration and production, 129-130, 134
general background, 127
Goulart, Joao, 127
industry organization, 133
international companies, 130
– cooperation, 131
liquefied natural gas, 136
National Mobilization Party, 127
natural gas consumption, 133-134
new developments, 134-135, 138
– government, 131
– pipeline developments, 135-136
oil, 127-129
Popular Socialist Party, 127
previous rounds, 132
promotion of natural gas, 134
sector organization, 131, 138-139
tax in Rio de Janeiro state, 131
unclear power, 139

Brunei, 143-148
- Amedeo Development Corporation, 143
- Ampa-Fairely Rationalization Profect, 147
- country, 147
- downstream, 146
- economic overview, 147
- electricity, 147
- energy overview, 148
- environmental overview, 148
- Exclusive Economic Zone, 145
- general background, 143-144
- natural gas, 146-147
- oil, 144-145
  - and gas industries, 148
- Sultan Hasan Bolkiah, 145

Burundi, Kenya, Rwanda, Tanzania, and Uganda, 365-374
- coal, 370
- Community of East and Southern African States, 365
- East Africa Community, 365
- electricity, 371-373
- energy overview, 367
- European Investment Bank, 372
- general background, 365-367
- Gross Domestic Product, 365
- Heavily Indebted Poor Countries, 366
- liquefied petroleum gas, 367
- natural gas, 370
- oil, 367-369
- pipelines, 369
- refining, 369
- renewable energy, 373-374
- South Africa Development Community, 366
- Tanzania Petroleum Development Corporation, 368
- Tanzania Zambia oil pipeline, 369

C

Cameroon, 149-156
- Bakassi Peninsula, 150-151
- Cameroon Oil Transport Company, 152
- Cameroon Petroleum Depot Company, 153
- Cameroon Societe Nationale de Raffinage, 153
- Chad-Cameroon pipeline, 151-153
- country overview, 154
- Democratic Republic of Congo, 153
- downstream, 153
- economic overview, 155
- electricity and natural gas, 153-154
- energy overview, 155
- environmental overview, 155-156
- general background, 149-150
- Heavily Indebted Poor Countries, 150
- International Court of Justice, 151
- International Monetary Fund, 149
- Memorandum of Understanding, 151
- Non-Government Organization, 153
- oil and gas industries, 150, 156
- recent upstream developments, 151

Canada, 157-172
- Aboriginal Pipeline Group, 167
- Alberta, 164-165
- Arctic, 166, 167-168
- Atlantic coast, 160-161, 165-167
- British Columbia, 165
- Canadian Association of Petroleum Producers, 159
- Chevron Texaco Corp., 162
- coal, 168
- coalbed methane, 166
- Coastal Zone Management Act, 167
- country overview, 170
- difficulties, 162-163
- economic overview, 170
- electricity, 168
- energy, 157-158, 170-171
- Energy and Utilities Board, 159, 164
- environment, 169
- environmental overview, 171-172
- exploration and production, 160, 164
- exports, 169
- Floating Production Storage and Off-loading, 160
- general background, 157
- Gross Domestic Product, 157
- imports, 169
- Internal Energy Agency, 172
- liquefied natural gas, 164, 166
- Maritimes and Northeast Pipeline, 166
- National Energy Board, 164, 169
- oil, 159
  - and gas industries, 172
  - exports, 159-160
  - sand pipelines, 163-164
- other developments, 162
- pacific coast, 161
- pipelines, 163, 166
- producers, 161-162
- Regional Transmission Organization, 169
- restructuring, 169
- Sable Offshore Energy Project, 165
- sector organization, 160
- synthetic crude oil, 161
- Trans Mountain Pipeline, 163
- Trillion Cubic Feet, 164

Western Canadian Sedimentary Basin, 160, 164

Caribbean fact sheet, 173-183
- additional trains, 177
- Barbados National Oil Company, 178
- British Petroleum, Trinidad, 175
- Cuba, 177
- deepwater exploration, 175
- Dominican republic, 180-182
- Economic Exclusive Zone, 178
- electricity, 180
- exploration and production, 175, 178
- exports to the united states, 180
- Federal Energy Regulatory Commission, 179
- general background, 173
- Jamaica, 182
- joint field development with Venezuela, 176
- liquefied natural gas, 173, 175
- LNG, 176-177
- new developments, 178
- new licensing round, 175-176
- oil and gas industries, 174
- other regional developments, 179
- pipelines, 176
- Puerto Rico, 183
- Puerto Rico Electrical Power Authority, 183
- recent developments, 175
- refining, 179
- renewable energy, 183
- storage, 179
- Trinidad and Tobago, 174-175

Caspian Pipeline Consortium, 188

Caspian Sea region, 184-191
- Baku Tiblisi Erzurum, 188
- environmental issues, 191
- European Bank for Reconstruction and Development, 189
- export issues, 188
- general background, 184-185
- Gross Domestic Products, 184
- natural gas, 186-188
- north for northwest, 190
- oil, 185-186
- regional conflicts, 190
- west, 188-189

Caucasus region, 192-207
- Armenia, 198-202
  - Azerbaijan Georgia, 202-207
  - coal, 200, 205
  - downstream refining, 203-204
  - electricity, 205-206
  - hydroelectric power, 201-202
  - hydropower projects, 207
  - Metsamor Nuclear Power Plant, 201
  - natural gas, 199-200, 204-205
  - oil, 199, 202-203
  - power outages, 206
  - privatization, 200, 206
- Azerbaijian International Operating Company, 195
- Baku-Ceyhan pipeline 194-195
- Baku-Erzurum pipeline, 196-197
- Baku-Novorossiisk pipeline, 195
- Baku-Supsa pipeline, 195
- European Union, 193
- general background, 192-193
- Georgian British Oil Company, 202
- Gross Domestic Product, 199
- International Finance Corporation, 197
- Iran-Armenia pipeline, 197-198
- main export pipeline, 194, 199
- National Petroleum Limited, 204
- natural gas transit, 196
- oil transit, 194
- other natural gas transit in the Caucasus, 198
- other oil transit, 196
- regional energy issues, 193-194
- Transport System Europe-Caucasus-Asia, 193
- Trillion Cubic Feet, 196

Central America, 674-686
- Belize, 682
- Costa Rica, 682
- electricity, 681
- energy, 676-677
- Environment, 686
- exploration, 678
- general background, 674-676
- Guatemala, 678, 683-684
- Honduras, 684-685
- natural gas, 780
- Nicaragua, 679, 685
- oil consumption, 679
- oil production, 677
- pipeline, 679-681
- plan Puebla Panama initiative for energy integration, 681-682
- privatization, 684
- rural electrification, 684
- Salvador, 683
- sector organization, 678, 682-683, 685

Central Asia, 208-211
- coal, 210
- downstream refining, 209-210
- electricity, 210-211
- general background, 208
- Gross Domestic Product, 208

natural gas, 210
oil, 208-209
Chad, 227-233
Cameroon Oil Transport, 229
Chad Oil Transport Company, 229
country overview, 231
downstream, 231
economics overview, 232
electricity and renewable energy, 231
energy overview, 232
environmental overview, 232, 233
general background, 227-228
Memorandum of understanding, 228
National Assembly, 228
oil and gas industries, 233, 228-231
Chile, 234-243
Chile National Energy, 237
coal, 239
country overview, 241
developments, 238
Downstream, 235
Economic Dispatching Central, 240
economic overview, 241-242
electricity, 239-240
Empreasa Nacional de Petroleo, 235, 243
energy overview, 234-235, 242
environment, 241
environmental overview, 242-243
Gas atacama, 238
general background, 234
import pipelines, 238
liquefied natural gas, 238
National Energy Commission, 243
natural gas, 237-238
new power bill, 240
North American Free Trade Agreement, 234
oil and gas industries, 235-243
oil trnsporation, 237
power projects, 240
transmission grids, 240
Trasandino, 237
upstream, 235-236
China, 244-255
China National Offshore Oil Corporation, 246
China National Oil Development Corporation, 255
China National Petroleum Corporation, 246
China National Power Industry Corporation, 255
China Oilfield Service Ltd, 247
cola, 250-251
country overview, 253-254
electricity, 251-252
energy industry, 255
energy overview, 254
environment, 253
environmental overview, 254-255
general background, 244-245
International Energy Agency, 255
Liquefied Natural Gas, 249
natural gas, 249-250
oil, 245-249
State Energy administration, 246
State Owned Enterprises, 245
Trillion Cubic Feet, 249
Water and Electric Corporation, 255
World Health Organization, 253
World Trade Organization, 244
Colombia, 256-265
Army National Liberation, 256
Billion Cubic Feet, 261
coal, 261-262
country overview, 263
country, 263
economic overview, 263
electricity, 262-263
energy overview, 264
energy overview, 265
environmental overview, 264-265
exploration and production, 258
Fuerzas Armada Revolcucionarias Colombia, 256
general background, 256-257
National Liberation Army, 260
natural gas pipelines, 261
natural gas, 260-261
oil, 257
– pipelines, 259
recent developments, 259
refining, 260
sector organization, 257-258, 261-263
security issues, 260
setback in exploration, 259
Congo-Brazzaville, 266-276
Africa, 269
country overview, 274
economic overview, 274
electricity, 273-274
energy overview, 275
environmental overview, 275
exports, 270
field developments and exploration, 270-271
Floating Production Unit, 270
Forces Democratiques Unies, 266
general background, 266-268
International Monetary Fund, 267

Magnesium Alloy Corporation, 273
Mer Tres Profonde Sud, 271
natural gas, 272
oil and gas industries, 268, 276
Parti Congolais du Travail, 266
Pointe Noire Grande Fonds, 269
Post-conflict Economic Rehabilitation Credit, 268
Poverty Reduction and Growth Facility, 267
production, 269-270
Refining and downstream, 272
Societe Nationale de Distribution, 274
Societe Nationale des Petroles du Congo, 268
Societe Nationale Electricite, 273

Cote d'ivoire, 277-284
Canadian Natural Resource, 279
Commonwealth Development Corporation, 282
Compagine Electricite Ivoirienne, 277
Economic Community of West African States, 278
economic overview, 283
Electricite De France, 281
electrity, 281-282
energy overview, 283
Energy Regulatory Commission, 282
Environmental overview, 283-284
Floating Production Storage and offloading, 279
general background, 277-278
Gross domestic Product, 277
International Monetary Fund, 278
Liquefied Petroleum Gas, 281
Memorandum Of understanding, 281
natural gas and LPG, 280-281
New Partnership for African Development, 282
oil and gas industries, 278-280, 284
Patriotic Movement for Cote d'Ivoire, 277
Production Sharing Contract, 279
Rally of Republicans, 277
refining and downstream oil activities, 280
Staff Monitored Program, 278
West Africa Gas Pipeline, 281
West African Economics and Monetary Union, 278

E

East Mediterranean Gas, 451

Economic Community of West African States, 285-300
common currency, 285
Compagnie Electrique Benin, 299
consumption, 295
Council of Ministers, 286
electricity, 297-299
energy overview, 286-289
exploration and production, 290-294
floating production and storage offloading, 290
general background, 285-286
Liberia Electricity Corporation, 298
Liquefied Natural Gas Corporation, 295
National Water and Electricy Company, 299
natural gas, 295-297
Nigerian National Petroleum Corporation, 290
Patriotic Movement For Cote d'Ivoire, 286
petroleum, 289
refining, 294
regional project, 299-300
Residual Catalytic Cracking, 294
Saltpond Offshore Producing Ltd, 290
Strategic Reserve Plant, 298
Tema Oil Refinery, 294
Trade Liberalization Scheme, 286
Volta River Authority, 297
West Africa Power Pool, 299
West African Gas Pipeline, 296
West African Monetary Union, 286
West African Refinery Company, 294

Ecuador, 301-310
Comunidad Andina de Naciones, 307
country overview, 308
economic overview, 308-309
electricity, 306-307
Energy Development Corporation, 307
energy development, 306
– overview, 309
environmental overview, 309-310
general background, 301-303
Instituto Ecuatoriano de Electrification, 307
International Monetary Fund, 301
Ishpingo Tambococha and Tiputini, 303
natural gas, 306
oil and gas industries, 303-305, 310
oil transporation, 305-306
Oleducto Crudos Pesados, 305
restructuring, 307-308
Value Added Tax, 302

Egypt, 311-320
Arab Petroleum Pipeline Corporation, 320

Build Own Operate and Transfer, 317
Compressed Natural Gas, 312
country overview, 318
economic overview, 318
Egypt Trans Gas Company, 320
Egyptian electricity Authority, 316
Egyptian General Petroleum Corporation, 312, 320
Egyptian Natural Gas Holding Company, 314
Electricite De France, 317
electricity, 316-317
energy overview, 318-319
environment, 317
environmental overview, 319-320
general background, 311-312
Gross Domestic Product, 311
International Egyptian Oil Company, 314
liquefied natural gas, 311, 313
natural gas, 314-316
oil and gas industries, 312-313, 320
refining, 314
State Owned Enterprises, 311
Suez canal/Sumed pipeline, 313-314

Empreasa Nacional de Electricidade (ENE), 27

Equatorial Guinea, 321-328
African Development Bank, 322
Atlantic Methanol Production, 325
Consumer price index, 322
country overview, 326-327
downstream, 325
economic overview, 327
electricity, 326
energy overview, 327
Enhanced Structural Adjustment Facility, 322
environmental overview, 327-328
floating production storage and offloading, 325
general background, 321-322
Gross Domestic Product, 321
International Court of Justice, 324
Liquefied Petroleum Gas, 325
Memorandum of Understanding, 326
natural gas, 325-326
oil and gas industries, 232, 328
Overseas Private Investment Corporation, 325
Production Sharing Contracts, 323
production, 324-325
reserves/exploration, 323-324

European Bank of Reconstruction and Development, 197

European Union, 687-693
energy, 689
– consumption, 689
– policy and the internal market, 689-693
– use and carbon emissions, 693
general background, 687-688

F

France, 329-339
Charbonnages de France, 333, 334
coal, 333
Commissariat a'I Energie Atomique, 337
country overview, 337-338
downstream, 331
economic overview, 338
Electricite De France, 330
electricity market privatization, 334-336
electricity, 334
energy overview, 338
energy, 330
environmental overview, 339
environmental, 337
European Union, 329
exploration and production, 331
French Electricity Regulation commission, 333
Gaz de France, 330, 332
General Agreement on Tariffs and Trade, 329
general background, 329-330
General du Travail, 336
International Atomic Energy Agency, 329
International Energy Agency, 329
liquefied natural gas, 332, 333
natural gas market privatization, 333
natural gas, 332
nuclear, 336-337
oil, 331
Publique et Industrielle, 336
Reseau de Transport Electrcite, 335
sector organization, 332, 334

G

Gabon, 340-347
Central Africa Economic and Monetary Community, 341
country overview, 345
economic overview, 345
electricity, 344
energy ovesrview, 345-346
environmental, 346
general background, 340-342
Gross Domestic product, 340
International Monetary Fund, 341
natural gas, 344

oil, 342-344
– and gas industries, 347
Parti Democratique Gabonais, 340
refining, 344
Societe'etd' Eaux du Gabon, 344

Germany, 348-346
coal, 354-355
coal, 357
consumption forecast, 351
country overview, 362
downstream, 351-352
economic overview, 362
electricity, 358
emden natural gas hub, 354
Emden Natural Gas Hub, 356-357
Energie Banden Wurttemberg Aktiengesellschaft, 360
energy, 349-350
energy industries, 364
energy overview, 362-363
Energy saving ordinance, 359
environmental overview, 363
environmental, 361
European union market liberalization requirements, 350
European Union, 348
general background, 348-349
German North Sea Consortium, 353
Gross Domestic Product, 348
imports, 358
Mittleplate, 350
monitoring report, 353, 355
natural gas, 352
North Atlantic Treaty Alliance, 348
north trans-gas pipeline, 354, 356
nuclear power, 360-361
oil, 350-351
other programs, 359-360
pipelines, 354, 356
production, 352-353, 355, 357-358
promotion of renewables and energy efficiecy, 359
regulatory agency for the electricity market, 360
– – – – German natural gas market, 353, 355
renewables, 358-359
sector organization, 353, 355, 360
Social Democrats, 349
Solar panel roof program, 359
Verbundnetz Gas, 353, 356

Greece, 375-383
Athens Stock Exchange, 376
Centre Renewable Energy Sources, 381
coal, 379-380
country overview, 381
downstream, 378
economic overview, 382
Economist Intelligence Unit, 375
electricity, 380, 381
Energy Photovoltacis, 381
environmental overview, 382-383
European Union, 375
Foreign Direct Investment, 375
general background, 375-377
Greece Public Gas Corporation, 376, 377, 378
Gross Domestic Product, 375
Hellenic Petroleum, 377
International Monetary Fund, 376
Interstate Oil Gas Transport to Europe, 379
liquefied natural gas, 379
Memorandum of Understanding, 379
natural gas, 378-379
North Aegean Petroleum Company, 377
oil and gas industries, 383
oil, 377-378
Public Power Corporation, 380

H

Horn of Africa, 384-395
Build Operate Transfer, 392
Build Own Operate, 387
Calub Gas Share Company, 391
Deadweight Tons, 384
Djibouti, 384-385
Dubai Ports Authority, 385
electricity, 387
Ente Nazional Energia Elettrica, 394
Eritrea, 387-390
Bab el-Mandeb, 388-389
downstream and refining, 389
electricity, 389-390
oil, 387-388
Eritrean Electricity Authority, 389
Ethiopia, 390-392
downstream, 391-392
electricity, 392
oil and natural gas, 391
Ethiopian Electric Power Corporation, 392
Ethiopian People Revolutionary Democratic, 390
Ethiopian Privatization Agency, 391
Front for the Restoration of Unity and Democracy, 384
Geothermal Development Associates, 387
Global Environmental Facility, 387

Gross Domestic Product, 385
Heavily Indebted Poor Countries, 391
Independent Power Project, 392
oil, 385-386
Production Sharing Contract, 387
Rift Valley Fever, 394
Somali National Movement, 393
Somalia, 393-395
electricity, 394-395
oil and natural gas, 394
Transitional National Government, 393
United Nations, 387

I

Independent Power Producers, 412
India, 396-406
Administered Pricing Mechanism, 397, 399
Bharat Petroleum, 399
Broken Hill Proprietary, 401
Coal India Ltd, 402
coal, 401-402
country overview, 404
Dabhol Power Corporation, 402
downstream refining, 398-399
economic overview, 404-405
electricity, 402-403
energy overview, 405
environment, 403-404
environmental overview, 405-406
Foreign Direct Investment, 397
Foreign Investment Promotion Board, 400
Gas Authority of India Limited, 400
general background, 396-397
Gross Domestic Product, 396
Hazira Bijaipur Jagdishpur, 400
Hindustan Petroleum, 399
Independent Power Producers, 402
industry restructuring and price deregulation, 399
Maharashtra State Electricity Board, 402
National Thermal Power Corporation, 400
natural gas, 399-401
New Exploration Licensing Policy, 398
oil, 397-398
Oil and Natural Gas Corporation, 398
State Electricity Boards, 402
Indonesia, 407-416
Broken Hill Proprietary, 412
China National Offshore Oil Corporation, 409
coal, 412
country overview, 413
economic overview, 413-414
electricity, 412-413
energy overview, 414
environment, 413
environmental, 415
general background, 407-408
Gross Domestic Product, 407
liquefied natural gas, 407, 410-412
oil, 408-410
– and gas industries, 415-416
– sector reforms, 410
Perum Gas Negara, 415
Perusahaan Gas Negara, 411
Perusahaan Listrik Negara, 412
refining, 410
International Energy Agency, 415
Iran, 417-431
Caspian sea region, 422
country overview, 428
crude swaps, 422-423
economic overview, 428-429
electricity, 425-426
energy overview, 429
Enhanced Oil Recovery, 419
environment, 428
environmental overview, 429-430
foreign investment buybacks, 420-421
Gas To Liquids, 424
general background, 417-418
Gross Domestic Product, 417
International Atomic Energy Agency, 425
Japan National Oil Company, 420
liquefied natural gas, 424
liquefied petroleum, 421
Ministry of Petroleum, 430
National Iranian Gas Company, 430
National Iranian Oil Company, 430
National Iranian Petrochemical Company, 430
National Iranian Tanker Company, 430
Natural Gas Liquids, 421
natural gas trade, 523-425
nuclear, 426
offshore development, 421
oil, 418-420
– and gas industries, 430-431
other natural gas development, 424-425
refining and transportation, 423
sanctions, 418
south Pars, 423-424
Iraq, 432-446
China National Petroleum Company, 439
Coalition Provisional Authority, 432

Council of International Coordination, 433
country overview,444
economic overview, 44
Economist Intelligence Unit, 433
electricity, 443-444
energy overview, 445
environmental overview, 446
export, 437
Gas Oil Separation Plant, 440
general background, 432-434
International Oil Companies, 436
Iraqi National Oil Company, 436
Iraq National Oil Company, 434
Iraq Pipeline Saudi Arabia, 441
Joint Logistics Centre, 436
liquid petroleum gas, 442
Middle East Economics Survey, 437
natural gas liquids, 442-443
North Oil Company, 434, 435
oil, 434-435
– and gas industries, 446
– export pipelines, 440-441
– terminals, 441-442
pre-war state of Iraq's oil sector, 437
production, 435-436
refining, 442
Sate Oil Marketing Organization, 434
South Oil Company, 434
status of oil development deals with foreign companies, 437-439
upstream oil sector current status, 439-440
Very Large Crude Carriers, 441

Israel, 447-458
coal, 453-454
country overview, 455-456
Crude refining capacity, 458
downstream, 450-451
economic overview, 456
Eilat Ashkelon Pipeline Company, 451
electricity, 454-455
energy, 449
– industries, 457-458
– overview, 456-457
environmental overview, 457
general background, 447-448
Independent Power Producers, 454
Israel Monopoly National Utility, 454
National Coal Supply Corporation, 453
natural gas, 451-453
oil, 449-450
Oil Refineries Ltd, 449

Italy, 459-469
Billion Kilowatt Hours, 465
coal, 465
country overview, 467
downstream, 462
economic overview, 467-468
Electrcite due France, 466
electricity, 465
energy Industry, 469
– overview, 468
Ente Nazionale Idrocarburi, 469
environment, 467
environmental overview, 468-469
European Union, 459
exploration and production, 461-462
general background, 459-461
Gross Domestic Product, 459
imports, 464
Independent power producer, 464
Initial Public Offering, 463, 465
International Energy Agency, 469
Libya Norway and Qatar, 464
liquefied natural gas, 464-465
natural gas, 462
North Atlantic Treaty Organization, 459
nuclear, 466-467
oil, 461
sector organization, 462-466

J

Japan, 470-479
Arabian Oil Company, 472
country overview, 477
economic overview, 477
Electric Power Development Co, 479
Electric Utilities Industry Law, 476
energy industries, 479
– overview, 471, 477-478
environmental overview, 478-479
Gross Domestic Product, 470
Japan National Oil Company, 472, 479
liquefied natural gas, 474
Ministry of Economy Trade and Industry, 472
natural gas, 474-477
coal, 474-475
electricity, 475-476
environment, 476-477
liquefied natural gas, 474
oil, 472-473
recent developments, 470-471
refining downstream, 473
Tokyo Electric Power, 474, 475

Jordan, 480-485
country overview, 483-484
economic overview, 484
Electricity Distribution company, 483

electricity, 483
energy industries, 485
– overview, 484
environment overview, 484-485
general background, 480
Gross Domestic Product, 480
Independent Power Producer, 483
Jordan Electric Power Company, 483
National Electric Power Corporation, 485
National Electric Power, 483
National Petroleum Co, 482
natural gas, 282-483
Natural Resources Authority, 482
oil, 481-482
– transportation, 482

K

Kazakhstan, 486-498
Atyrau Samara, 490
Caspian pipeline consortium, 490
coal, 492-493
country overview, 495-496
downstream refining, 490
economic overview, 496
electricity, 493-494
energy industries, 498
– overview, 496-497
environmental overview, 497
general background, 486-487
Gross Domestic Product, 486
industry organization, 495
Karachaganak Integrated Organization, 488-489
Kashagan, 489
Kazakhstan Electricity Grid Operating Company, 495
natural gas distribution,, 491-492
nuclear power, 495
oil, 487-488
– exports, 489-490
Tengiz, 488
transmission and distribution, 494-495
Kuwait, 499-509
Arabian Oil Company, 504
country overview, 507
Crude Distillation Units, 505
Crude oil exports, 503
current oil production, 502
economic overview, 507-508
electricity, 507
energy overview. 508
environmental overview, 508-509
foreign downstream operations, 505
– upstream operations, 504-505
Gathering Center, 502
general background, 499-500
Gross Domestic Product, 499
Gulf Cooperation Council, 500
International Energy Agency, 509
International Oil Companies, 503
Kuwait Foreign Petroleum Company, 504
Kuwait Gulf Oil Company, 504
Kuwait National Petroleum Company, 505
Kuwait National Petroleum Company, 509
Kuwait Oil Tanker corporation, 509
Kuwait Petrochemical Industries Company, 505
Kuwait Petroleum Corporation, 502, 509
Ministry of Electricity and Water, 507
natural gas, 506-507
neutral zone, 504
oil, 501-502
– and gas industries, 509
Petrochemical Industries Company, 509
petrochemicals, 505-506
plans to expand oil production, 503-504
Production Sharing Agreements, 500
refining, 505
Supreme Petroleum Council, 504

L

Libya, 510-521
country overview, 519
economic overview, 519-520
electricity, 518-519
energy overview, 520
environmental overview, 520-521
exploration and development, 515-516
general background, 510-512
Great Man Mad River, 512
Gross Domestic Product, 510
Iran Libya Sanctions Act, 511
Joint Oil Company, 513
liquefied natural gas, 515
liquefied petroleum gas, 517
National Oil Corporation, 512
natural gas, 517-518
oil, 512-513
– and gas industries, 521
production exports and reserves, 514-515
refining marketing, 516-517
Sirte Oil Company, 515
Waha Oil Company, 515
Zueitina Oil Company, 515

M

Malaysia, 522-529
country overview, 526-527
economic overview, 527

electricity, 525-526
energy overview, 527-528
environmental overview, 528
general background, 522
natural gas, 524-525
oil, 523-524
– and gas industries, 529
refining and downstream, 524
Mexico, 530-542
energy overview, 540
coal, 536-537
country overview, 539-540
downstream, 533-534
economic overview, 540
electricity, 537-538
energy industries, 541-542
environment overview, 539, 541
general background, 530-531
infrastructure, 539
liquefied natural gas, 536
Mexico oil exports and OPEC, 533
natural gas, 534
oil, 531
reserves and production, 532-533
sector organization, 531-532, 536, 538-539
transportation and distribution, 535-536
upstream, 534-535
Monthly energy chronology
1996, 937-977
1997, 978-1032
1998, 1033-1082
1999, 1083-1112
2000, 1113-1138
200, 1139-1178
2002, 1179-1223
2003, 1224-1264
2004, 1265-1274

N

Nicaragua, 679, 685
Nigeria, 543-555
border dispute, 549
country overview, 553
crime and ethnic unrest, 546-547
economic overview, 553
electricity, 552
energy overview, 554
environment, 552-553
environmental overview, 554-555
exploration and field development, 547-549
exports, 546
general background, 543-544
natural gas, 550-552
oil, 545
– and gas industries, 555
production, 545-546
refining and downstream, 549-550
North Atlantic Treaty Alliance, 348
North Atlantic Treaty Organization, 459
North Central Europe, 556-567
coal, 563-564
downstream, 560-561
electricity, 565
general background, 556-558
imports, 562
natural gas, 561-562
– – transit, 559-560
nuclear, 566-567
oil, 560
– transit, 558-559
Poland imports Denmark and Norway, 563
– – Russia, 562
regional integration, 560
restructuring, 564-565
sector liberalization, 565-566
North Korea, 568-573
country overview, 571-572
domestic energy resources, 570
economic overview, 572
energy overview, 572
environmental, 572-573
general background, 568
natural gas transportation, 571
North Korean South Korean relations, 569
nuclear, 571
oil, 570
United States economics sanctions, 569
North Sea, 574-582
Denmark and Germany, 581
exploration and production, 577
general background, 574-576
government promotional programs, 577
independent producers, 577
licensing, 576
natural gas infrastructure, 581-582
natural gas, 579-580
Netherlands, 580
Norway, 576, 580-581
and the UK joint development, 576
oil infrastructure, 576-579
other North sea oil producers, 577-578
United Kingdom, 577, 580
Norway, 576, 580-581, 583-595
coal, 591-592
country overview, 592-593
economic overview, 493
electricity, 592
energy overview, 593-594
environment overview, 592, 594-595
general background, 583-584

natural gas, 588-589
exports, 589-590
production, 590-591
North Sea oil and natural gas, 584-585
oil, 585-586
production, 586-588
sector restructuring, 586

O

Oman, 596-602
country overview, 600
economic overview, 601
electricity, 600
energy overview, 601
environmental overview, 601-602
general background, 596-597
liquefied natural gas exports, 599-600
natural gas, 599
oil, 597-598
– and gas industries, 602
refining and petrochemicals, 598

P

Pakistan, 603-611
coal, 608
economic overview, 609-610
electricity, 607-608
energy industries, 611
– overview, 610
environment, 608-609
environmental overview, 610-611
general background, 603-604
natural gas, 605-607
oil, 604-605
refining downstream, 605
Panama, 612-618
canal expansion and modernization, 616
– traffic, 615
country overview, 616-617
economic overview, 617
energy, 613-614
– overview, 617
environmental overview, 617-618
general background, 612-613
Panama canal, 614
Trans-Panama pipeline, 615-616
treaties, 614-615
U.S. petroleum imports and the Panama canal, 616
Paraguay, 619-624
country overview, 622
economic overview, 622
electricity, 620-622
energy industries, 624
energy overview, 622-623
environmental overview, 623
general background, 619
natural gas, 620
oil, 619-620
Persian gulf oil and gas exports fact sheet, 625-632
general background, 625-626
Offshore Persian Gulf Oil Development, 627-628
Offshore Persian Gulf Natural Gas Development, 628-629
oil and gas reserves production capacity, 626627
oil flows
Bab al Mandab,630
oil imports from the Persian gulf, 631-632
other export routes, 631
strait of Hormuz, 629-630
Suez Sumed complex, 631
Peru, 633-643
Aguaytia project, 639
Comisea project, 637-639
country overview, 641
economic overview, 641
electricity, 639
energy overview, 641-642
environment overview, 642
general background, 633-635
natural gas, 637
oil, 635
– and gas industries, 643
pipeline, 636
power grid, 640
production and exploration, 635-636
refining downstream, 637
sector organization, 639-640
Philippines, 644-652
coal, 647
country overview, 650
economic overview, 650
electricity, 648-549
energy overview, 651
environmental overview, 651-652
general background, 644
natural gas, 646-647
oil, 645-646
– and gas industries, 652
refining and downstream, 646
renewables, 649-650
Portugal, 653-660
country overview, 658
economic overview, 658-659
electricity, 656
energy, 653-654
– overview, 659
– sector restructuring, 654-655
environmental overview, 659-660
exploration and production, 655, 656
general background, 653
Iberian integration, 657
natural gas, 655-656
nuclear power, 658
oil, 655
refining downstream, 655

sector organization, 657-658
storage, 656
Production Sharing Agreement, 223
Puerto Rico Fact Sheet, 661-665
coal, 663
electricity, 663-664
energy overview, 665
energy, 662
environment, 664
general background, 661-662
natural gas, 662
oil, 662
political and economic overview, 664
renewable energy, 664

Q

Qatar, 666-673
country overview, 671
downstream, 669
economic overview, 671-672
electricity, 671
energy overview, 672
environmental overview, 672-673
general background, 666-667
natural gas, 669-671
offshore development, 668-669
oil, 667-668
– and gas industries, 673
onshore development, 668

R

Russia, 694-709
Adria reversal project, 699
Angarsk Daqing, 700
Angarsk Nakhodka, 700
Baltic pipeline system, 688
BP, 696
China and South Korea, 704
Coall, 704-705
country overview, 707
downstream refining, 700
eastern oil pipeline options, 699
economic overview, 707-708
electricity, 705-706
– exports, 706
– industry structure, 706
energy overview, 708
environment, 706-707
environmental overview, 708-709
export markets, 702
gas industry structure, 701-702
general background, 694-695
Murmansk, 688-699
natural gas, 701
new export pipelines, 702-703
North Trans gas pipeline, 703
Odessa Brody reversal project, 699
oil, 695
– exports, 697-698
– industry structure, 696
Sakhalin, 704
Yukos Sibneft, 696-697

S

Saudi Arabia, 710-722
country overview, 720
economic overview, 720-721
electricity, 718-720
energy overview, 721
environment, 720
environmental overview, 721-722
general background, 710-712
natural gas, 717-718
oil and gas industries, 722
oil, 712-715
ports and pipelines, 715-716
refining, 716
Singapore, 723-728
country overview, 726
economic overview, 726-727
electricity, 726
energy industries, 728
– overview, 727
environmental overview, 727-728
general background, 723
natural gas, 725-726
oil, 723-724
petrochemicals, 724-725
South Africa, 729-739
coal, 729-730
consumption, 732
country overview, 737
economic overview, 737-738
electricity, 736
energy industries, 739
– overview, 738
environment, 736-737, 738-739
exploration, 733-734
exports, 731-732
general background, 729
oil and natural gas, 733
production, 730-731, 734-735
refining and downstream oil activities, 735-736
synthetic fuels, 732-733
South Asia, 740-748
bio-mass non-commercial fuels, 746
coal, 745
electricity, 746-748
energy, 740-742
general background, 740
natural gas, 744-745
oil, 742-743
refining, 743-744
South China, 749-755
general background, 749-750
liquefied natural gas, 754
natural gas, 754
oil, 753
piracy, 755

regional conflict and resolution, 752-753
shipping, 754-755
South China sea territorial issues, 750-751
U.N. law of the sea, 751-752
South Korea, 756-763
coal, 759
country overview, 761
economic overview, 761
electricity, 759-760
energy industries, 763
– overview, 761-762
environment, 760-761
environmental overview, 762
general background, 756-757
natural gas, 758-759
oil, 757-758
Southeastern Europe, 764-775
Burgas Alexandropoulis pipeline, 768-769
Burgas Vlore pipeline, 769
coal, 771-773
Constanta Trieste, 769-770
electricity, 773-775
general background, 764-766
natural gas, 770-771
– – transit, 771
oil, 766-768
– transit, 768
regional energy issues, 766
Southern African development community, 776-789
coal, 785-787
consumption, 782-783
economic overview, 777-778
electricity, 787-788
energy, 778-779
exploration and production, 780-782
general background, 776-777
natural gas, 783-785
oil integration, 783
petroleum, 779-780
refining, 782
renewable energy, 788-789
Spain, 790-800
coal, 796
country overview, 799
economic overview, 799
electricity, 796-798
energy overview, 799-800
environmental overview, 800
general background, 790-792
liquefied natural gas, 795-796
natural gas, 793-795
nuclear power, 798
oil, 792-793
refining, 793
Sudan, 801-809
country overview, 807
economic overview, 808
electricity, 806-807
energy overview, 808
environmental overview, 808-809
general background, 801-803
greater nile oil project, 803-804
oil, 803
production, 804-805
refining and downstream, 805-806
Syria, 810-818
country overview, 816
economic overview, 816-817
electricity, 815-816
energy overview, 817
environmental overview, 817-818
natural gas, 814-815
oil, 811-912
– and gas industries, 818
recent development, 810-811
refining downstream, 812-813

T

Taiwan, 819-825
coal, 821
country overview, 824
economic overview, 823-824
electricity, 821-822
energy industries, 825
energy overview, 824
environment overview, 823-825
general background, 819
natural gas, 821
oil, 820-821
Tajikistan energy sector, 212-215
coal, 214
electricity, 214-215
general background, 212
natural gas, 213
oil, 212-213
Turkmenistan energy sector, 216-220
Thailand, 826-832
country overview, 830
electricity, 829-830
energy overview, 831, 832
environment, 930
environmental overview, 831-832
general background, 826
joint development area, 828-829
natural gas, 828
oil, 826-927
refining downstream, 827-828
Tobago, 174-175
Trinidad, 174-175
Turkey, 833-845
coal, 835-836, 840
country overview, 843
economic overview, 843-844
electricity, 841-842
energy overview, 835, 844
environment, 848-843
environmental overview, 844-845
natural gas, 837-839
blue stream pipeline, 838-839
consumption and production, 837-838

oil and gas industries, 845
other natural gas import deals, 839-840
--- issues, 840
pipeline, 836
recent development, 833-843
refining downstream, 836-837
Turkmenistan energy sector: coal, 219
downstream refining, 218
electricity, 219-220
general background, 216-217
natural gas, 218-219
oil, 217-218
Trillion Cubic Feet, 218

U

Ukraine, 846-855
coal, 851
country overview, 853
economic overview, 853-854
electricity, 851
energy overview, 854
environment, 852-853
environmental overview, 854-855
general background, 846-847
natural gas, 849
nuclear, 852
Odessa Brody reversal project, 847-848
oil, 847
– transit, 847
refining downstream, 848-849
transit gas to Europe, 849-851
United Arab Emirates, 856-864
country overview, 861-862
dolphin project, 860
economic overview, 862
electricity, 860-861
energy overview, 862-963
environmental overview, 863
foreign affairs, 857
– downstream operation, 859
general background, 856
government structure, 856
natural gas, 859
oil, 857-858
– and gas industries, 863-864
other industry, 856-857
projects, 859
refining, 858
supplying Dubai, 860
United Kingdom, 865-878
coal, 872-973
country overview, 876
downstream, 869-870
economic overview, 876
electricity, 873-875
energy overview, 876-877
environment overview, 877-878
environmental, 875-876
general background, 865-866
industry structure, 869, 871-872
natural gas, 870-871
north sea oil and gas, 866-867
nuclear, 875
oil, 867-868
production, 868-869
United States of America, 879-900
coal, 891-892
consumption marketing, 883
country overview, 897
domestic and import pipelines, 889-890
economic overview, 897-898
electricity, 893-894
energy industries, 900
– overview, 898-899
environment, 896-897
environmental overview, 899-900
general background, 879-880
hydroelectricity renewable, 894-896
imports exports, 883
natural gas demand, 888-889
– – merges, acquisitions, bankruptcies, 890-891
– – production and storage, 887-888
– nuclear, 894
oil, 880-883
– mergers and acquisitions, 885-887
refining downstream, 884
strategic petroleum reserve, 884-885
U.S. energy sanctions issues, 883-884
Uruguay, 901-907
country overview, 905
economic overview, 905-906
electricity, 904-905
energy industries, 907
– overview, 906
environmental overview, 906-907
general background, 901-903
natural gas, 903-904
oil, 903
Uzbekistan energy sector, 221-226
coal, 225
downstream refining, 223
electricity, 225-226
general background, 221
natural gas, 223-224
oil, 221-223

V

Venezuela, 908-920
coal, 915-916
country overview, 918
economic overview, 918
electricity, 916
energy overview, 918-919
environment, 917-918
environmental overview, 929-920

exploration and production, 912, 915
extra heavy crude oil, 913-914
general background, 908-909
generation disruptions, 917
LNG, 915
natural gas, 914-915
oil, 909-911
– and gas industries, 920
PdVSA, 911-912
political role, 911
recent strike, 911
resumption of operation, 911-912
pipeline, 915
recent development, 912-913, 915
refining, 914
sector organization, 915, 916-917
Venezuela and CPEC, 912
Vietnam, 920-930
coal, 926
country overview, 928
economic overview, 928
electricity, 926-927
energy overview, 928-929
environmental overview, 929
general background, 920
liquefied petroleum gas, 925-926
natural gas, 924-925
oil, 920-924
– and gas industries, 929-930
refining, 924

Y

Yemen, 931-936
country overview, 934-935
economic overview, 935
electricity, 934
energy overview, 935
environmental overview, 935-936
general background, 931-932
natural gas, 933-934
oil, 932-933
– and gas industries, 936
refining, 933